Biology of Women

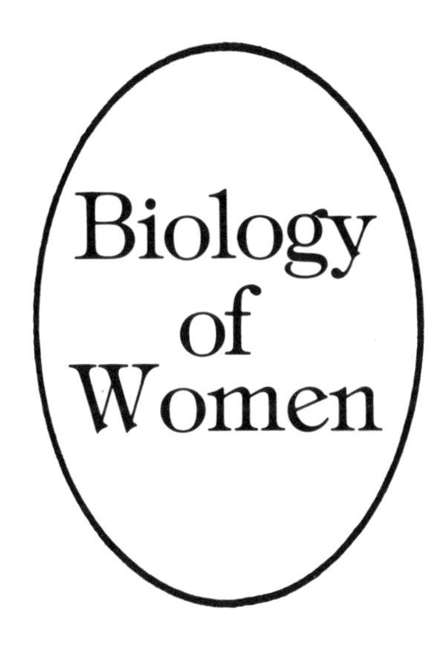

Biology of Women

by

Eileen S. Gersh, Ph.D.
Lecturer in Biology
Department of Biology
University of Pennsylvania

and

Isidore Gersh, Ph.D.
Professor Emeritus of Anatomy
Department of Anatomy
University of Pennsylvania

illustrations by
Mary Jo Larsen

University Park Press
Baltimore

UNIVERSITY PARK PRESS
International Publishers in Science, Medicine, and Education
300 North Charles Street
Baltimore, Maryland 21201

Typeset by Action Comp Co. Inc.
Manufactured in the United States of America by The Maple Press Company.

Library of Congress Cataloging in Publication Data
Gersh, Eileen S
Biology of women.
Bibliography: p.
Includes index.
1. Women—Physiology. 2. Human biology.
I. Gersh, Isidore, 1907- joint author.
II. Title.
QP34.5.G47 612′.62 80-25534
ISBN 0-8391-1622-5

Contents

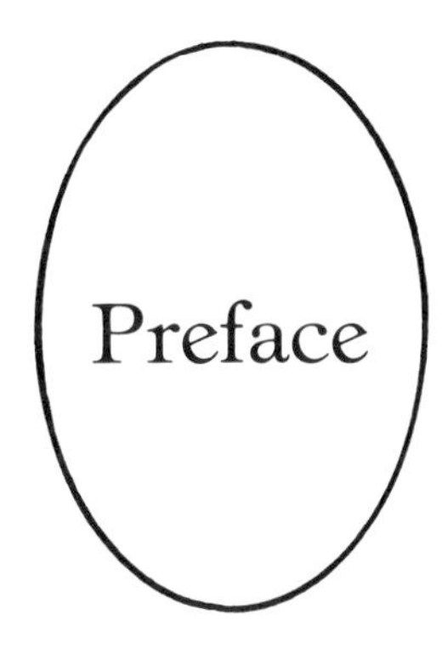

Preface

This book is based on a course on the biology of women that has been taught for five years in the Fine Arts College at the University of Pennsylvania. The students are all undergraduates, and may be majoring in science or in the humanities.

The purpose of the course is expository—that is, it lays out the biologic bases of being a woman. This would seem to be a simple matter, but it is not. The main reasons for the difficulty are the following: 1) the biologic bases have certain theoretical and factual underpinnings that are not readily acquired in the course of undirected reading; 2) they are obscured by long-standing taboos and myths; 3) in many areas, information on important points is still lacking because of insufficient research; and 4) the old problems of nature versus nurture, or genes versus culture, plague and bias discussion of the facts.

Although the catch phrase, "anatomy is destiny," is based on the writings of Freud (1925), the underlying ideology goes back very much further. Many people believe that women's role has always been largely confined to the bearing and rearing of children and the performance of household chores. Others argue for a primitive communal stage in which women as well as men played an important role as providers. In any case, the stereotyped role of woman as wife, mother, and homemaker looks increasingly obsolete in view of the scientific and technologic developments of our advanced industrial age, the fact that nearly half the women in this country are employed, and the increasing spread of education and sophistication among women. The contradictions between what women are capable of doing in this age, and the limited opportunities that society provides for them, have given rise to the "second wave" of the women's movement. Recognizing the limited advantages conferred by the vote, women have gone into the streets to demonstrate and have entered the halls of Congress as elected officials and lobbyists, to struggle for constitutional equality, the right to control their own reproductive functions and other demands for expanded opportunities and new lifestyles.

These demands have been met with resistance. We are still told that a woman's place is the home, yet there have also been subtle shifts in the line of attack. The simple statement that woman is destined by her biology to bear and care for children was not enough. It was proclaimed that their "raging" hormones must exclude women from a sphere where men (controlled, no doubt, entirely by cerebral processes) must naturally hold a monopoly. The Darwinian "struggle for existence" and the discovery of a hormonal influence on aggressiveness in male animals has also been used to justify the dominance of men and the exclusion of women from the competitive world of free enterprise. The focus of the new science of sociobiology on the genetic component in determination of human behavior likewise tends to encourage the view that gender-related differences in nonsexual behavior are "natural" and inevitable. The latest argument is based on differences in mental ability (verbal versus mathematical and spatial skills) between the sexes that develop at the time of puberty, and which, according to some (as yet insufficient) evidence may be based on innate, prenatally determined, differences in the brain. A good discussion of the "biology is destiny" arguments has been presented by Lambert (1978).

The subject of sex is frequently associated with emotional overtones, particularly of fear and anxiety. We hope to avoid invoking such feelings by presenting the biologic facts objectively in the same way that we would describe any other bodily system, such as the circulation or the brain. We hope also to stimulate thought and discussion as to how the biologic facts should be interpreted. Important contributions to these interpretations must come from the disciplines of psychology, sociology, history, and anthropology; but they have to start from an understanding of the biologic facts.

We believe that a clear understanding of the situation should facilitate rational decision-making about the roles women are to play in society and how they are to play them. Also the numerous taboos and myths that surround sexual topics might be seen for what they are, and their recognition will help women view their bodies realistically and thus help to do away with the alienation from their own bodies that most women have experienced through generations past. Finally, the sifting of the facts from the myths may cast an illuminating light on very real problems that are at present obscured by the propagation of myths, thus making it easier for women to establish satisfactory relationships with men (and vice versa), with other women, and with society.

"There goes another myth!", students have periodically exclaimed throughout our course. We believe that we must find the facts and base our conclusions on them. Myths are created when understanding is wanting. Without the facts, women have tended to create new myths to suit themselves. We are as strongly opposed to feminist myths as to sexist myths, and we deal with one or two of the former (Chapters 2 and 4). At best, these myths are an obstacle to acquiring knowledge and real understanding: at worst, they can lead to errors in practice and to disillusionment.

All over the world, women tend to be shorter than men (on the average). This is a statement of fact, not a value judgement. If we make this statement, we are not putting women down. As a matter of fact, the reader can certainly think of circumstances in which it would be an advantage to be small, as well as others in which it is an advantage to be tall. There are more color-blind men than color-blind women. This, too, is a neutral statement of fact. It does not put men down, even though it is hard to imagine conditions in which color-blindness would be an advantage. We should not shut our eyes to these real differences and thereby build a myth that there are no differences.

In general, biologic arguments against feminism ignore the remarkable plasticity of the human animal, the extent to which we respond to our environment. The nature versus nurture controversy is a sterile one as long as we cannot do controlled experiments in which one of the variables can be isolated and studied, others being kept constant (as we discuss in Chapter 7). Yet a heavy emphasis on either nature or nurture can lead us to formulate quite different practical policies. Emphasis on what is "natural," or simply on what *is*, tends toward support of the status quo, and where dissatisfaction arises, to the pessimistic view that nothing can be done about it. Emphasis on "nurture," on the other hand, while it can open the way to progressive change, can also open the way for manipulation.

For instance, although boys tend to excel in some cognitive skills and girls in others, if these skills can be learned, this can be used educationally to open up new opportunities for both sexes; on the other hand, it can be used negatively, to "track" boys and girls into different academic disciplines.

A word about the terms *natural* and *normal.* Nature creates freaks, so they are a part of nature. A mutant type arising from a genetic change may become the norm for the population in the course of evolution. We use the terms *normal* and *abnormal* without passing judgement.

Normal signifies inclusion in a continuous range around the average for the population (in considering height, for instance, this would exclude pituitary dwarfs), or inclusion in a group that comprises all but a small minority (for instance, females with two X chromosomes as opposed to those with one or three). We use the term *abnormal* to describe rare exceptions, knowing that it is sometimes used in a pejorative sense, because any other term we might use (exceptional, anomalous, or whatever) is equally susceptible to being given a pejorative twist.

This book differs from others of a similar kind in several ways. 1) Its primary point of view is biologic rather than psychologic, behavioral, or social. Two main groups of topics are covered: the genetic aspects of sex, and the female genital system, including its biologic controls. These topics are not treated together comprehensively in any other single book. We have found it necessary to refer our students to specialized books used by medical and other graduate students. In these books the ideas, and even the terminology, present problems to the uninitiated. We also deal with some parts of the male genital system to provide a better understanding of the female system. 2) It is basic, and in most topics, discusses the principles involved before embarking on the specific aspect, whether genetic, physiologic, or biochemical. Because some of the disciplines involved (e.g., genetics, cytology, endocrinology, anatomy) have theoretical aspects that are very broad indeed, we have selected and emphasized those parts that pertain to the biology of women. 3) The information from many specialized sources is integrated. In addition to covering structure, function, and control mechanisms, of the reproductive system, indications for the study of behavior, psychology, and physical performance are discussed. This is not a practical book on sex, but we do consider matters having to do with menstrual and menopausal variations, pregnancy, contraception, abortion, exercise, sports, and drugs. We hope we have introduced these topics in a way that will lead to self-understanding without encouraging medical self-treatment.

This book is written primarily for those with no more than a high-school training in biology and chemistry, with enough added material to make the book interesting to college-trained men and women who are better prepared. We hope the book will be useful in college courses on the biology, psychology, and behavior of women and to high school teachers of biology as well as in courses on nursing and on physical education. We hope it will be of interest also to general readers of both sexes regardless of age and education.

We are thankful to Mr. Howard Deck for his courteous, efficient, and invaluable library assistance. We wish to express our thanks to Ms. Joyce Duncan, production editor at University Park Press, for her sensitive and precise editorial treatment of the text. Finally, we are grateful for the encouragement and support of successive coordinators of Women's Studies at the University of Pennsylvania, Dr. Cynthia Secor, Dr. Elsa Greene, and especially Dr. Ann Beuf.

REFERENCES

Berman, E. 1970. New York Post, August 3.

Freud, S. 1925. Some psychological consequences of the anatomical distinction between the sexes. In J. Strachey (ed.), The Standard Edition of the Complete Psychological Works of Sigmund Freud, Vol. XIX. Hogarth Press, London, pp. 248–258.

Lambert, H. H. 1978. Biology and equality: A perspective on sex differences. Signs 4:97–117.

Biology of Women

1 Gross Anatomy of the Female and Male Genital Systems

TERMINOLOGY

Anatomic Terms

Some general terms should be defined at the outset because they are used throughout the book. In Figures 1.1A and 1.1B basic terms are applied to the quadripedal position, a position comparable to that of a cat. *Anterior* refers to the front (or head) part of the body and *posterior* to the hind (or tail) part. *Dorsal* refers to the backbone side of the body and *ventral* to the opposite side. This terminology is used regardless of the position of the body—whether on all fours, upright, prone, or kneeling. If the body is cut lengthwise dorsoventrally in the midline, the plane of section is called *midsagittal* (Figure 1.1C). A *parasagittal* plane is parallel to the midsagittal plane. *Lateral* refers to the part away from the midsagittal plane. When a plane of section is a cross section and at right angles to a sagittal plane, it is called *transverse.*

Some parts are referred to as being *proximal* or *distal*. The term *proximal* generally means closer to the midline or to the attachment or point of origin. The term *distal* generally means away from, or peripheral to the midline or to the attachment or point of origin. For example: the shoulder is proximal compared to the hand, which is distal; or the wrist is proximal to the fingers, which are distal.

Ducts are tubes that conduct a fluid from its point of origin to a surface. The space that conducts the fluid is called the *lumen*. The ducts connected with the organs or structures that make the fluids are called *secretory* or *glandular* parts. Those glands that secrete their products in water solution or suspension or as oil droplets into the ducts are called *externally secreting* or *exocrine glands*, because their ducts empty on a surface such as the intestine or the skin. But some glands do not have ducts, and secrete their products more or less directly

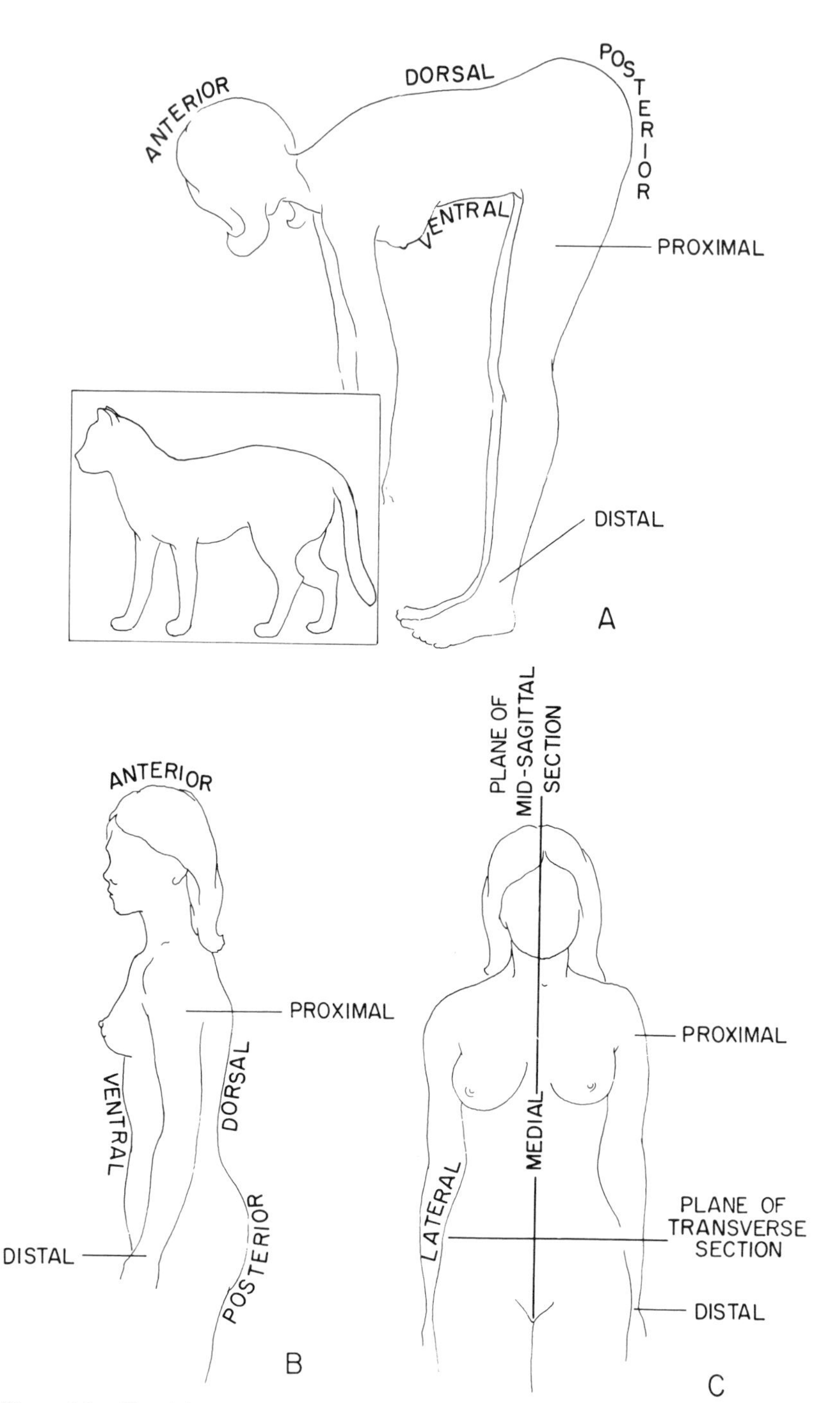

Figure 1.1. Sketch illustrates use of anatomic terms as defined in the text. A) Terms describe person in the quadrupedal position illustrated by the sketch of a cat (in box). B) The terms are the same regardless of the position of the body (standing, lying, etc.) C) The planes of sectioning originated from dissection of cadavers, in which separation and examination of both sides are necessary.

into the blood vessels. These are called *ductless* or *internally secreting* or *endocrine glands*.

Many terms are commonly used in connection with blood vessels or vascular channels. The vessels that carry blood from the heart to the periphery are called *arteries*, and they may be large or small arterial vessels. After dividing into microscopic blood vessels (or *capillaries*), the blood vessels form *veins*, which join together and become progressively larger venous blood vessels usually leading back to the heart. The capillaries almost always form an extensive network, and sometimes the veins form a similar network. Such a network is called a *plexus*, and is characteristic of erectile tissue and of many parts of the genital tract and of endocrine organs.

Units of Measurement

The units of measurement used are mostly metric because they are used by most countries now, and are likely to become official in the United States in the foreseeable future. They are given in Table 1.1.

FEMALE GENITAL ORGANS

The relations of the various genital organs to each other are sometimes difficult to appreciate. The relationships are simplified in Figure 1.2. If all the genital organs were removed in one piece from a cadaver, and laid out flat on a table, and then sectioned lengthwise, the cut surface would resemble the diagram in Figure 1.2A, of which Figure 1.2C is a cross section. If the same flattened, but intact, genital tract were sectioned in a lengthwise plane at right angles to that in Figure 1.2A, the cut surface would resemble the diagram in Figure 1.2B, of which Figure 1.2D is a cross section.

The genital organs are divided into those which are external and

Table 1.1. Equivalent measurements in the metric and other systems

1 meter (m)	= 100 centimeters (cm)
1 cm	= 10 millimeters (mm)
1 mm	= 1000 micrometer (μm)
1 μm	= 1000 nanometers (nm)
1 nm	= 10 Ångströms (Å)
1 μm	= 10,000 Å
1 inch (in)	= 2.54 cm = 25.4 mm
1 gram (g)	= 1000 milligrams (mg)
1 mg	= 1000 micrograms (μg)
1 ounce (oz)	= 28.35 g

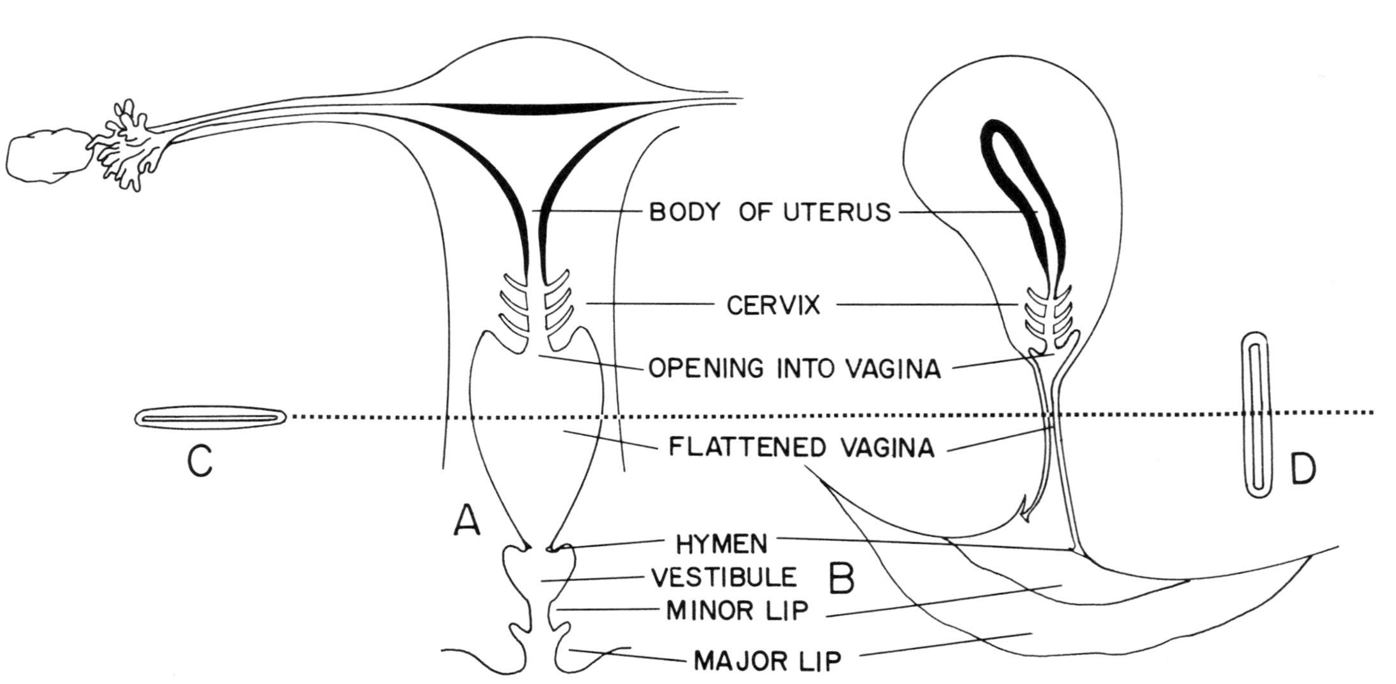

Figure 1.2. The vagina is mostly flattened except during sexual intercourse, when it is distended and tubular. In A, the genital organs of a cadaver are sectioned lengthwise after being laid out flat. The vagina is collapsed, and when sectioned transversely as in C, appears as a flattened tube. When the vagina is sectioned longitudinally in a plane 90° to that in A, it appears as in B, with the walls closely applied. When this is sectioned transversely, it appears as in D, which is the same as C but in a different plane of sectioning.

can be seen by simple inspection, and those which are internal. The chief genital structures are listed in Table 1.2, and are described systematically in the remainder of this chapter.

Major and Minor Lips and Mons (Labia Majora and Minora, Mons Veneris)

In viewing the external genitalia of a woman when she is standing, sitting, or lying in a prone position, little can be seen except the major lips, which touch each other (Figure 1.3). The slit between them is called the *pudendal cleft*. Each lip is a prominent rounded fold of skin that is customarily dry and covered with hair. The major lips extend lengthwise and fuse as they pass anteriorly to form a raised fullness which lies over the symphysis pubis, where the pubic bones are fused to each other. The raised eminence is called the *mons veneris*. The raised, rounded appearance of the mons and major lip is caused by the fat and loose connective tissues that underlie them.

More is exposed when a woman spreads her thighs (Figure 1.4). The lesser lips are now visible, as they also run lengthwise. A part of each passes over the *glans* of the clitoris, forming the *prepuce* or *hood*; another part is attached to the underside of the glans, and is known as the *frenulum*. Largely devoid of fat tissue, the minor lip is thinner.

Table 1.2. Parts of the genital systems of men and women and their equivalents

Female	Male
External Genital Organs	
Labium majus (external or major lip)	Scrotum
Labium minus (internal or minor lip)	Floor of urethra of penis
Vestibule	Part of urethra of penis
Hymen	?
Corpora cavernosa clitoridis; Glans clitoridis	Corpora cavernosa penis
Vestibular bulbs	Corpus spongiosum penis; Glans penis
Internal Genital Organs	
Vagina (part)	Part of prostatic utriculus
Urethral glands	Prostate gland
Vestibular glands	Bulbourethral glands
Uterus	Appendix testis (a rudiment)
Ovaries	Testes
Part of epoophoron (a rudiment)	Duct of epididymis; Vas deferens; Ejaculatory duct; Seminal vesicle

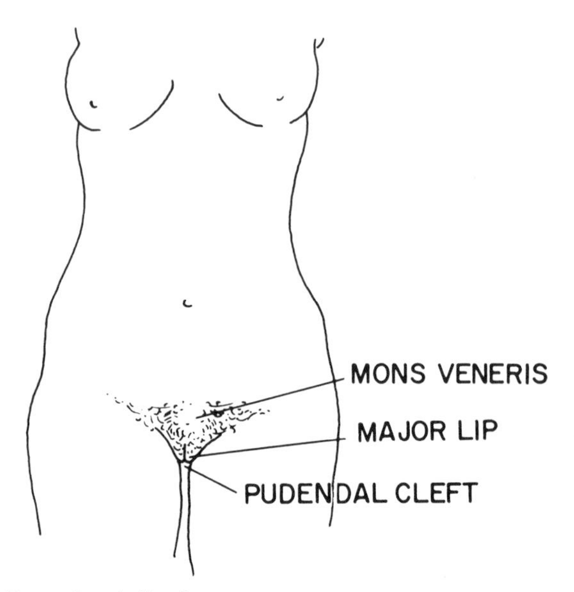

Figure 1.3. External genitalia of woman.

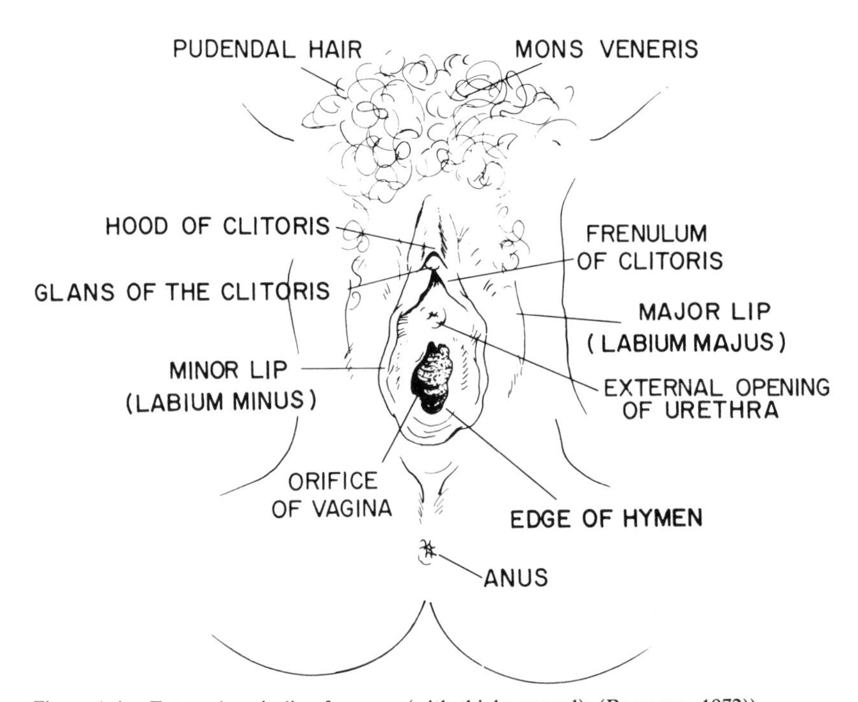

Figure 1.4. External genitalia of woman (with thighs spread). (Romanes, 1972)).

Vestibule of Vagina

When the lesser lips are separated, the vestibule of the vagina is exposed. It is separated partially from the inner part of the vagina by the *hymen* (which blocks the vagina incompletely) and by the tissues associated with the opening of the *urethra*. The urethra of women is relatively simple as compared with that of men, though the function in both sexes is to act as a duct or conduit for the urine from the urinary bladder to the outside. Also opening into the vestibule are the ducts that conduct the secretion, such as it is, of the greater vestibular glands to the genital canal. The hymen is a thin pliable membrane, usually perforated to a greater or lesser degree (and obliterated in many women), which stretches loosely between the vestibule and the rest of the vagina (Figures 1.4 and 1.5). The hymen may be stretched and torn by digital manipulation or the insertion of tampons, but not by vigorous exercise such as horse riding.

Erectile Organs

The clitoris is also visible externally, and like the penis, is mostly composed of erectile tissue (Figure 1.6). It is about 2-2½ cm long from base to tip, and has no other relation with the urethra than to be near the urethral opening. The size at rest varies greatly from person to person. The clitoris is derived from and continuous with the *crus* of the clitoris on each side (Figure 1.6). The crus also contains some erectile tissue. Both crura (or legs) are closely attached to the medial edges of the pubic arch. They fuse anteriorly just beneath the symphysis pubis to form the clitoris, which bends forward and ends as the glans of the clitoris. The

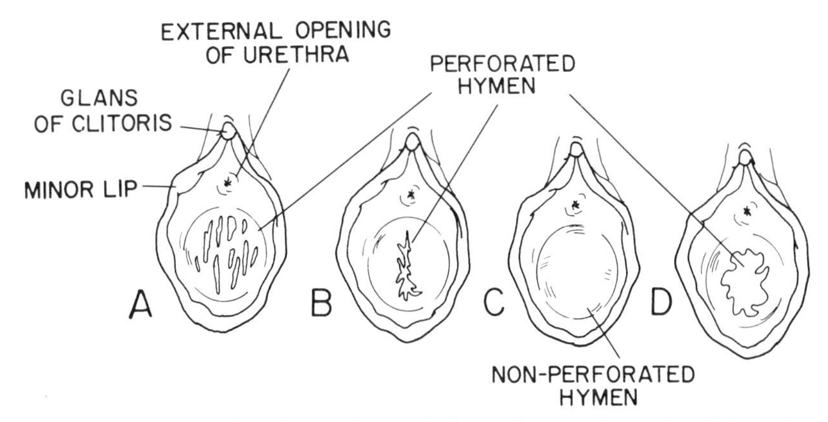

Figure 1.5. The hymen is a thin membrane which stretches over the opening of the vagina. Sometimes it is not perforated, and must be ruptured surgically to allow passage of menstrual fluid. The kinds of perforations vary greatly in virgins. (After Stedman, 1961).

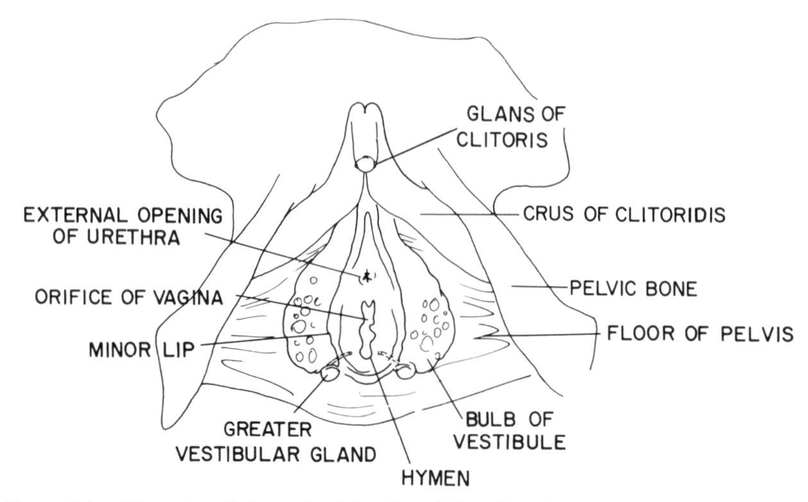

Figure 1.6. Dissection of the pudendal region. (After Cunningham (Romanes, 1972)).

nerve supply to the glans is exquisitely sensitive (see Chapter 12). The glans is nearly covered by a thin fold of skin (the *prepuce*) and is anchored by another fold beneath it, both folds being anterior extensions of the minor lips. Bilaterally beneath the surface, somewhat dorsally and posteriorly, is another pair of erectile structures called the *bulbs* of the vestibule. These are on each side of the midline around the external opening of the vagina and urethra, and extend anteriorly to reach the glans. Further details will be found in Chapter 12, which deals with the orgasm.

Vagina

The inner portion of the vagina is a thin-walled tube about 8-11 cm long that is ordinarily collapsed and flattened; it appears as a slit for most of its course, with a somewhat more open part where the *cervix* of the uterus dips into the vagina (Figures 1.2 and 1.7). The more open channel which surrounds the vaginal portion of the cervix is called the *fornix*. The dorsal portion is functionally significant, in that it is the basis of a "strategic" pool where semen may collect (depending on position) before being transferred to the cervix and uterus, as discussed in Chapter 12. The length and potential diameter when distended vary greatly from person to person, depending in part on the number of pregnancies. The vagina is somewhat arched, and accordingly its dorsal wall is slightly longer than the ventral wall. The wall of the vagina is moist and has many small folds which encircle it. This surface layer is surrounded by a muscular layer. Between these two layers is a layer of loose connective tissue that contains a very rich network of inter-

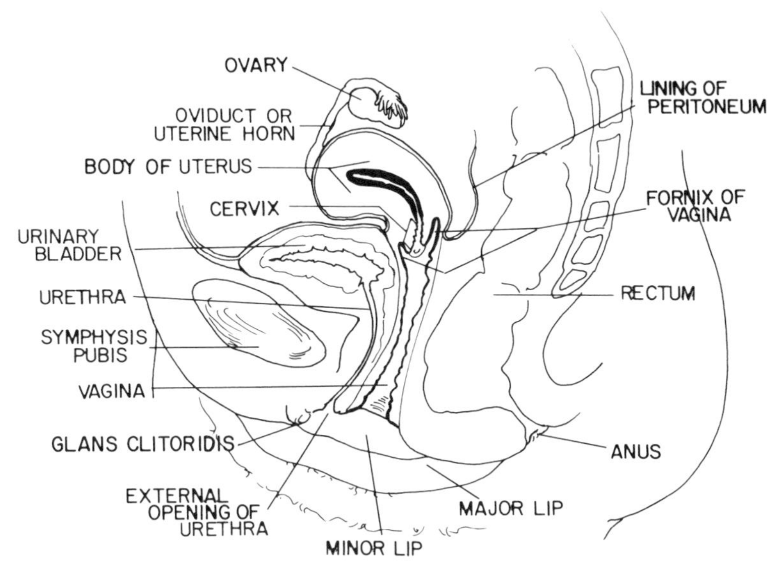

Figure 1.7. A mostly midsagittal section of the pelvic organs of a woman, with the oviduct and ovary and the major and minor lips brought into the same plane for didactic purposes only. (After Cunningham (Romanes, 1972)).

connected, branching, distensible veins. These are so numerous as to resemble erectile tissue. Numerous larger veins also form a close net on the outer side of the muscular layer (Figure 1.8). The uterus is even more vascular than the vagina. The veins communicate with the very extensive network in the floor of the pelvis. The tortuous form of the blood vessels, and the rich and extensive anastomoses (or interconnections) are reflections of the distensibility and cyclicity of the genital organs.

Uterus

The uterus extends anteriorly from the vagina, and makes a bend with it of about 100°, to lie in the pelvis between the urinary bladder and rectum (Figure 1.7). The uterus appears like an upside-down pear, the large part (about two-thirds) being the body of the uterus, and the narrow part the cervix. Both are characterized by their thick muscular walls and their narrow internal cavities, which, at their largest, in the nonpregnant state, are no larger than a pencil. The opening of the cervix into the vagina is flattened and slitlike (Figure 1.9). The inner lining of the cervix is characterized by a very unusual sequence of folds that slope towards the vaginal opening (Figure 1.10). The cervix is about 2 cm long and tapers slightly where it is attached to the vagina. The

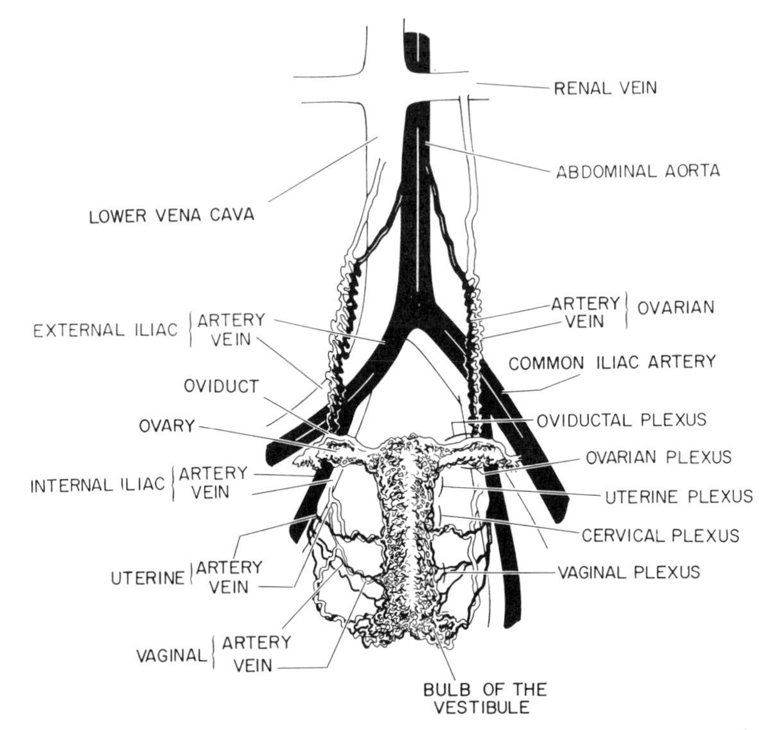

Figure 1.8. The genital tract is characterized by an enormously rich vascularity, here represented by the rich, anastamotic venous network on the outside of the vagina, uterus, oviduct, and ovary. (After Sherfey, 1972).

body of the uterus is about 5 cm long and about 4 to 5 cm at its widest, being flattened dorsoventrally. The wall is about 1 cm thick. All of these dimensions vary from person to person, and all of these are always larger after pregnancy. More details are given in Chapter 11.

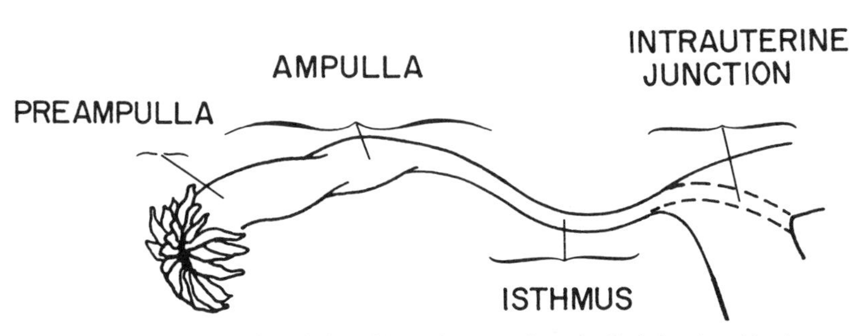

Figure 1.9. Outline drawing of the oviduct of women from its fimbriated end in the preampulla to the junction with the uterus and its passage through the uterine wall. (After Hafez and Blandau, 1969).

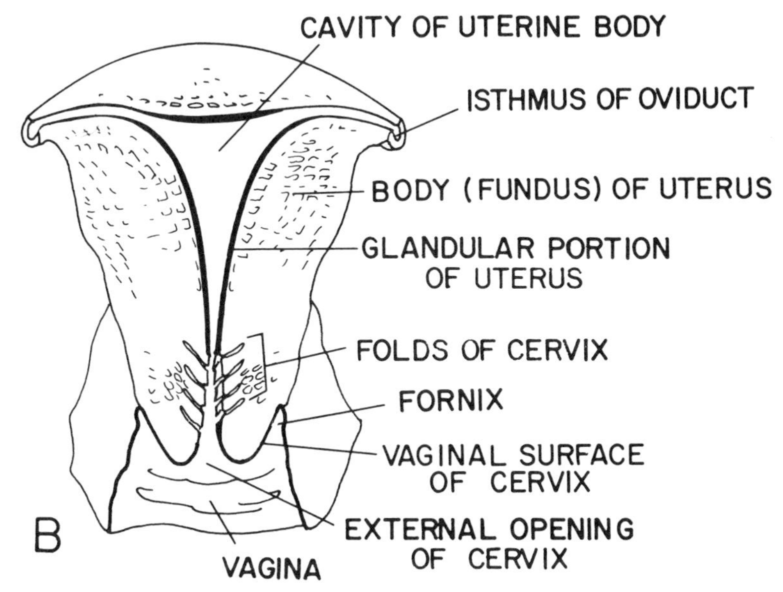

Figure 1.10. A more detailed view of a plane of sectioning of uterus and vagina. (After Cunningham (Romanes, 1972)).

Oviducts

At about the maximum diameter of the dome-shaped top of the uterus, it extends laterally into the paired oviducts (or fallopian tubes) (Figures 1.2, 1.9, 1.11). These are about 1 cm in diameter, and about 10 cm long, and extend laterally to cover partially the ovaries. Near their origin, in the region called the *isthmus*, the internal diameter of the uterine tube is less than 5 mm. Toward the other end, the *ampulla* of the uterine horn is somewhat wider. Terminally, it expands to form the fimbriated portion, where the surface is furrowed, corrugated, branched, and fringed. The inner canal terminates in this expanded fashion by opening directly into the body (peritoneal) cavity or space. There is thus a small gap between the uterine tube and the ovary, though there seems to be some contact between the ovary and one or more terminal fringes of the oviduct.

Ovaries

The ovaries are paired, one on each side of the pelvis (Figure 1.11). They are 3 cm or more long, about 2 cm wide and about 1 cm thick, and are shaped somewhat like an almond. Their surface markings and colors change during the menstrual period. The ovaries are mostly dense, except for the egg follicles, especially the ripening ones that appear monthly in fertile women. The blood vessels are very prominent.

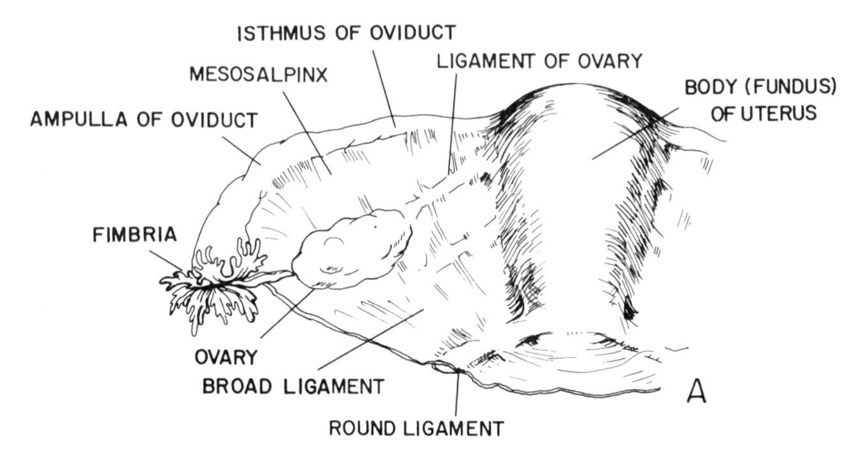

Figure 1.11. Ligaments covered with peritoneum (except for the fimbria) hold the various parts of the genital tract in their proper relations, and prevent kinks and other improper positions. (After Cunningham (Romanes, 1972)).

Periodically the ovaries of adult women produce ripe egg follicles that release their eggs into the oviducts. Each ruptured egg follicle is replaced by a *corpus luteum* (from Latin, meaning "yellow body"). The cells of these two structures, perhaps also with the stromal cells between them in the remainder of the ovary, synthesize and secrete three types of steroid hormones: estrogens, progesterones, and androgens (discussed in Chapters 7 and 11). The release of the hormones is controlled by feedback mechanisms that involve the anterior lobe of the hypophysis and the hypothalamus (discussed in Chapter 7).

Ligaments

The various parts of the female genital tract, from the cells on up to the organs, are held together in functional relations by connective tissue. This also serves to hold the various parts of the genital tract in proper relationship with the urinary tract, the rectum, and all other structures in the pelvis including blood vessels and nerves. A delicate sheet of connective tissue, the *peritoneum*, extends over the bladder and rectum and encloses the uterus between them. Some of the connective tissue strands and bands, which are essential for maintaining relations of structures in the pelvis, are very dense and tough, and are clear anatomic or dissectable entities. These are known as *ligaments* (Figure 1.11). These include the following:

1. The broad ligaments of the uterus (one on each side) that attach the uterus to the pelvis and also maintain the uterine tubes in their proper relations.

2. The round ligament (one on each side) is a special band of the broad ligament that passes to the back of the pelvis from the region above the origin of the uterine tubes. It circles around the inside of the pelvis to pass through the inguinal canal into the deep layer of the skin and ends by attaching to the major lip.

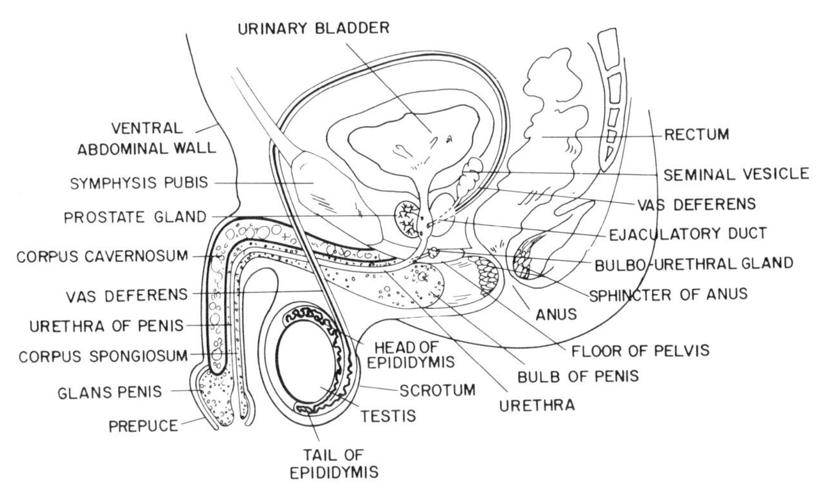

Figure 1.12. A nearly midsagittal section of the male pelvis, with some parts brought into the plane of section (vas deferens to its emptying point in the urethra, testis, epididymis, and seminal vesicle) and some parts sectioned just lateral to the midsagittal plane and placed as if in that plane of section (corpus cavernosum). (After Masters and Johnson, 1966).

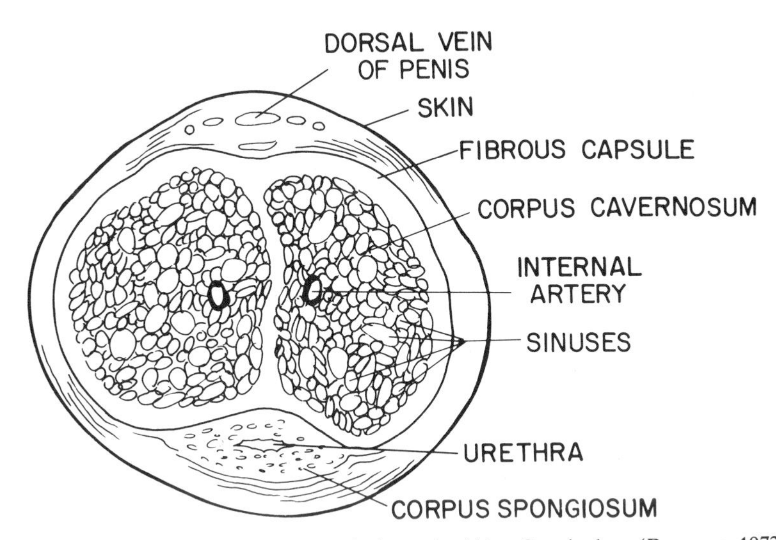

Figure 1.13. Transverse section through the penis. (After Cunningham (Romanes, 1972)).

3. The ovarian ligament (one on each side) is a special thickening of the broad ligament that helps to maintain the ovary in its correct relations.
4. The infundibulopelvic ligament helps to maintain the relations between the uterine tube and the major blood vessels of the region.

Plasticity of the Female Genital Tract

The formal anatomy of the female genital tract gives little idea of the variability and plasticity built into it. For example, the ovary is sufficiently adaptable and responsive to a whole sequence of hormones to permit one or more eggs to grow, mature, and be released every month, without any permanent scar such as would form in any other part of

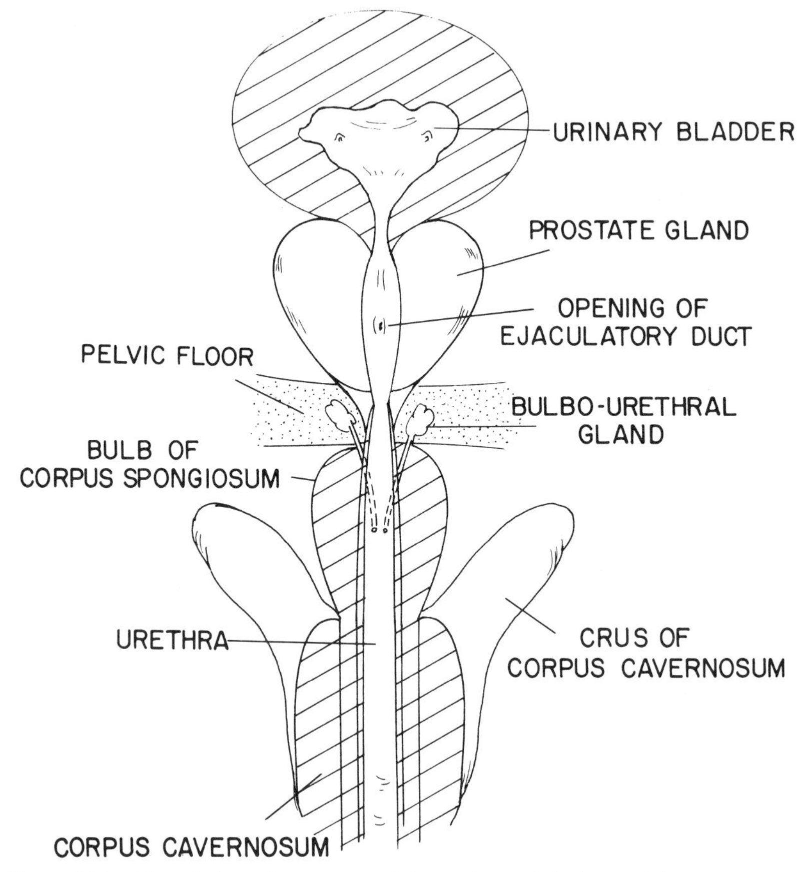

Figure 1.14. The relations of some parts of the urethra and the bases of the corpora of the penis. The preparation is of the various organs removed in one piece, laid out flat, and sliced lengthwise. (After Cunningham (Romanes, 1972)).

the body subject to monthly rupture of, say, a sterile abcess of 1.5 cm in diameter. The development and growth of a corpus luteum every month (see Chapter 11) and its subsequent regression are equally remarkable. The uterus similarly undergoes precisely controlled cell losses

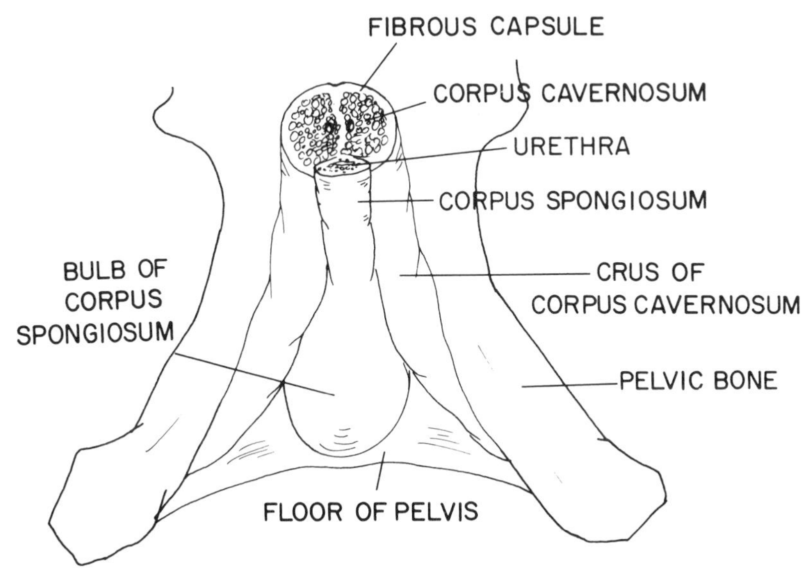

Figure 1.15. Dissection of the base of the penis. (After Cunningham (Romanes, 1972)).

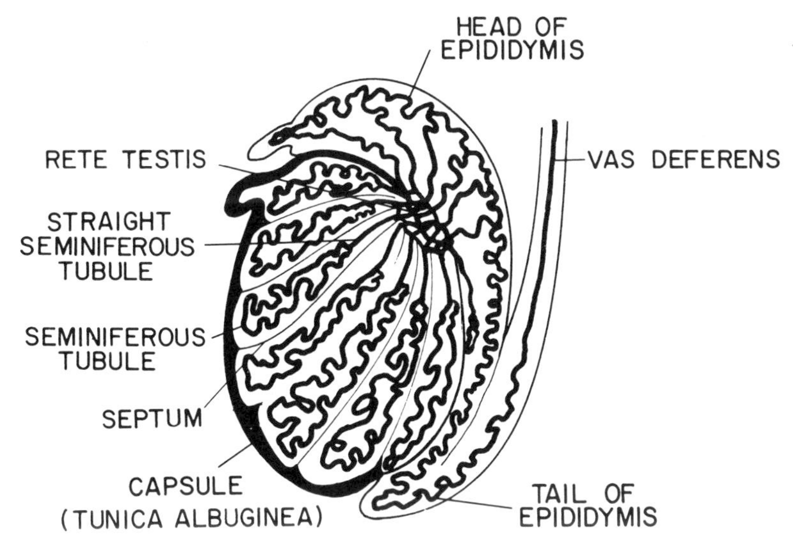

Figure 1.16. The relations of the testis and epididymis. (After Bloom and Fawcett, 1975).

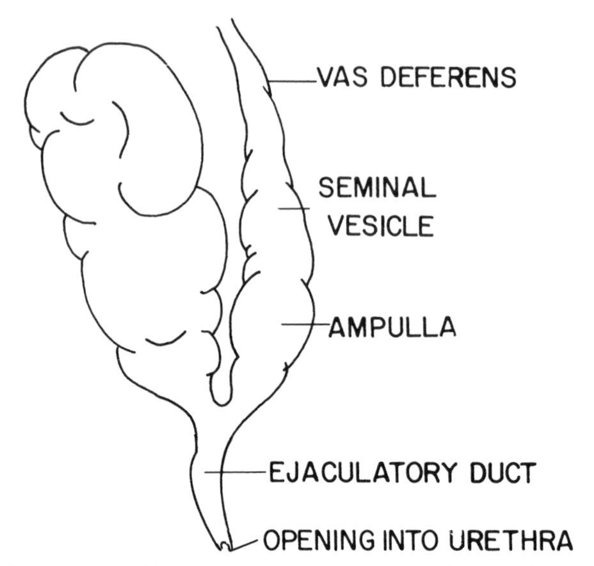

Figure 1.17. The seminal vesicle and enlarged termination of the vas deferens (the ampulla). As viewed from the outside, the duct of the seminal vesicle empties together with the vas deferens to form the short ejaculatory duct that empties on the utricle into the prostatic portion of the urethra. (After Cunningham (Romanes, 1972)).

that in other parts of the body could affect the health of the woman disastrously; yet these are replaced with great facility. And there is no change in any organ of the body as major and dramatic as that which takes place in the uterus during and immediately after pregnancy. The clitoris and vagina continue even into old age to be capable of going through the movements and vascular changes associated with the orgasm. The cells lining the internal surface of the vagina go through a series of cyclical changes related to the hormonal environment. The reader is advised to consider the formal anatomic description of the female genital organs as, in a manner of speaking, a baseline that is sufficiently variable to encompass maturation of the genital system, ovulation and menstruation, orgasm, pregnancy and parturition, and atrophy in aging women.

MALE GENITAL ORGANS

These extend in three dimensions in a more complicated manner, since the sperm arising externally in the scrotum have to be conducted through the peritoneal cavity by a series of complicated passages and then to the outside via the penis. Thus the genital system is intimately integrated with the urinary system, since the sperm utilize a part of the urinary system, the urethra, to reach the outside. And finally, the male system is characterized by glands whose secretions comprise the major part of the semen and serve to protect and eject the stored sperm. The various parts are described by means of drawings only (Figures 1.12–1.17). These serve to identify the parts mentioned in later chapters. The drawings also serve as a basis of comparison of female and male equivalent structures summarized in Table 1.2.

REFERENCES

Anderson, J. E. 1978. Grant's Atlas of Anatomy. 7th Ed. Williams & Wilkins Co., Baltimore.

Basmajian, J. V. 1975. Grant's Method of Anatomy. 9th Ed. Williams & Wilkins Co., Baltimore.

Bloom, W., and D. W. Fawcett. 1975. A Textbook of Histology. 10th Ed. W. B. Saunders Co., Philadelphia.

Clemente, C. D. 1975. Anatomy: A Regional Atlas of the Human Body. Lea & Febiger, Philadelphia.

Goss, C. M. (ed.). 1966. Gray's Anatomy of the Human Body. 28th Ed. Lea & Febiger, Philadelphia.

Hafez, E. S. E., and R. J. Blandau (eds.). 1969. The Mammalian Oviduct. University of Chicago Press, Chicago.

Masters, W. H., and V. E. Johnson. 1966. Human Sexual Response. Little, Brown & Co., Boston.

Netter, F. H. 1965. The CIBA Collection of Medical Illustrations. Vol. 2. Reproductive System. Ciba, Summit, N.J.

Romanes, G. J. 1972. Cunningham's Textbook of Anatomy. 11th Ed. Oxford University Press, London.

Sherfey, M. J. 1972. The Nature and Evolution of Female Sexuality. Random House, New York.

Sobotta, J. 1963. Atlas of Human Anatomy. F. H. Figge (ed.). 9th English Ed. Vol. 2. Urban & Schwarzenberg, Baltimore.

Stedman's Medical Dictionary. 1961. 20th Ed. Williams & Wilkins Co., Baltimore.

Woodburne, R. T. 1978. Essentials of Human Anatomy. 6th Ed. Oxford University Press, New York.

Zuckerman, S. 1961. A New System of Anatomy. Oxford University Press, London.

2 Development of Female and Male Sex Organs

The main aim of this chapter is to give the biologic basis for the high degree of variation in the genital structures of women. To begin with, females differ from males from the time of fertilization in their sex chromosomes (the X and Y chromosomes), the cells of females normally having two X chromosomes and 44 others (the autosomes), while cells of males normally have one X and a Y plus 44 autosomes. The genetic expression in the adult woman is not necessarily all or none. There is a basic range of variation that is considered normal: the clitoris (or vagina, or uterus, etc.) can be small, large, or in-between. Superimposed on this may be unusual genetic or environmental effects, due to certain hormones that can be synthesized in the embryo or fetus in abnormal amounts or which may be given to the mother as drugs (e.g., to prevent spontaneous abortions). These substances can be metabolized to form androgenic substances and may affect the normal sequence of maturational stages of the genital tract early in embryonic life, interfering with its step-by-step development. The resulting variations can be minor or extreme, but supplement and in some cases overshadow the expected normal degree of variation.

DEVELOPMENT OF EMBRYONIC GENITAL SYSTEMS

For reasons that are discussed later, it is necessary to outline the development of the genital system of embryos of both sexes to appreciate at least some of the changes that occur in female embryos.

At 4 weeks after fertilization, the sex organs begin to develop. A month later, all the beginnings are in place, and thereafter development of the potential organs takes place. The actual size of the embryo at 4 weeks is shown by the marker in Figure 2.1A, its actual appearance when magnified is shown beside the marker. At 2 months, the embryo

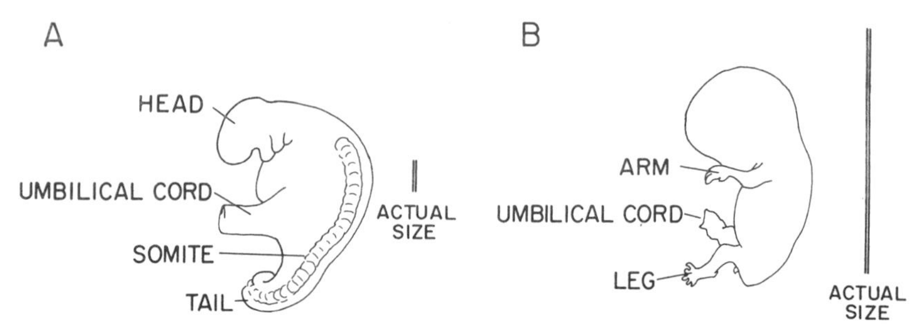

Figure 2.1. Left side of human embryos, enlarged. Actual length is indicated by the vertical bar to the right of the drawing. A) Embryo of 25 somites or body segments, about 5 mm long, approximately 28 days old. B) Embryo about 8 weeks old, crown-rump (CR) length about 30 mm. (After Langman, 1975).

has grown considerably in size, and its appearance is different (Figure 2.1B).

At first, there is no ovary or testis (Figure 2.2). There is only a common genital mass along what will later be the back. This is the same in male embryos. The sex cells wander into this genital mass from their point of origin in the yolk sac (a total distance of less than 1 mm), and colonize the genital mass (Figure 2.3). They multiply furiously, and in the end give rise to millions and billions of potential egg cells and sperm cells, respectively.

Two sets of tubes are laid down in the common genital mass (Fig-

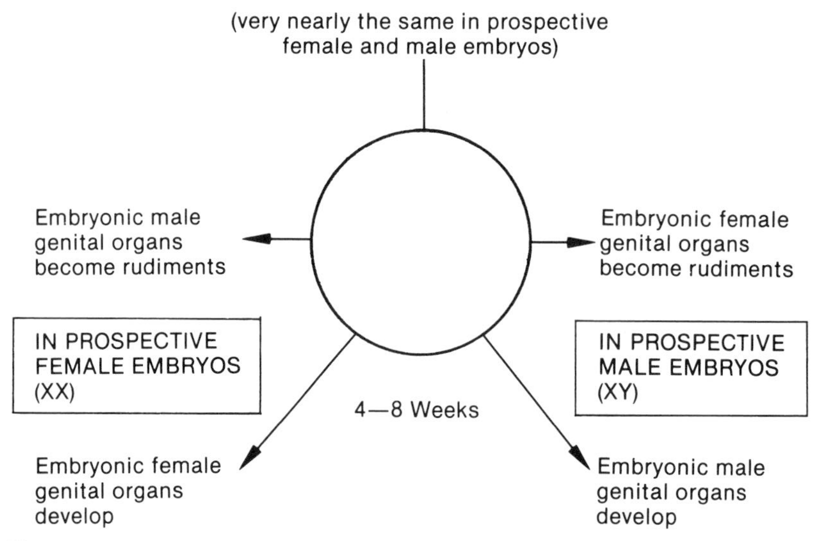

Figure 2.2. General diagram of the common genital mass in the very early embryo and how it develops in female (XX) and male (XY) embryos. (After Jost, 1972.)

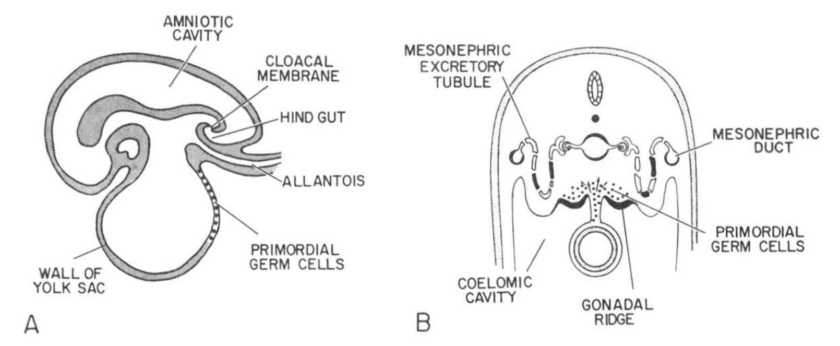

Figure 2.3. Schematic drawings of the relations of the urogenital mass in human embryo of about 4 weeks (about 5 mm long), before even the first germ cells appear in the genital regions. (After Langman, 1975). A) Longitudinal section of an embryo even earlier than that shown in Figure 1.2A. The primordial germ cells originate in the wall of the yolk sac, and migrate dorsally to reach the wall of the hind gut. B) Transverse section of an embryo about 5 weeks old.

ure 2.2), known as the *Müllerian ducts* and *Wolffian ducts*, after the names of the distinguished embryologists who discovered them about a century ago. In genetic females (44 autosomes + XX), the Wolffian duct and structures related to it degenerate, while the Müllerian ducts develop into the oviduct, uterus, cervix, and inner part of the vagina. In genetic males (44 autosomes + XY), the Müllerian duct and structures degenerate, while the Wolffian duct develops into the epididymis, the vas deferens, the seminal vesicle, the ejaculatory duct, and the prostate gland (see later, Figures 2.6B, and 2.7A). External swellings in the genetic female give rise first to an anal fold, and then to the cloaca, the bulbourethral glands, and the major and minor lips. External swellings in the genetic male (indistinguishable from those of the genetic female except cytologically) give rise to an anal fold, then to the penis, scrotum, and the terminal portion of the urethra (Figures 2.4 and 2.5).

Some of the major similarities (as well as major differences) in the origin and development of the various parts of the male and female genital systems are shown in Table 2.1. The structures, which are stated above to degenerate, do not disappear altogether. Small bits remain, and for reasons which are not known, they develop in a small percentage of adults as abnormal growths that may have to be removed surgically.

In the genetic females, the genital ridge becomes circumscribed as the ovary, and it is in the outer part of it (the cortex) that the sex cells develop (Figure 2.6). In the genetic male, the cortex is replaced by dense connective tissue and the sex cells are lodged in the seminiferous tubules that form at about this time (Figure 2.7). Completion of some of the main stages in the development of the female genital tract is shown in Figure 2.6; descent of the testis into the scrotum is illustrated in Figures 2.7C and 2.8.

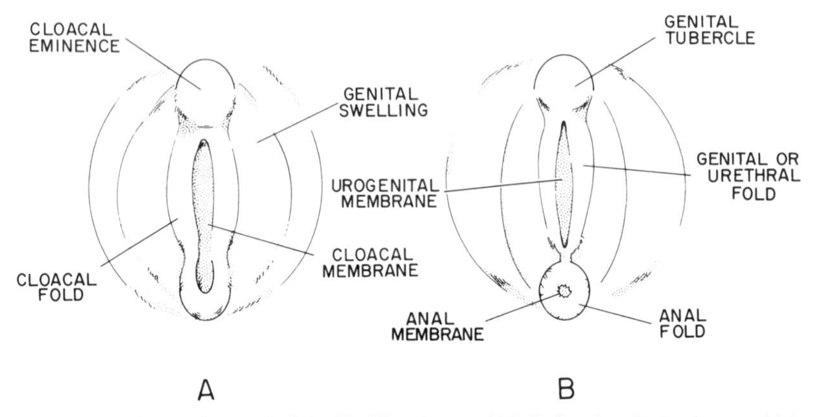

Figure 2.4. At the distal end of the Wolffian duct and Müllerian duct is the cloaca, which also is continuous with the hind gut. A) Viewed from the surface of the 4-week-old embryo, the cloaca is closed by a cloacal membrane. The proliferation of underlying tissue results in the raising of the cloacal folds and eminence, themselves superimposed on a moderately raised portion, the genital swelling. B) The hind gut is separated by a wall of tissue from the urogenital sinus as the external genital tissues enlarge (After Langman, 1975).

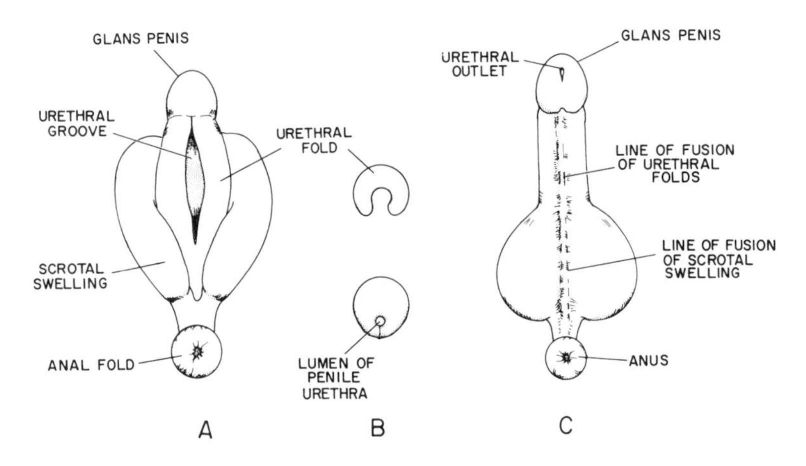

Figure 2.5. Stages in the development of the external genitalia in the male human embryo and fetus. A,B) At about 10 weeks the tissues in the genital tubercle of Fig. 2.4B enlarge to form a plate that curls over on itself until the edges meet and fuse, thereby forming the penile urethra. The line of fusion of the urethral folds remains permanently, and is continuous with the line formed by the fused scrotal foldings derived from the genital folds of earlier stages. (After Langman, 1975).

FACTORS INFLUENCING VARIATION IN SIZE AND FUNCTIONAL STATE OF GENITAL SYSTEMS

A very important aspect of the development of these organs is the timing. If the timing is off, and some organ stops development or slows down,

Table 2.1. Major embryological equivalents of the female and male genital systems

Site of origin	Female	Male
Genital ridge	Ovary	Testes
Wolffian duct		Epididymis Vas deferens Ejaculatory duct Seminal vesicle
Müllerian duct	Uterus, Uterine tubes, Part of vagina	
Urogenital sinus	Part of vagina, Most of urethra, Vestibule of vagina, Greater and lesser vestibular glands	Prostate gland, Prostatic urethra, Part of penile urethra, Penile urethral glands, Bulbo-urethral glands
Genital swelling	Major lips Glans of clitoris Vestibular bulbs Corpus cavernosum	Scrotum Glans of penis Corpus spongiosum Corpus cavernosum
Urethral folds	Minor lips	Floor of urethra of penis

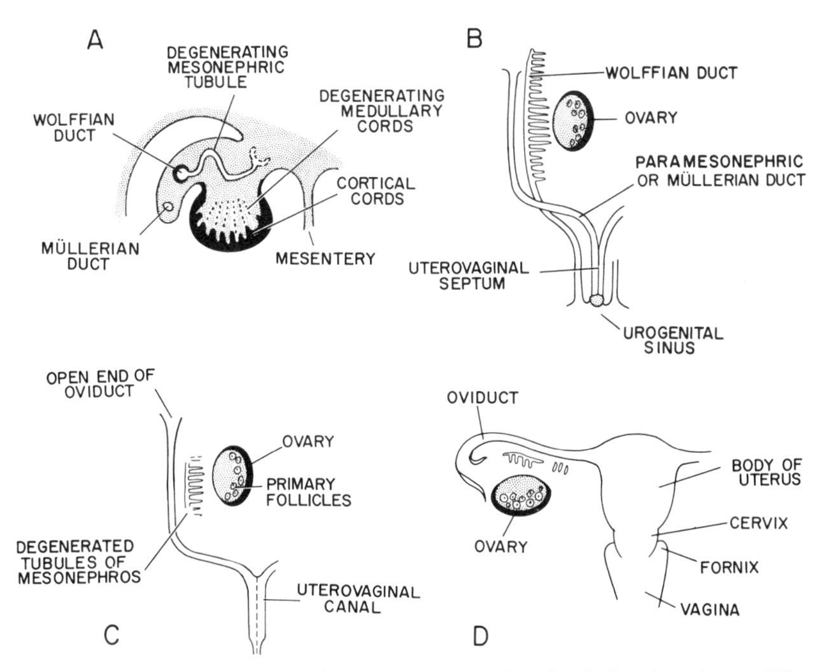

Figure 2.6. Formation of the genital tract, uterus and vagina in female embryos. (After Langman, 1975).

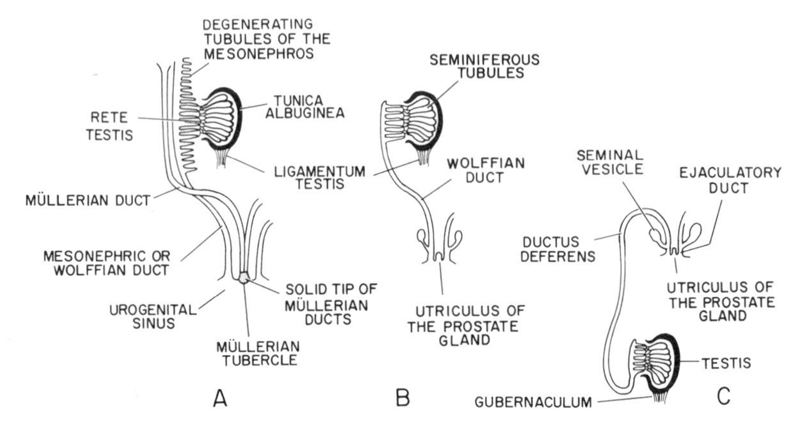

Figure 2.7. Stages in the development of the genital system in the male human embryo and fetus. (After Langman, 1975).

it may not develop fully at all, or may develop abnormally. Another important factor is that all the organs at all stages of development are affected by hormones or substances similar to hormones circulating in the mother's blood stream, and by hormones produced by the embryonic testis, adrenal cortex, and perhaps by other hormone-secreting glands. These are the bases for several kinds of intersex states (rare events), and for variations in size and functional capacity of both external and internal sex organs. These causes of variation supplement the genetic factors, and together they determine the range of variation of the sexual organs.

EFFECTS OF HORMONES ON DEVELOPMENT OF THE GENITAL SYSTEMS IN FEMALES AND MALES

On the whole, the male genital system develops earlier than the female genital system of embryos and fetuses. The earlier development of the male genital system can be attributed to androgenic substances. These stimulate the development of certain prospective genital organs (certain testicular structures and derivatives of the Wolffian duct) in the male embryos, while inhibiting the formation of other prospective organs (ovarian tissues and derivatives of the Müllerian duct). In general, the critical time for the full development of the genital tract in human embryos is when they are 1.25–2.5 cm (1/2 to 1 inch) long, or approximately 50 to 60 days after fertilization. At this time the androgenic substances seem to be synthesized by the testis, especially the steroid hormone testosterone.

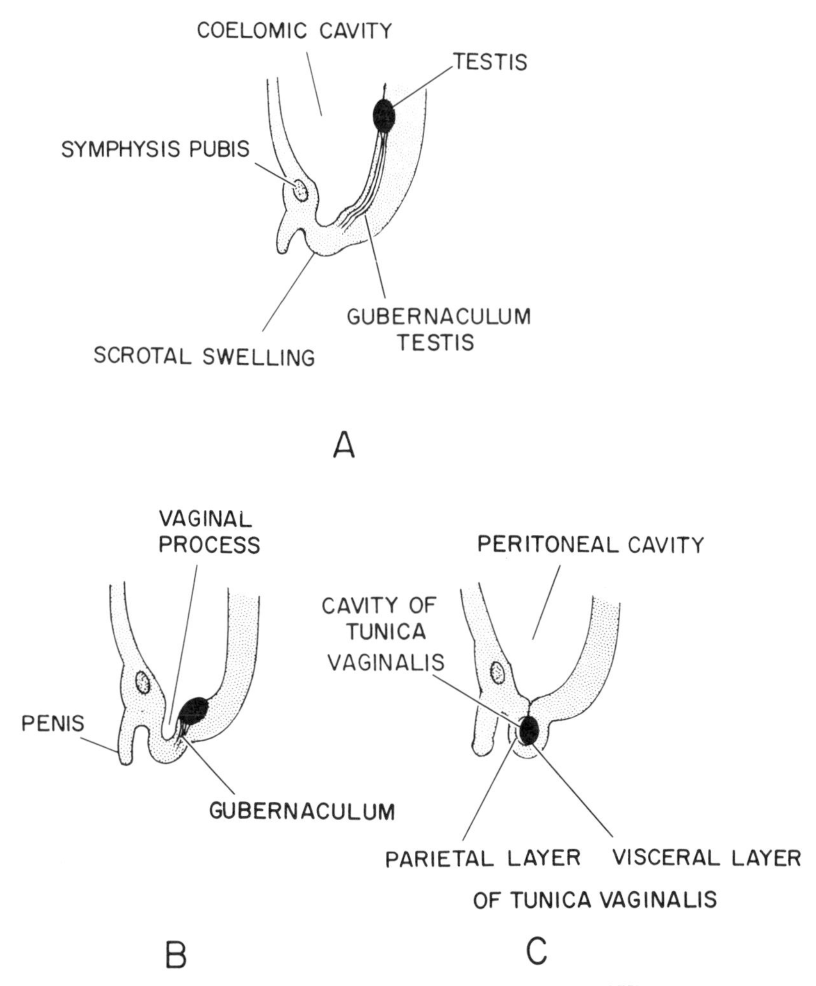

Figure 2.8. Stages in the descent of the testis. (After Langman, 1975).

Numerous drugs are or were administered to avoid spontaneous abortion and to prevent miscarriage. Some of these were found to be metabolized by the body to androgenic substances. It is believed that these substances affected the genital system in genetically female embryos and caused the more extensive development of the genital swelling and possibly of some of the derivatives of the Wolffian duct system. Sometimes the same abnormalities may develop when excessive amounts of androgens are secreted during sensitive periods, notably by the maternal or fetal adrenal cortex.

SOME INTERSEX STATES

Such variations are uncommon (about one in one thousand) and are presented here very briefly.

Genetic Females

Abnormal Development of Gonads The ovaries are vestigial and have no ova in the adult. The uterus and oviducts are small. There is no testis, and no major derivatives of the Wolffian duct persist in the adult. In most females, the ovary is genetically programmed so that it matures slowly, and begins to ovulate at puberty. In the individuals described here, it seems that the ovary was not programmed genetically for persistence, and the remainder of the genital system matured, but only slowly and incompletely. This is part of Turner's syndrome.

Adrenogenital Syndrome The clitoris is sometimes enlarged, and the labial folds fused, resulting in abnormalities of the vagina when this syndrome is present. The ovaries of the adult appear immature. In general, derivatives of the Müllerian duct are present, though underdeveloped, while derivatives of the Wolffian duct are atrophied. In such individuals, the inner part of the adrenal cortex is markedly enlarged. It is thought that the cells of this part of the cortex secrete excessive amounts of androgenic substances, and these cause masculinization of the external genitalia. The excessive secretion of androgens perhaps takes place after the sensitive or critical period for the development of the Wolffian duct, for its derivatives are not enlarged.

Maternal Ingestion of Androgenic Hormones or Other Drugs If such substances are administered to the mother early in pregnancy, they pass through the placenta and act on the embryonic genital tissues. The derivatives of the Wolffian duct may be fully developed from the epididymis to the external genitalia. The ovaries may be normal or almost normal, and the derivatives of the Müllerian duct may be normal, though the vagina may be abnormal. The age of the embryo when the drug is administered, the duration of drug treatment, and the dosage of the drug are important factors that influence the extent of abnormality.

Genetic Males

Male Hermaphroditism (Type 1) The testes are abnormal, producing few or no sperm except when they have descended into the scrotum. A vagina is usually present, but a uterus may be absent. The clitoris may be enlarged, and the major lips may be fused. There may be a testis on one side, and an ovary on the other. The speculation is based on the theory that androgens have two hormonal actions: 1) stimulation of the Wolffian duct and its derivatives, and 2) inhibition of the Müllerian duct and its derivatives. The second action in these individuals seems to be defective, and allows the Müllerian duct derivatives to develop to varying degrees, depending on hormonal levels, sensitivity of receptor sites, etc.

Male Hermaphroditism (Type 2) In this syndrome, sometimes called *androgen insensitivity* or *testicular feminization*, the testes appear normal, though undescended, and contain no sperm. The external genitalia appear female, but the vagina may be shallow, or end as a blind pouch. The uterus and tubes are rudimentary. It is thought that the androgenic substance(s) secreted by the developing testis is normal in amount, and that the second action (no derivatives of the Müllerian duct) is not fully active. The androgenic action (though sufficient in amount) is ineffective in these embryos because the affected organs are insensitive to the androgen. In other words, the cell surface lacks androgen receptors (see p. 173) and is "unaware" that the stimulus (androgen) is present in the circulating blood.

VARIABILITY

The normal and abnormal development of the genital system has been discussed at sufficient length to emphasize at the earliest possible time the interplay between purely genetic factors and environmental factors. It is not possible to quantitate these effects—not now, and perhaps never. But it is well to be aware of the great variability in the genital tracts of women and men, including those extremes in both sexes that are described above.

A MYTH IN THE MAKING?

Sherfey (1972) refers first to the story of Adam and Eve and then to Freud's view of female sexuality, according to which the clitoris is a rudimentary penis and therefore the woman's true sexuality is centered not in the clitoris (which is rudimentary) but in the vagina (which is not). According to Freud's theory, the female starts out like the male but has to develop her own specific form of sexuality, just as Eve was reshaped from a rib of Adam.

During the first few weeks of a human embryo's life, there is no way of telling whether it will develop as a female or a male, except by its chromosomes. What determines development as a male rather than a female is the production of androgens, which is certainly determined by the chromosomes.

Sherfey uses these facts to turn the Adam and Eve story upside down. She proposes that because without androgens males develop as females, we all start as females and the male is a derivative form.

But is this a proper way to look at the development of the reproductive system? In biology, we have to think in terms of process, interactions, and change, and to consider all the factors involved. In the course of female development, certain genes are activated and produce their products: these products react with other substances in the cells, resulting in change and differentiation in the female direction. In the course of male development, a different series of changes take place. Androgens are produced by the embryo in amounts sufficient to cause one set of ducts (Wolffian) to develop and the other (Müllerian) to atrophy. And the stage is set for the subsequent development of maleness. In the earlier stages, up to the fourth week, the embryo with its two sets of ducts is already potentially a female or potentially a male, although it is not yet fully determined. We cannot say that these embryos are the same, or that they are both essentially female. The embryos are already different because in the one case, the egg has been fertilized by an X-bearing sperm and in the other by a Y-bearing sperm. This is a dialectical way of thinking, as opposed to Sherfey's method of comparing human embryos at a particular time when, isolated from their antecedents, female and male seem to be alike.

Sherfey is setting up a new myth in place of the male-oriented myth of Adam and Eve. Davis (1972) goes even further when, in *The First Sex* she speculates that at some point early in time there were only females, who reproduced parthenogenetically, i.e., without fertilization (see Chapter 5), and that somehow males were derived from females later in the evolution of the human species.

REFERENCES

Arcy, L. B. 1954. Developmental Anatomy. 6th Ed. W. B. Saunders Co., Philadelphia.

Corliss, C. E. 1976. Patten's Human Embryology. McGraw Hill Book Co., Philadelphia.

Davis, E. G. 1972. The First Sex. Penguin Books, Inc., Baltimore.

Flanagan, G. 1962. The First Nine Months of Life. Simon & Schuster, New York.

Hamilton, W. J., J. D. Boyd, and H. W. Mossman. 1972. Human Embryology. Williams & Wilkins Co., Baltimore.

Jirasek, J. E. 1971. Development of the Genital System and Male Pseudohermaphroditism. Johns Hopkins University Press, Baltimore.

Jones, H. W., Jr., and W. W. Scott. 1971. Hermaphroditism, Genital Anomalies and Related Endocrine Disturbances. Williams & Wilkins Co., Baltimore.

Jost, A. 1972. A new look at the mechanism controlling sex differentiation in mammals. Johns Hopkins Med. J. 130:38–53.

Langman, J. 1975. Medical Embryology. 3rd Ed. Williams & Wilkins Co., Baltimore.

Money, J., and A. A. Ehrhardt 1972. Man and Woman, Boy and Girl. Johns Hopkins University Press, Baltimore.

Patten, B. M. 1968. Human Embryology. 3rd Ed. McGraw Hill Book Co., New York.
Patten, B. M., and B. M. Carlson. 1974. Foundations of Embryology. McGraw Hill Book Co., New York.
Sherfey, M. J. 1972. The Nature and Evolution of Female Sexuality. Random House, New York.

3 Genes, Chromosomes, and Mutations

GENES AND CHROMOSOMES

Our bodies are made up of cells and intercellular substances, and every cell has a nucleus at some stage of its life. In that nucleus are two sets of 23 *chromosomes*, one set inherited from our mother and the other from our father. Each chromosome of the maternal set has a counterpart in the paternal set. So our 46 chromosomes comprise 23 pairs. In these chromosomes are the *genes*, which are arranged linearly. Genes determine, on the one hand, the general nature of the human species and our resemblances to parents and other relatives, and on the other, some of the individual peculiarities that distinguish us from each other. Each chromosome set normally has a complete set of genes, or *genome*. The role of the genome is to program important functions of the body—growth and development of the embryo and fetus, growth and maturation of the infant to the adult, and ultimately, maintenance of all activities of the normal body.

Distribution of Genes to New Cells

All the cells of the body, as different as they are, have their origin from a single fertilized egg cell. The differences are due, in large part, to the action of different groups of genes that determine the development and activity of the different types of cells—such as those of the skin, brain, and liver. Therefore, the right group of genes must be present in a given type of cell. This is ensured by a process that delivers two complete sets of chromosomes and genes to every new cell, followed by a process of selective activation of genes, which determines in large part the cell's subsequent differentiation and function.

Mitosis

The first process, by which chromosomes and genes are distributed regularly to new cells, is called *mitosis*. A mitotic division of the nucleus occurs each time a cell divides. Cell division makes growth and development possible. Cell division also provides for replacement of cells that die or are worn off, like the cells of the skin. Cells that divide are immature. Recent work suggests that at each division cells become restricted in their potentialities. As cells mature, they stop dividing and become differentiated from other cells, or specialized for a particular function.

THE CELL CYCLE

Cells that divide go through a cell cycle, or mitotic cycle (Figure 3.1) that may take minutes, or hours, or even days to complete. Although changes may occur continuously and one phase merges with another, the cycle has been divided into five stages: interphase, prophase, metaphase, anaphase, and telophase (Figure 3.2).

Interphase is divided into substages, and during interphase a new set of genes and chromosomes is formed from each old set. Very little is known about what triggers this process, or about what causes cells to divide. A description of mitosis can be found in one of the genetics texts listed at the end of this chapter.

CHEMICAL NATURE OF GENES AND CHROMOSOMES

During the 1940s evidence began to accumulate that genes are made of nucleic acid, and in 1953, Watson and Crick proposed a model for the

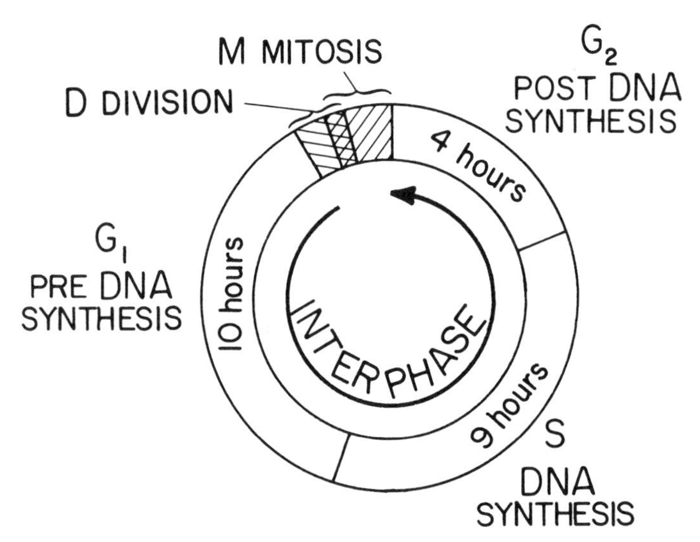

Figure 3.1. The mitotic cycle. The times are characteristic of cultured mammalian cells. The interphase, especially G_1, is usually longer in vivo. (After Dyson, 1974)

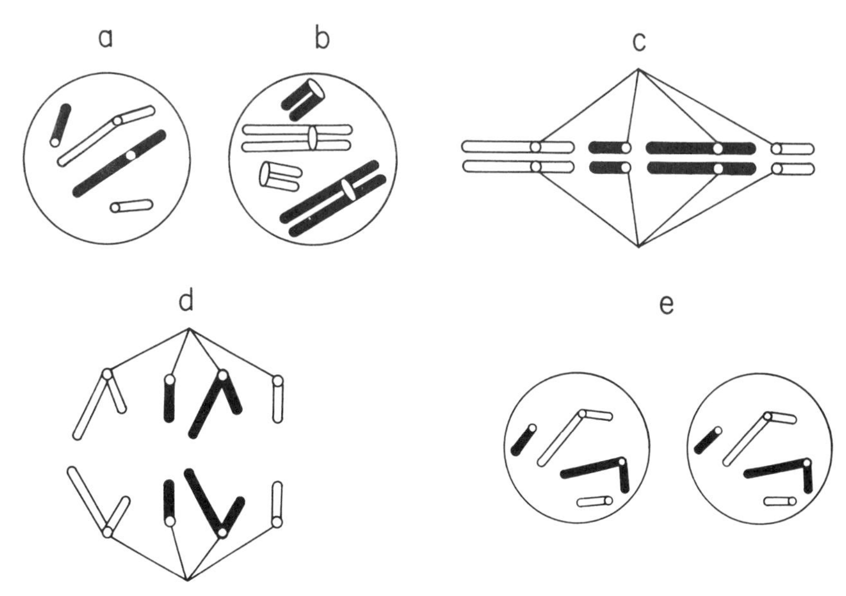

Figure 3.2. Mitosis. a and b) early and late prophase; c) metaphase; d) anaphase; and e) telophase. Only two pairs of chromosomes, a long pair and a short pair, are shown. Centromeres are indicated by open circles. (After Srb, Owen, and Edgar, 1965)

structure of the deoxyribonucleic acid (DNA) molecule. Human genes, and those of all animals and plants, are composed of DNA.

The DNA molecule consists of two long strands wound around each other to form a double helix (Figure 3.3). Each strand is a polymer, that is, it is made up of many units (called *nucleotides*). A nucleotide is made from a nitrogenous base, a sugar, and phosphoric acid. There are four different kinds of base in DNA: adenine (A), guanine (G), thymine (T) and cytosine (C). Each base in a strand is linked to a base in the opposite strand, but the pairs of linked bases are of only two kinds—adenine paired with thymine (AT) and guanine paired with cytosine (GC).

The order in which the bases follow one another in a single DNA molecule is tremendously variable. There are over 1000 different possible base arrangements in a sequence only five nucleotides long. There would be 4^{1000} (four to the power 1000) possible sequences for a molecule 1000 nucleotides long, a reasonable size for a gene. It is the specific nucleotide sequence of a gene which gives it its characteristic properties.

Probably there is one continuous giant DNA molecule running the length of each chromosome. (At any rate, there is at least one.) This molecule comprises a sequence of genes, perhaps interspersed with spacer DNA (nongenic DNA whose function is unknown) and controlling DNA elements (whose function is to control the activity of other genes). A chromosome contains many genes, ranging from tens

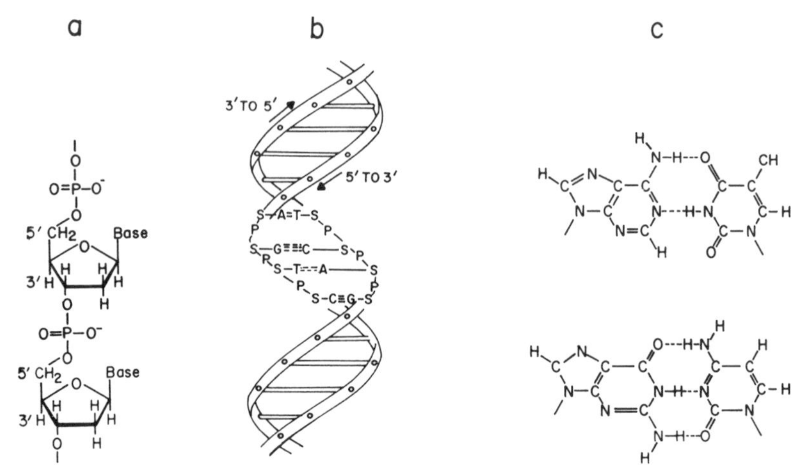

Figure 3.3. a) Two nucleotides linked together. The phosphate-sugar-phosphate linkages form the DNA backbone. b) Part of a double-stranded DNA molecule showing, in the center region, the phosphate-sugar backbones (S-P-S-P-), and the bases (A, T, C, G) extended towards one another inside the helix. c) Diagrams of the hydrogen-bonded base pairs, AT and GC. The purines (adenine and guanine), have two rings and are larger than the pyrimidines (thymine and cytosine) (After Dyson, 1974).

of genes in small chromosomes to hundreds, or perhaps even thousands, in the largest chromosomes. Besides DNA the chromosome contains other substances, notably proteins. The formation of two complete DNA molecules from one is called *DNA replication*. Each strand of the old DNA serves as a mold or template. Because of the limited pairing of bases (AT or GC), the order of bases in a template strand determines the order of bases in the new complementary strand being formed, so that the two strands in the new molecule are similar to those of the old. DNA replication takes place during interphase, at the DNA synthesis stages (Figure 3.1). This stage is usually relatively close to the beginning of the next mitosis, with a large gap (G_1) between the previous telophase and the S stage, and a smaller gap (G_2) between S and the next prophase.

PROTEIN PRODUCTION

The structure of the DNA molecule and the nature of the enzymes involved in DNA replication provide a basis for exact replication of the gene, with its unique sequence of bases. Mitosis provides a mechanism for distributing the copies of genes to new cells. Reproducing itself is the first function of the gene. The second function is to code for a specific molecule of RNA (ribonucleic acid), which in turn codes for the production of a polypeptide and of a protein.

Different kinds of RNA are involved in one way or another in the synthesis of proteins, which is carried on mainly in the cytoplasm. Along with nucleic acids, proteins are the most important chemical constituents of living cells. There are many different kinds of protein. Some of them are structural components of cells—for instance, the keratin of skin and its derivatives and the protein tubulin in the mitotic spindle. Other proteins, enzymes, facilitate chemical reactions that could not occur in the cell without them. Still others are antibodies protecting us against foreign substances. The specificity of a protein is determined, like that of DNA, by the unique sequence of building blocks of which it is made. The building blocks of DNA are the four nucleotides with their four different bases and the building blocks of proteins are twenty different amino acids. Amino acids can be linked together in a chain which is called a *polypeptide*. A polypeptide may consist of as few as three amino acids or as many as several hundred. A protein may consist of a single polypeptide, of two or more identical polypeptides, or of two or more different polypeptides.

How does a gene with its specific nucleotide sequence, located in the nucleus, code for a polypeptide with a specific amino acid sequence, which is synthesized in the cytoplasm? An intermediary, called *messenger RNA* (*mRNA*) is synthesized by each gene that is responsible for the synthesis of a polypeptide and the messenger RNA passes out from the nucleus to the cytoplasm. RNA synthesis is called *transcription*. A new RNA molecule is formed in essentially the same way that a new strand of DNA is formed, using a strand of the DNA as a template or mold. The mRNA cannot, however, serve directly as a template for polypeptide synthesis, for there is no affinity between nucleotides or nucleotide sequences and amino acids. The assembly of amino acids into polypeptides is a very complex process involving ribosomes and some other kinds of RNA besides mRNA. The most abundant, known as *ribosomal RNA* (*rRNA*), makes up 80% of the total RNA of cells. It is found in the ribosomes, submicroscopic bodies frequently located in the endoplasmic reticulum of the cytoplasm.

Another kind of RNA is represented by small molecules known as *transfer RNA* (*tRNA*), each of which has an affinity for a particular amino acid.

All kinds of RNA are produced in the nucleus except small amounts made in two kinds of cell organelles called *mitochondria* and *plastids*.

In the cytoplasm, a molecule of mRNA becomes attached to a ribosome. In turn, different tRNAs bring in their specific amino acids, each of which corresponds to a sequence of three bases (a codon) in the message.

When DNA is used as a template for RNA synthesis in the process called *transcription*, the building blocks of template and product are essentially similar. The "language" is the same. When polypeptides are formed, on the other hand, the amino acid building blocks do not make direct contact with the RNA template, but are assembled through the mediation of tRNA molecules. The amino acid sequence in the polypeptide is like a different language from the codon sequence of the mRNA. This process is called *translation*. A good description of transcription and translation is provided in Watson (1970).

MUTATIONS

If the proper amino acid sequence is changed by accidentally substituting one amino acid for another, the effect may or may not be serious, depending on whether active sites are altered. This kind of accident sometimes happens in the course of assembly of amino acids into a polypeptide on the ribosome. It can also occur if a change occurs in the base sequence of a DNA molecule. This can change the code so that every molecule of mRNA produced by the gene thereafter will be translated to a polypeptide with an altered amino acid sequence. Such a change in the gene itself is called a *mutation* and the changed gene is called a *mutant allele*. If a mutation occurs in a cell that subsequently divides, the gene will be replicated in its mutant form and transmitted to the daughter cells. If the cell lineage gives rise to germ cells and gametes, the mutant gene will be heritable.

Genotype and Phenotype

Without the occurrence of mutations, we would never have known that genes exist. If genes were invariable, all individuals would be genetically alike. Different hair colors, eye colors and blood types are caused by mutant genes. Thus, although the initial mutational change is in the nucleotide sequence of DNA and would be hard to detect and identify, it can result in mutant forms of mRNA and polypeptides, and consequently in changed function of a protein. This in turn may lead to an alteration distinguishable to the naked eye. The change in *genotype* (genetic make-up) may result in an altered *phenotype*.

Spontaneous and Induced Mutations

Mutations occur spontaneously. This is a way of saying that we do not know the cause in a given instance. The rate of spontaneous mutation in our species is about one in a million gene replications, or less, depending on the gene. Mutation can also be induced by agents that are mutagens, such as radiation and chemicals.

Although "safe" limits are cited for mutagens such as X-rays, even the smallest dose is capable of producing a genetic change. In some cases, the precise mode of action of a mutagen is known. Although specific mutagens produce specific types of change in DNA, their place of action in the genome is random, so that there is no way to produce a change in a particular gene at will in human beings.

It has been estimated that in the genome as a whole, a mutation occurs in one out of every ten germ cells (Muller, 1950).

Natural Selection

Not all mutations are deleterious, otherwise evolution would have been a downhill process if it had occurred at all. An example of a mutant gene that confers an advantage on individuals heterozygous for it (i.e., with one mutant and one normal copy, or allele, of the gene) is that which determines sickle cell anemia. Although the mutant gene is lethal in homozygotes (i.e., those who inherit two similar mutant alleles), it confers resistance to malaria when heterozygous. In malaria-ridden countries, therefore, there would be positive selection for heterozygotes who carry the gene for resistance, even though homozygotes for the mutant allele would die before reaching the reproductive stage and thus be negatively selected against. Because of the positive selection, the mutant gene would be preserved at a relatively high frequency in the population. On the other hand, in areas where malaria is not endemic there will be no selection of the heterozygotes. From this it is clear that one cannot say offhand that a mutation is either good or bad. Its effects will depend on the interaction with particular factors in the environment of an individual who carries the mutant gene.

REFERENCES

Dyson, R. D. 1974. Cell Biology: A Molecular Approach. Allyn & Bacon, Inc., Boston.

Muller, H. J. 1950. Our load of mutations. Am. J. Hum. Genet. 2:111-176.

Roberts, J. A. F. 1970. Introduction to Medical Genetics. 5th Ed. Oxford University Press, New York.

Srb, A. M., R. D. Owen, and R. S. Edgar, 1965. General Genetics. 2nd Ed. W. H. Freeman, San Francisco.

Stern, C. 1960. Principles of Human Genetics. 2nd Ed. W. H. Freeman, San Francisco.

Watson, J. D. 1970. Molecular Biology of the Gene. 3rd Ed. Chapters 11-13, pp. 330-434. W. A. Benjamin, Inc., Menlo Park, California.

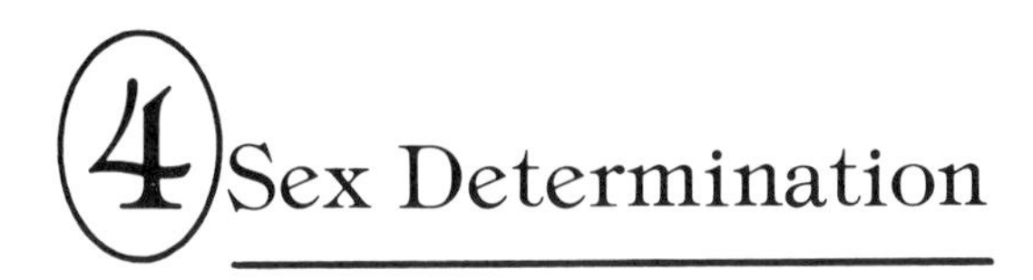

4 Sex Determination

The Role of Sex Chromosomes

HUMAN KARYOTYPES

Individuals of the human species, *Homo sapiens*, normally have 46 chromosomes, two of which are called sex chromosomes, because they play an important part in determining the sex of the individual. The human sex chromosomes are known as the X chromosome and the Y chromosome. The X is one of the largest chromosomes, the Y one of the smallest. Each of them is divided by a centromere into two unequal arms, long and short. Females generally have two X chromosomes and no Y: males generally have one X and one Y. The chromosome constitution, or *karyotype*, of a normal human female is designated as 46,XX, that of a normal males as 46,XY; 46 represents the total number of chromosomes, and is followed by the sex chromosome constitution, XX or XY. The 44 chromosomes other than the X and Y are called autosomes.

Haploid and Diploid Chromosome Numbers

An individual is formed by the union of two sex cells or gametes, the egg (ovum) from the mother, and sperm (spermatozoon[1]) from the father. A normal human gamete has one complete set of 23 chromosomes, which is known as the *haploid* chromosome number. The zygote or fertilized egg cell, resulting from the union of two gametes, has the *diploid* number of 46 chromosomes (that is, two complete sets) and, as discussed in Chapter 3, this number is maintained constant throughout the development of the fetus and from birth to death by the process of mitosis. The only exceptions are the sex cells (which are haploid) and some body cells (in the liver, for instance) which become *polyploid* by doubling their chromosome number one or more times.

[1] In most biologic terms, where two *o*'s come together, both are pronounced—thus spermatozō'on, ō'ocyte, etc.

Meiosis and Gamete Formation

Gametes are formed in the gonads—the ovaries or the testes. During this process, a special type of nuclear division occurs in which the chromosome number is reduced from 46 to 23. This reduction occurs in two steps or divisions and is called *meiosis*. In the previous interphase (stage 1 in Figure 4.1A and B), the usual DNA replication occurs. Early in the first meiotic division (prophase), each chromosome inherited from the mother lines up with its homologous partner inherited from the father (stage 2 of Figure 4.1A and B). Chromatids (half-chromosomes) of the two paired chromosomes may physically exchange segments. These exchanges, known as crossovers, can result in recombination of genes inherited from father and mother. Then (at metaphase), instead of 46 separate chromosomes arranged on the equatorial plate, as in mitosis (Figure 3.2), there are 23 *pairs*. Each chromosome of a pair has two half-chromosomes or chromatids, joined at the centromere. The centromeres, which are responsible for chromosome movement, remain single and at anaphase (stage 3 of Figure 4.1A and B), the two whole chromosomes of each pair separate to opposite poles. One chromosome of each pair is thus included in each daughter nucleus, so that each nucleus has one complete chromosome set. All the chromosomes are double (with two chromatids) except at the centromeres (stage 4). A second step is required to produce gametes with one set of single chromosomes in each. A second division takes place without the usual intervening replication of DNA (stage 5). At metaphase, the single set of chromosomes assembles on the equatorial plate. Only now do the centromeres split, and the two chromatids of each chromosome become separate and independent chromosomes and are distributed to separate daughter nuclei.

There are two striking differences between the process of gamete formation in the male and that in the female (stages 5 and 6). First, in the male, all four nuclei produced by the two meiotic divisions become sperm nuclei; but in the female, only one of the nuclei becomes a large nutritive egg nucleus, while the second nucleus from each division degenerates. Second, because the two X chromosomes of the female separate at meiosis, all eggs contain a single X chromosome; but in the male, where there is a single X with a Y for a partner, the result of their separation is that half the sperm contain an X chromosome and the other half contain a Y chromosome. The sex of the offspring is determined at fertilization: it depends on whether the sperm contains an X or a Y chromosome. Thus, if there are equal chances for an egg to be fertilized by an X-bearing sperm or a Y-bearing sperm, the primary sex ratio (the ratio of male to female zygotes at conception) should be 1:1. Sex ratios will be considered further in Chapter 5.

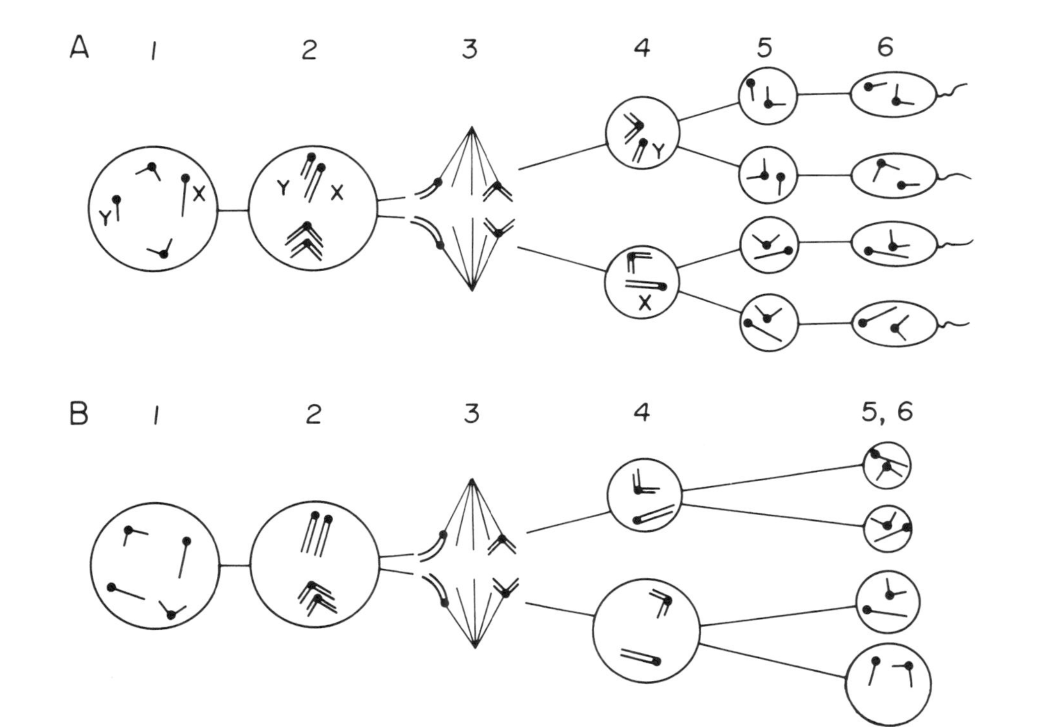

Figure 4.1. Meiosis. A) sperm formation. The separate chromosomes (1) replicate themselves and pair (2). The four-stranded pairs assemble on the metaphase plate and two-stranded chromosomes separate at anaphase (3) to form two nuclei (4). At the second division the two chromatids of each chromosome separate and four nuclei are formed (5), each of which subsequently becomes a sperm nucleus (6). B) The stages of egg formation are the same. Only the large nucleus (5) with its surrounding cytoplasm, develops into an egg. The others are polar bodies. (After Srb, Owen, and Edgar, 1965).

In the female, gamete formation has already started at birth. The meiotic divisions of the maturing egg (oocyte), which are then under way, are not completed until fertilization takes place. This cannot occur before puberty because the eggs are not completed and released until then. In the male, meiosis (and gamete formation) does not begin until puberty.

CHROMOSOMAL ACCIDENTS

Sometimes accidents happen to chromosomes, and aberrant gametes are formed. One such accident, which may occur during meiosis, can result in nondisjunction, or failure of a pair of chromosomes (at the first division) or chromatids (at the second division) to disjoin, or separate into the two new daughter nuclei.

Failure of the sex chromosomes to disjoin leads to a variety of abnormal gametes and abnormal karyotypes, according to whether the failure happens in the male or the female, and whether it occurs in the first or second meiotic division. The possibilities are shown in Table 4.1. In the female, nondisjunction of X chromosomes at either the first or second division may give rise to an egg which has an extra X chromosome—(24,XX), or an egg which has no X chromosome—(22,O) (where O represents absence of an X chromosome). If the first type of egg is fertilized by an X-bearing sperm, the zygote will have three X chromosomes (47,XXX), and if it is fertilized by a Y-bearing sperm the zygote will have two X's and a Y chromosome (47,XXY). Individuals with these sex chromosome anomalies have been identified. The second type of egg can give rise to 45,XO and 45,OY zygotes. Only individuals of the former type

Table 4.1. Inheritance of sex chromosomes following nondisjunction at meiosis

Division	Ova	Sperm	Zygote	Syndrome
	Abnormal	Normal		
First or Second	24,XX	23,X	47,XXX	Triple X
	24,XX	23,Y	47,XXY	Klinefelter
	22,O	23,X	45,XO	Turner
	22,O	23,Y	45,YO	? Inviable
	Normal	Abnormal		
First	23,X	24,XY	47,XXY	Klinefelter
	23,X	23,O	45,XO	Turner
Second	23,X	24,XX	47,XXX	Triple X
	23,X	22,O	45,XO	Turner
	23,X	24,YY	47,XYY	
	23,X	22,O	45,XO	Turner

are known. The latter, lacking any X chromosomes, has never been found and presumably it is inviable and does not develop beyond the zygote or a very early embryonic stage.

Nondisjunction of X and Y at the first meiotic division in the male will result in sperm that have both X and Y (24,XY) and sperm that have neither (22,O). These will give rise to anomalies which are also produced by nondisjunction in the female. Second division nondisjunction in the male, besides producing sperm lacking a sex chromosome, can result in a sperm with two X chromosomes or one with two Y's. The latter can give rise to a distinctive, viable type of zygote (47,XYY) (see Table 4.1).

Individuals who have extra chromosomes may result from these meiotic accidents, which are called primary nondisjunctions. During gamete formation in these individuals there will be secondary nondisjunctions. Thus, in an XYY individual, the Y chromosomes may go to the same daughter nucleus at the first meiotic division, instead of separating. This too results in a gamete with two Y chromosomes.

Other types of chromosomal accidents are deletion, which results in the loss of part of a chromosome; inversion, in which a segment of a chromosome is turned around; and translocation, which involves the transfer of a segment from one chromosome to another.

The Barr Body

In 1949 Barr and Bertram, observing cells of cats, found that in some cells of the female there was a sizeable DNA-containing body lying against the inner surface of the nuclear membrane. This *Barr body*, or sex chromatin, was not present in the corresponding cells of the male. A Barr body is also characteristic of the nuclei of women and may be seen in cells scraped from the inside of the cheek and suitably prepared for microscopic study. It can also be seen in cells derived from a fetus developing in the uterus (see "Amniocentesis," below). Lyon (1961) developed the hypothesis that the Barr body is an inactivated X chromosome, and that in the female as in the male, there is only one active X. This hypothesis seems to be correct, with qualifications. It is apparently a matter of chance which X (the maternal or paternal) remains active: in fact, in a given individual, one X may be active in some cells and the other X in other cells. Some of the evidence that the Barr body represents at least part of an X chromosome comes from individuals with irregular numbers of X chromosomes. If there is only one active X, the single X in a 45,XO individual should be active and there should be no Barr body. On the other hand, a 47,XXX individual would have two supernumerary X chromosomes and should have two Barr bodies. These expectations have been verified so thoroughly that the number of Barr bodies can now be used to determine the number of X chromosomes a person has.

There is now also a simple way in which the number of Y chromosomes can usually be ascertained (the exceptions being when a particular segment of the Y has been deleted). A fluorescent dye, quinacrine, has been found to stain the Y chromosome and produce a particularly bright fluorescent spot at the tip of its short arm. A Y chromosome stained with quinacrine also produces a bright spot in interphase nuclei. When nuclei of known XYY individuals are studied, in about 30% of them, two widely separated fluorescent spots can be seen.

Amniocentesis

Withdrawal of fluid from the uterus during pregnancy has been used since 1966 as a means of obtaining information about the chromosomal and genetic constitution of the developing fetus. The fetus is surrounded by amniotic fluid within an intrauterine sac called the *amnion*. Somewhat as cells from our skin rub off, fetal cells are detached in the surrounding fluid. By the 13th to 15th week of pregnancy, the volume of amniotic fluid is large enough so that a sample of amniotic fluid can be removed with a hypodermic needle with relative safety in a process called *amniocentesis*. The cells in this sample can be cultured, and after about 3 weeks analyzed for karyotype and also for the level of activity of certain substances such as enzymes. Amniocentesis can be used when it is known that parents have a high risk of having a child with an abnormality that can be diagnosed in the fetal stage. High-risk parents might have had a previous child or a relative with such an abnormality. Also, older women are at relatively high risk of producing aneuploid germ cells (cells with irregular chromosome numbers). Over eighty genetic diseases can be diagnosed in this way, including galactosemia, Tay-Sach's disease, and Lesch-Nyhan syndrome (Milunsky, 1976). The chromosomal abnormalities include those to be described below. If the fetus is diagnosed as seriously abnormal, the parents may want to have a therapeutic abortion. If the diagnosis is normal the parents can be relieved of their anxiety on this score for the rest of the pregnancy. A report from the National Institutes of Health (NICHD, 1976) covered over 1000 instances of amniocentesis and concluded that the dangers of amniocentesis to the physical welfare of mother and fetus are small. Another study of 3000 amniocenteses led to the same conclusion (Golbus, Loughman, and Epstein, 1979).

A third study, by the British Medical Research Council (Working Party on Amniocentesis, 1978) indicated that the risks may be greater than the other studies suggest. An increased frequency of miscarriages (1.0–1.5% greater than for controls) and of perinatal deaths was reported, and also a higher frequency of respiratory difficulties and postural abnormalities in the infant.

Other Diagnostic Techniques

There are other techniques that can be used to diagnose some genetic defects at early stages or to identify some deleterious mutant genes in would-be parents. These include blood tests for parents, the use of fetoscopy or ultrasonography for fetal diagnosis, and testing of babies at birth. The practice, and even the advocacy of such techniques can raise some ethical questions (Neel, 1971; Carter, 1979). People who might avail themselves of these techniques should become familiar with their uses and risks and the attitudes prevalent in our society concerning these techniques, so that they can make informed decisions.

SYNDROMES CAUSED BY SEX CHROMOSOME ANOMALIES

For a long time, physicians have recognized complexes of abnormal characteristics or syndromes. Since 1956, with the development of improved cytologic techniques for chromosome studies, it has been found that some of these syndromes are related to changes in number of specific chromosomes or parts of chromosomes.

Turner's Syndrome

One example, Turner's syndrome, can result from the presence of only one X chromosome (45,XO). Individuals with this syndrome are female. The full syndrome includes a large number of abnormalities, most of which are highly variable in their expression from one affected individual to another. The one abnormality invariably present, though expressed to differing degrees, is gonadal dysgenesis—deterioration of the gonads. In the 3-month-old fetus, the ovaries are normally developed for a female of that age; but by the time of birth they have degenerated and few germ cells, if any, are left. At the age of puberty, even fewer 45,XO women have any recognizable ovarian tissue, and because of the lack of ovarian hormones, the usual pubertal changes, including menstruation, do not occur. A common feature of Turner's syndrome is short stature, apparent even at the beginning of puberty, but accentuated by the lack of the usual growth spurt during puberty. The height attained at maturity is usually between 4 ½ and 5 feet. Other characteristics that may or may not occur are webbed neck and abnormalities of internal organs. Contrary to some reports (Hunt, 1972), which were based on an inadequate number of cases, mental retardation is not a characteristic. IQ follows a normal distribution (see Chapter 6), according to Ehrhardt (1976) and Simpson (1976). Like many females, 45,XO females score low in the ability to discriminate in matters of space and shape (see Chapter 10). In the absence of externally visible abnormalities, the XO condition is often not discovered until puberty. Although hormone treatment at puberty can correct some of the abnormalities that would otherwise occur, no treatment can restore the normal functions of the ovaries.

What if only part of a sex chromosome is missing? The full syndrome can appear in women who lack only a part of the second X chromosome. It seems that the full syndrome follows deletion of the short arm of the X, while deletion of the long arm causes gonadal dysgenesis without other symptoms. A person with the short arm of the Y deleted may also manifest gonadal dysgenesis. The effect of loss of the long arm of the Y is not known for certain. It is clear, though, that it would be incorrect to refer to the small Y chromosome, as Davis did, as a "deformed X" from which part has been broken off (Davis, 1972). Part of the Y chromosome is unique, containing genetic elements that determine maleness, activating many genes (on other chromosomes) that are inactive in females.

The abnormality of the 45,XO phenotype raises a question. If only one X is active in the normal female, as the Lyon hypothesis assumes, why should the absence of a second X result in abnormality? Different answers have been proposed. One is that the extra X is not inactivated in the zygote, but remains active during at least part of the embryonic development, and may then be inactivated in some tissues earlier than in others. A second possible answer is that not all of the second X is inactivated: in fact, different amounts and different parts of it might become inactive in different tissues and at different times. There is evidence for both of these limitations on the inactivation of all but one X. Some degree of inactivation explains, however, why the organism is so tolerant of supernumerary X chromosomes (see next paragraph), whereas individuals with additional autosomes (which are not inactivated) are apparently inviable.

Additional X Chromosomes in the Female

Additional X chromosomes do not produce a recognizable syndrome, probably because of inactivation. Women with 47,XXX and 48,XXXX karyotypes are identifiable by the presence in the nucleus of two and three Barr bodies, respectively. Menstruation may be irregular, fertility may be reduced because of underdeveloped ovaries, and mental retardation may be present but none of these features appears regularly. These women may give birth to 47,XXX daughters and 47,XXY sons, in addition to normal daughters and sons, because of secondary nondisjunction of the X chromosomes at meiosis.

Klinefelter's Syndrome

The chromosome complement 47,XXY is associated with a syndrome that includes small testes, sterility, and small penis. The production of androgen is often reduced and the production of gonadotropins usually high. There is sometimes enlargement of the breasts, and mental retardation is more common than it is in the population as a whole. Individuals with more than one additional X chromosome—48,XXXY and 48,XXXXY—have also been identified. They show the above characteristics, often more markedly than those with only one extra chromosome.

XYY Anomaly

A very thorough study of the incidence and characteristics of 47,XYY individuals is necessary before firm conclusions can be drawn about the effect of the extra Y. Studies of men in prison hospitals for the mentally abnormal have revealed frequencies of about 2% of this karyotype, whereas the frequency found in samples of the population at large was less than 0.2 percent. Most 47,XYY individuals fall within the normal range of intelligence and a large majority are not criminally inclined, however, so that it would be naive to suppose that the extra Y of itself makes men violent, retarded, or criminally inclined. One alternative that has been put forward was suggested by the fact that in some mental and penal institutions, the incidence of 47,XYY karyotypes was especially high in samples of men selected for being at least 6 feet tall. This suggests that men with an extra Y tend to be tall. It has also been suggested that when police arrive on the scene of a disturbance they may apprehend large men preferentially. If the extra Y is associated with a tendency to increase in height, this would increase the frequency of arrest of 47,XYY individuals. This suggestion has not been validated, although it is still sometimes repeated. Cautious and reasonable statements about the 47,XYY data have been made by Hook (1973) and Goldstein (1974).

Starting in 1968, a research team at Harvard Medical School began to collect karyotypes of all babies born at the Boston Hospital for women. Of the 15,000 infants, there were some boys with an extra Y. A controversy arose in 1974 about this research (Culliton, 1974). It centered mainly on the following question: if a baby boy is found to have an extra Y and the parents are informed of this, are they not likely to develop an expectation that their child will develop criminal behavior, and will not this expectation tend to become self-fulfilling? One group of scientists demanded that this research be discontinued. The research group thought that information on the effects of chromosome abnormalities is important, and wanted to carry on; nevertheless, the program was terminated in 1975 (Culliton, 1975).

Meanwhile, the real relationships between the extra Y, and criminality and mental retardation remain obscure.

Mosaics

The accident of chromosome nondisjunction is not confined to the meiotic divisions. During mitosis, the two daughter chromosomes derived from the chromatids of a single chromosome may fail to separate at anaphase to opposite poles, so that one daughter cell gets an extra chromosome and the other lacks one chromosome of a pair. Chromosome loss may also occur because a chromosome fails to arrive at a pole and is not included in one of the daughter nuclei. The resulting cells with aberrant karyotypes may divide and give rise to clones of cells (see Chapter 5) from which whole organs or parts of organs may be derived. Thus women and men may be what is called *mosaic* for different chromosome constitutions such as XO/XXX, or XO/XX/XXX (in women). Other mosaics that have been discovered are XO/XY/XXY, and XXXY/XXXXY.

When the mosaic is of a female/male combination such as XO/XY or XX/XXY, both ovarian and testicular tissue may develop. In a study (Polani, 1972) of 91 individuals in whom both types of tissues were present, 30% of them were chromosomal mosaics. When the mosaic is of a female/female or male/male combination such as XO/XX or XXY/XY, the presence of tissues with the normal chromosome number tends to normalize the development of the individual, as compared with one whose cells are all of the aberrant type. Nearly 11% of females with gonadal dysgenesis are XO/XX, and may have some normal germ

Table 4.2. Frequencies of sex chromosome anomalies

Chromosome Constitution	Designation	Frequency
45,XO	Turner syndrome	1 in 2500 female births[a]
	Gonadal dysgenesis	1 in 5000 infants[b]
46,XX	(Male)	Less than 1 in 10,000 male births[b]
47,XXX	Triple X	1 in 800 female births[a]
		1 in 1250 female births[b]
47,XXY	Klinefelter syndrome	1 in 400 male births[a]
		1 in 700 male births[b]
47,XYY		1 in 700 male births[b]
46,XXY/47,XY	Mosaic	1 in 1400 male births[b]

[a], Data from Roberts (1963).
[b], Data from Polani (1972).

cells and be fertile. About 10% of males with Klinefelter's syndrome are XY/XXY (Polani, 1972).

Frequency of Aneuploids and Aneuploid Mosaics

What are the chances that a person will have one of the unusual constitutions considered above? Table 4.2 shows that none of them seems to occur with a frequency greater than 1 in 400 at birth. A study made in Denver during the 1960s discovered 69 sex chromosome aneuploids in a total of 38,000 newborn babies—a frequency of 1 in 550 for all sex chromosome aneuploids combined. Although Turner's syndrome appears with much less frequency than this, it is estimated that the XO karyotype constitutes 0.75 percent of fertilized eggs (1 out of 133) but that prenatal mortality of this karyotype (i.e., spontaneous abortion) may be as high as 97%.

DETERMINATION OF SEX AND ASSOCIATED CHARACTERS

The Rule

Individuals with a Y chromosome are generally male, while those without a Y are generally female; both groups include individuals who are sexually immature and/or sterile. This generalization is questionable in some cases.

At this point, we must distinguish between sex as determined chromosomally, or by hormone differences and the differences that they induce, and the concept of gender. Most individuals with two X chromosomes and no Y are perceived as females and assigned a feminine gender role—that is, they are expected to behave as females. They usually think of themselves as females and behave more or less appropriately—i.e., they have feminine gender identity. Most individuals with an X and a Y are correspondingly assigned a masculine gender role and assume a masculine gender identity (Money and Ehrhardt, 1972). The concept of gender comprises attitudes and types of behavior that are not associated in a hard and fast way with differentiation by sex. In our culture, for example, men are supposed to be more aggressive and women more nurturant. The characteristics attributed to the masculine and feminine gender differ from one culture to another, however, and the characteristics acquired by males and females are culturally influenced.

The Exceptions

Among those of constitution 46,XX, prenatal exposure to abnormally high concentrations of hormones (progesterone) can result in masculinizing of external genitalia. Deviation beyond the normal range for females may also be caused by a gene mutation which affects the metabolism of the adrenal cortex and is responsible for the adrenogenital syndrome (Chapter 2). Individuals affected in either of these ways have sometimes been identified, and reared, as boys: they have assumed a gender role opposite to that indicated by their sex chromosome constitution (Money and Ehrhardt, 1972). Also rarely, 46,XX individuals, female in all physical respects, may feel that they are really males. These transsexuals show that gender identity is not necessarily determined by either the chromosome complement, the gonads or secondary sex characteristics. Finally, there are instances of individuals with the 46,XX karyotype who are anatomically males, though with abnormalities resembling those of Klinefelter's syndrome. It is possible, however, that in these cases a part of a Y chromosome has been translocated to one of the X chromosomes, in which case they would not be exceptions to the general rule.

Among individuals with a Y chromosome, those with an extra X may have some feminized secondary sex characteristics (breast development, pubic hair) and may be sterile. When only one arm of the Y is lacking, a female with gonadal dysgenesis can result. Transsexuals also occur among 46,XY individuals; although they are anatomic males, they feel that they should really be females. There is also a mutant gene which, in males, results in androgen insensitivity, also known as testicular feminization, so that 46,XY individuals with this gene (probably located in the X chromosome) have immature testes and completely feminized external genitalia and secondary sex characteristics (see Chapter 2). Children showing the effects of this mutant gene, though chromosomally male, are usually identified as and brought up as females.

The normal sex chromosome complements are necessary, but not sufficient, for normal male or female sexual development. The norm may be disturbed by chromosomal or gene mutations or by environmental factors. Besides, the evidence from transsexuals shows that gender identity is not necessarily determined by the chromosome complement, the gonads, or secondary sex characteristics, but can in fact be established independently and in apparent contradiction to the biologic determinants.

REFERENCES

Barr, M. L., and E. G. Bertram 1949. A morphological distinction between neurones of the male and female, and the behavior of the nuclear satellite during accelerated nucleoprotein synthesis. Nature 163:676–677.

Carter, J. L. 1979. The anatomy of controversy: Freedom and responsibility for teaching. BioScience 29:478-481.
Culliton, B. J. 1974. Patients' rights: Harvard is site of battle over X and Y chromosomes. Science 186:715-717.
Culliton, B. J. 1975. XYY: Harvard researcher under fire stops newborn screening. Science 188:1284-1285.
Ehrhardt, A. A. 1976. Lecture given at University of Pennsylvania, May 20, 1976.
Golbus, M. S., W. D. Loughman, and C. J. Epstein. 1979. Prenatal diagnosis in 3000 amniocenteses. New Eng. J. Med. 300:157-163.
Goldstein, M. 1974. Brain research and violent behavior. Arch. Neurol. 30:1-35.
Herskowitz, I. H. 1965. Genetics. 2nd Ed. Little, Brown & Co., Boston.
Hook, E. B. 1973. Behavioral implications of the XYY genotype. Science 179: 139-150.
Hunt, C. 1972. Males and Females. Penguin, Harmondsworth, England.
Lyon, M. F. 1962. Sex chromatin and gene action in the mammalian X-chromosome. Am. J. Hum. Genet. 14:135-148.
Milunsky, A. 1976. Prenatal diagnosis of genetic disorders. New Eng. J. Med. 295:377-380.
Neel, J. V. 1971. Ethical issues resulting from prenatal diagnosis. *In* M. Harris (ed.), Early Diagnosis of Human Genetic Defects: Scientific and Ethical Considerations. Fogarty International Center Proceedings, No. 6, Government Printing Office, Washington, D.C., pp. 219-229.
NICHD Study Group, 1976. Midtrimester amniocentesis for prenatal diagnosis: Safety and accuracy. JAMA 236:1471-1476.
Polani, P. E. 1972. Errors of sex determination and sex chromosome anomalies. *In* C. Ounsted and D. C. Taylor (eds.), Gender Differences: Their Ontogeny and Significance. Churchill Livingstone, Edinburgh, pp. 13-39.
Roberts, J. A. Fraser. 1963. An Introduction to Medical Genetics. Oxford University Press, London.
Simpson, J. L. 1976. Disorders of Sexual Differentiation. Academic Press, New York.
Srb, A. M., R. D. Owen, and R. S. Edgar. 1965. General Genetics. 2nd Ed. W. H. Freeman, San Francisco.
Williamson, N. E. 1976. Sex preferences, sex control and the status of women. Signs 1:847-862.
Working Party on Amniocentesis, Medical Research Council. 1978. An assessment of the hazards of amniocentesis. Br. J. Obstetr. Gynecol. 85 (Suppl. 2):1-41.

5 Sex-Linked and Sex-Limited Forms of Inheritance

AUTOSOMAL INHERITANCE

Long before the chemical nature of the gene was known, even years before genes were so named, Mendel recognized that individual traits were determined by hereditary factors and discovered some fundamental laws with regard to their inheritance. The first of these was the principle of segregation of alternative factors, or *alleles*, as we now call them.

Suppose that in an experiment a female fruitfly with normal wing gene (+) in both chromosomes of a pair is mated with a male that has tiny vestigial wings because it has a mutant allele of that gene (*vg*) in both chromosomes. Each offspring will inherit the normal allele from one parent and the mutant allele from the other. The offspring are heterozygous. In this case, the normal gene is dominant over the mutant gene, and all offspring have normal wings. The vestigial trait does not appear: it is recessive. Now, if heterozygous male and female offspring are mated together, the vestigial trait will reappear in about one-quarter of their offspring. The mutant gene has remained intact in the heterozygote and segregates from the normal allele at meiosis, so that germ cells are equally likely to receive the + or the *vg* allele. Figure 5.1 shows the different ways in which the two types of female and the two types of male gametes, + and *vg*, can combine, giving a ratio of three zygotes with a dominant allele to one that is homozygous recessive (*vg vg*). The three-quarters to one-quarter ratio is statistical, like the chance of getting heads or tails when one tosses a coin: in any particular trial, one will get an approximation of the expected proportions.

Some normal genes are only partially dominant over mutant alleles, and the heterozygotes are intermediate between the normal and the homozygous mutant form. In this case too, however, both alleles remain intact even though the heterozygote seems to be a blend of the two traits. As can be seen from Figure 5.1, the result of mating two of these

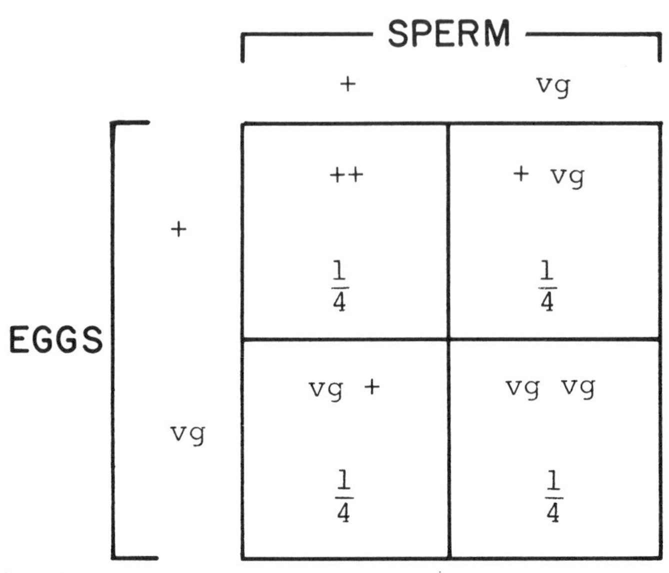

Figure 5.1. Genotypes expected from a mating of two heterozygotes for alternative alleles + and *vg*.

heterozygotes will be approximately one-quarter normal and about one-quarter homozygous mutant, while the remainder (about one-half) will be of the intermediate heterozygous type.

SEX-LINKED INHERITANCE

X Chromosome

Genes that are located in the X chromosome are called *sex-linked* genes. One of the earliest mutants found in the fruitfly, a white-eyed type, turned out to have a sex-linked mutant gene. Figure 5.2 shows how inheritance of such a sex-linked trait differs from that of an autosomally determined trait. When normal (X^+) males are mated with white-eyed females ($X^w X^w$), the female offspring are all wild type, with red eyes, but their brothers are all white-eyed. This result is typical for genes that are sex-linked, that is, carried on the X chromosome. The males are hemizygous because they inherit only one X chromosome, that from their mother, and with it they inherit their mother's sex-linked mutant characters. The daughters, who receive an X chromosome from their father as well as from their mother, are heterozygous for the wild type (+) and mutant (*w*) alleles, as their fathers had the wild type allele and the mutant gene was carried by their mothers. They are phenotypically wild type because the mutant character is recessive to the wild type (Figure 5.2A). Figure 5.2B-D shows the expected results of other matings involving sex-linked genes.

If the mating is made the other way—i.e., normal female and white-eyed male—all the offspring will be normal in appearance but the daughters will carry the mutant allele (Figure 5.2B). If a female heterozygous for the white eye gene is mated with a normal male, half the sons will have red eyes and half white eyes. All the daughters will have red eyes, but half will be heterozygous carriers of the white eye allele (Figure 5.2C). Finally, if a heterozygous female is mated with a white-eyed male, half the daughters and half the sons will have red eyes (these daughters will be heterozygous carriers) and the other half will have white eyes (Figure 5.2D).

As is clear from the diagrams, males can only inherit an X-linked gene from their mother, and can only transmit such a mutant to their daughters. This is known, therefore, as *X-linked* (or *criss-cross*) *inheritance.*

Human Sex-Linked Genes Over 100 known human genes are sex-linked (McKusick, 1978) and there are probably hundreds more. They are subject to the conditions discussed above. They tend to be inherited together, but may recombine (see Chapter 6). They show certain peculiarities of inheritance and expression. As illustrated above, because males have a single X chromosome, inherited from their mothers, they can inherit X-linked genes only from their mothers, and they transmit them only to their daughters. A second peculiarity, known only in mammals, is due to the Lyon effect—that is, inactivation of at least part of one X chromosome in the female—which, as we shall see below, affects the expression of recessive mutant alleles in heterozygous females.

Sex-Linked Lethals Some diseases with a genetic basis may cause death. An example is hemophilia, which is due to a recessive mutation of a sex-linked gene responsible for making a blood-clotting factor. Males hemizygous for this mutant gene and females who are homozygous for it lack the blood-clotting factor, and are hemophiliacs, or bleeders. If they begin to bleed, even from a small wound, they may lose so much blood that they die. This can be prevented, however, by treatment, and nowadays hemophiliacs can use emergency self-treatment. Females heterozygous for the mutant gene, because of the Lyon effect, will produce the factor in some cells but not in others; and depending on the proportion of normal and mutant cells, may exhibit varying degrees of hemophilia ranging from none to mild to fairly severe.

Although the hemophilia gene can be lethal in hemizygotes and homozygotes, it only causes death if bleeding occurs and if treatment is not available. It is therefore called a *conditional lethal.*

There are also genes which are *unconditional lethals*. The sickle-cell gene is an example, and so is the mutation causing the infantile form of

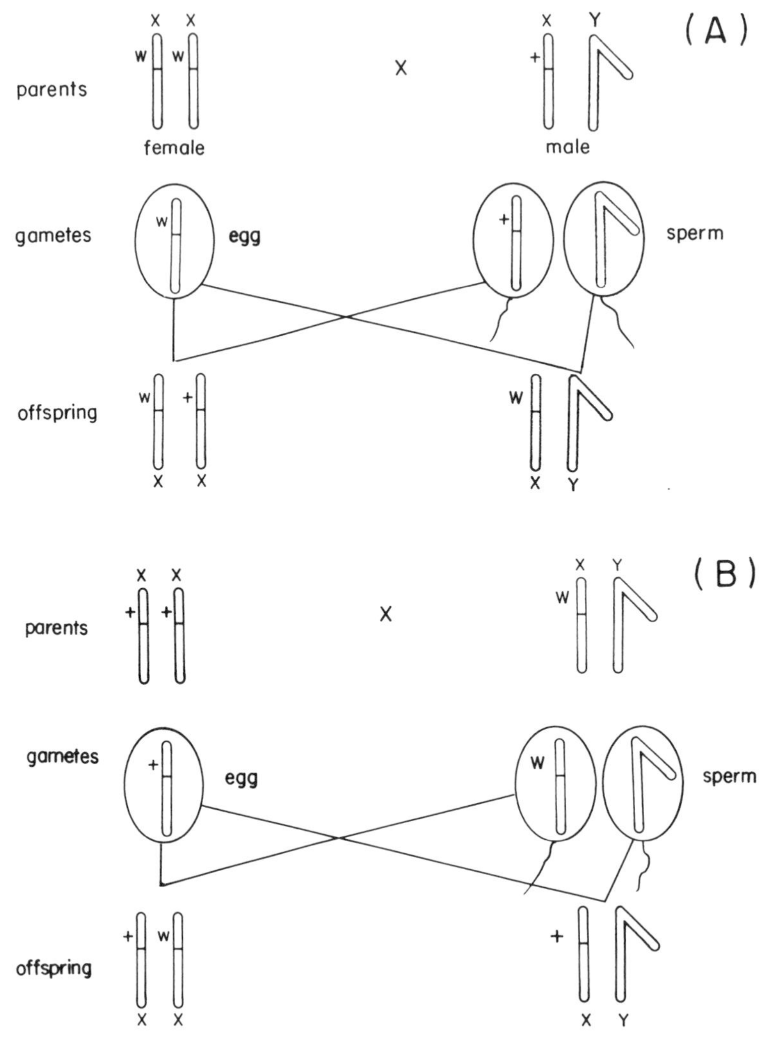

Figure 5.2. Inheritance of a sex-linked gene from four different types of mating. A) Mutant ♀ × wild-type ♂; B) wild-type (homozygous) ♀ × mutant ♂; C) wild-type (heterozygous) ♀ × wild-type ♂; D) wild-type (heterozygous) ♀ × mutant ♂. The alleles carried by parents, their gametes, and their ♂ and ♀ offspring are shown for each mating. (After Srb, Owen, and Edgar, 1965).

Gaucher's disease. In the former, deaths occur during childhood or adolescence. In the latter, which is due to an enzyme deficiency in nerve cells, children die when 1 or 2 years old. These are not sex-linked mutations: they affect males and females alike and with equal frequency. But there are sex-linked mutations that are outright lethal and manifested in males

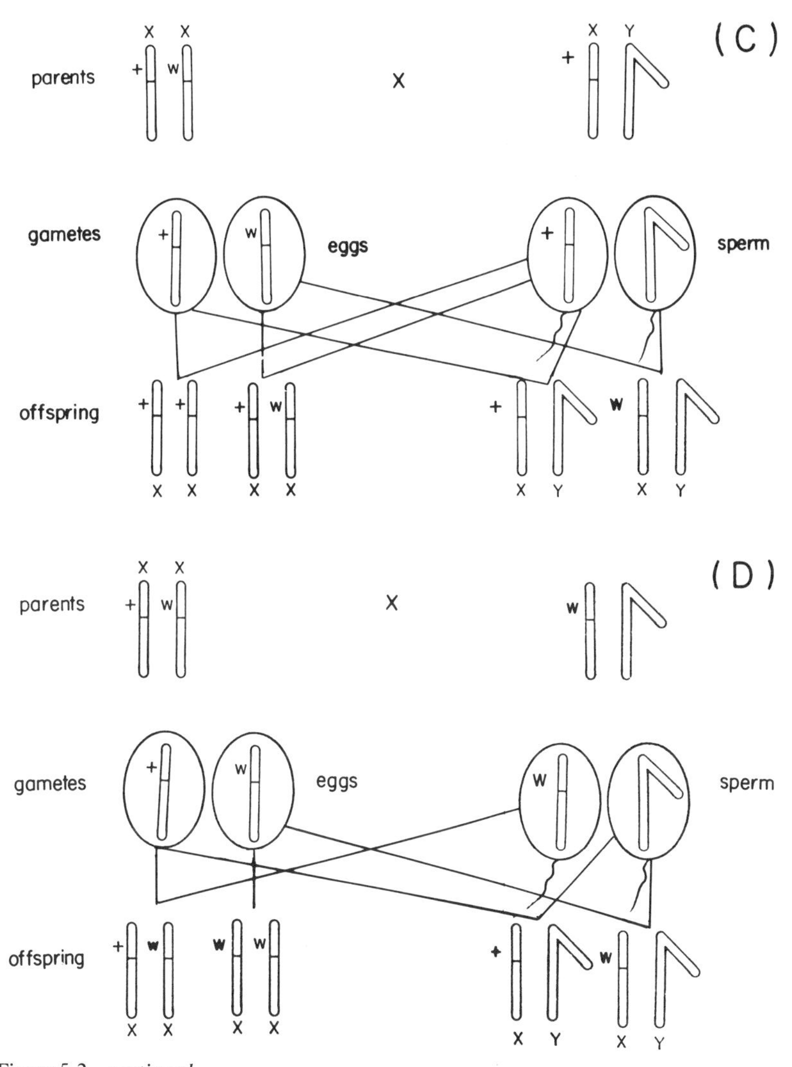

Figure 5.2 *continued.*

more often than in females. Examples are the mutant genes causing Aldrich syndrome, Duchenne-type muscular dystrophy, Fabry's disease, and hypochromic anemia. An illustration of the difference in frequency is provided by the sex-linked (nonlethal) form of color blindness. Eight percent of men, but only 0.5% of women have this disability.

Sex Ratios and Differential Mortality It has often been assumed that deleterious sex-linked genes are responsible for an excess of male deaths over female deaths in prenatal stages as well as in childhood.

Another common assumption used to be that there was an equal probability of an egg being fertilized by either an X- or a Y-bearing sperm, so that the sex ratio of males to females at conception (which is called the *primary sex ratio*) would be 1:1. If both of these assumptions were correct, we would expect the *secondary sex ratio* (that at birth) to show a deficit of males. On the contrary, the secondary sex ratio for white North Americans is 1.06 males to 1 female, and in other groups adequate samplings also show an excess of male births. If the first assumption is true, therefore, the second (about the primary sex ratio) must be incorrect, and vice versa. In fact, there is little evidence for either assumption.

Prenatal Mortality Of course, we cannot determine the primary sex ratio directly. Until methods were discovered for identifying X and Y chromosomes in cell nuclei, it was impossible to determine the sex of a fetus less than seven weeks old with any degree of reliability (Stevenson, 1966). Although we are beginning to get some information on the sex of stillborn fetuses and infants, which tells us a little about the ratio of male to female deaths *in utero*, these figures do not include deaths at very early embryonic stages. Many pregnancies terminated this early would not be detectable even by a missed menstrual period. Ounsted (1972) thinks there may be a considerable loss of embryos between fertilization and implantation in the uterus, with an excess of females lost at this stage. She suggests the possibility that antigens may be involved. Difficulties in implantation might account for the fact that figures from spontaneous abortions show an excess of female deaths, in the earliest stages available for diagnosis. In order to affect the sex ratio, a sex-linked antigen would be required. The suggestion that such an antigen affects implantation of the embryo is at present purely speculative.

In data from nine different studies of spontaneous abortion, each showed an overall excess of male over female deaths (Stevenson, 1966), with the ratios ranging from 1.19:1 to 3.29:1. This excess was consistent in each study for each month of gestation from the second to the sixth. Only in the first month, which was included in five of the nine studies, was there a departure from this situation. Three of the five series with data for the first month showed an excess of female deaths.

It is known that 20% of aborted fetuses have abnormal chromosome numbers, which presumably affect males and females equally when autosomes are involved. There is also a high prenatal mortality of XO fetuses, perhaps as high as 97% (which might include some deaths due to sex-linked lethal mutations). This mortality is certainly not enough to account for the excess of males at birth, beginning with a primary sex ratio of 1:1, for the frequency of XO individuals at conception is estimated at only 0.75%.

The current consensus is that there is a great excess of male zygotes at conception. On the other hand, it is known that the primary sex ratio for mice and rabbits is 1:1 (Melander, 1962; Vickers, 1967)—i.e., there is no excess of males—although this is not true of the hamster (Sundell, 1962). If it should also be proved that the human primary sex ratio is 1:1, some hypothesis, such as Ounsted's, to explain the excess of males at birth would be needed.

Postnatal Mortality Neonatal and postnatal death rates show a consistent picture of an excess of male deaths in early childhood. In the first month after birth, as indeed in the first year, the mortality rate is high—the rates for both sexes combined in the U.S. (1970 figures) were 15.1 per 1000 live births in the first month and 20.0 per 1000 live births in the first year. The first month figures for boy and girl babies were 17.0 and 13.1 per 1000 live births, respectively. For the whole first year, the corresponding figures are 24.1 male deaths and 18.6 female deaths per 1000 children under one year old.

For the next age group (those between 1 and 4 years old) the mortality rate was down to 84.5 per 100,000 of the population in this age group, with the rate for boys at 93.2 per 100,000 and that for girls at 75.4 per 100,000. More than half of the difference between boys and girls at this stage is due to the difference in the rate of deaths from accidents, which were responsible for 36.8 deaths per 100,000 in boys and 25.9 per 100,000 in girls. The next greatest causes of differential deaths were influenza and pneumonia, which accounted for 16% of the total difference between boys and girls (U.S. Public Health Service, 1974).

Clearly, after birth, the major differential causes of death are not due to deleterious mutations. Although we know much less about the causes of prenatal deaths, it seems possible that here, too, sex-linked lethals are of minor importance.

Mortality and Lifestyle The high contribution of accidents to the excess of male over female deaths is continued from early childhood through adolescence and into early adulthood. While men continue to be more subject to fatal accidents than women throughout their lives, this cause of death becomes of lesser importance after age 44, yielding to heart disease, which is responsible for nearly three times as many male deaths as female between the ages of 45 and 65 (U.S. Public Health Service, 1974). In a study of differential mortality, Waldron (1976) has concluded that one-third of the sex differential in mortality is due to causes which can be related to differences in lifestyle. The "coronary-prone behavior pattern" is a label for the high-stress lifestyle of people who display such traits as ambitiousness, aggressiveness, and competitiveness. It is more common among men than women, and men who show this pattern are twice as likely to develop coronary heart disease as others. Smoking, drinking alcohol, hunting, suppressing emotions such as fear, have also been more acceptable or more encouraged in males than in females, and contribute to the causes of differential mortality. Between 1950 and 1970, however, the difference in mortality rate from some of these causes has decreased, in part owing to an increase of female deaths from automobile accidents, respiratory disease, suicide, cirrhosis of the liver, and heart disease.

Y Chromosome

Genes on the Y-chromosome should show a different pattern of sex-linkage from that of X-linked genes, for the Y chromosome is always passed on from father to son. There seem to be few, if any, characteristics controlled by Y-linked genes though, other than the whole complex of maleness (as contrasted with femaleness).

SEX-LIMITED INHERITANCE

Some traits appear only in males and may seem as if they were inherited from father to son, such as baldness. If baldness were due to a Y-linked gene, it would be passed exclusively from father to son. The son of a nonbald father would not be bald except as a result of a new mutation, which is a rare event. This hereditary sequence is quite common, however, and in reality, baldness is due to an autosomal gene that is dominant in males but recessive and rarely expressed in females. The gene can be inherited from the mother as well as the father, and the pattern of baldness at temples and crown of the head may follow either the mother's side of the family or the father's. The gene's expression in heterozygotes is limited to males. In this case, as in many others, the cause of sex-limitation is obscure. Sex-limitation may be complete or partial: Many diseases are more common in one sex than the other.

Parthenogenesis

Parthenogenesis, or virgin birth, results when an egg develops without fertilization. This occurs regularly in some animals (for instance, in some insects and rotifers, a class of minute aquatic invertebrates). Obviously, this is a situation in which the hereditary characters of the offspring are determined entirely by the genes of the mother. If the unfertilized egg develops completely normally, the resulting individual will have a single set of chromosomes instead of two sets. More commonly, chromosome replication gets one step ahead of cell division so that the double (diploid) chromosome number is restored.

Parthenogenesis has been induced in some higher animals such as frogs, mice, and rabbits, although parthenogenesis in such mammals as the rabbit is more often than not arrested at an embryonic stage (Graham, 1970).

Several cases of supposed human parthenogenesis have been reported in the twentieth century (Stern, 1960). If they were genuine virgin births, the resulting individuals should be either 23,X or 46,XX, and in either case, because they would lack a Y chromosome, they would be females. Two kinds of tests have been made to determine whether supposedly parthenogenetic females were really parthenogenetic. First, the blood group was ascertained. If the child differed from the mother by any blood group factor, this must be due either to a gene inherited from a male parent or to mutation of a maternally inherited gene. The probability of two such rare events as mutation and parthenogenesis coinciding would be very small. The second test is skin grafting. Individuals usually differ in genes that cause antigen-antibody reactions when skin grafts are made from one to another. The grafted skin tends to be sloughed off. If a graft were made between a parthenogenetic child and her mother, there should be no such reaction, because they would be genetically identical. All but one of the supposed parthenogenetic females gave evidence of having blood group genes other than those of their mother. The remaining one did not pass the skin-graft test for parthenogenetic origin.

REFERENCES

Graham, C. F. 1970. Parthenogenetic mouse blastocysts. Nature 226: 165-167.

McKusick, V. A. 1978. Mendelian Inheritance in Man. 5th Ed. Johns Hopkins University Press, Baltimore.

Melander, Y. 1962. Chromosomal behavior during the origin of sex chromatin in the rabbit. Hereditas 48:645-661.

Ounsted, M. 1972. Gender and intra-uterine growth: With a note on the use of the sex proband as a research tool. *In* C. Ounsted and D. C. Taylor (eds.), Gender Differences: Their Ontogeny and Significance. Churchill Livingstone, Edinburgh, pp. 177-201.

Stern, C. 1960. Principles of Human Genetics. 2nd Ed. W. H. Freeman San Francisco.

Stevenson, A. C. 1966. Sex chromatin and the sex ratio in man. *In* K. L. Moore (ed.), The Sex Chromatin. W. B. Saunders Co., Philadelphia, pp. 263-276.

Sundel, G. 1962. The sex ratio before uterine implantation in the golden hamster. J. Embryol. Exp. Morphol. 10:58-63.

U.S. Public Health Service. 1974. Facts of Life and Death. U.S. Dept. HEW Pub. no. (HRA) 74-1222.

Vickers, A. D. 1967. A direct measurement of sex-ratio in mouse blastocysts. J. Reprod. Fert. 13:375-376.

Waldron, I. 1976. Why do women live longer than men? Part I. J. Hum. Stress 2:2-13.

Waldron, I., and S. Johnston. 1976. Why do women live longer than men? Part II. J. Hum. Stress 2:19-31.

6 Sources of Variation and Individual Differences

"VARIETY IS THE VERY SPICE OF LIFE"

In an introduction to her book *Male and Female*, Mead (1949) makes the point that our highly mechanized society has a strong tendency toward homogeneity. Not only are commodities mass-produced in vast numbers of identical copies, but we are supposed to follow fashions in dress and furniture styles and to approximate certain norms of appearance and behavior. If we insist on sexual equality on the basis that there are no essential differences between men and women other than in reproductive function, are we really getting rid of undesirable restrictions on women and men? Or are we denying and suppressing some real and valuable differences that could and should be made use of and that would contribute a welcome element of variety to our society? In other words, is the "unisex" outlook one which impoverishes society and harms the individual by "going against nature"?

The answer to this question must take into account the available variation in the human population. This chapter deals primarily with genetic variation and only touches on environmentally caused variation.

MUTATION

All the genetic variation in a species arises by mutation. If a mutation does not cause a genetic death before puberty, it can be transmitted to the next generation. Some individuals will be heterozygous, with the mutant allele in one chromosome and the normal allele in the other chromosome of a pair. As is discussed in Chapter 3, some of these mutations may be beneficial, some more or less detrimental.

GENES IN DIFFERENT CHROMOSOMES

Independent Assortment of Genes

Although half our chromosomes come from our mother and half from our father, each pair of chromosomes behaves independently of every other pair at meiosis. The maternal chromosomes of some pairs will be distributed to the same gamete as the paternal chromosomes of other pairs, providing new assortments of genes (Figure 6.1). This rule was formulated by Mendel as the law of independent assortment.

The Test Cross

If, using an experimental organism such as the tomato, guinea pig, or fruitfly, we make a test cross, in which the heterozygous individual, such as that in Figure 6.1, is mated with one homozygous for the recessive mutants *a* and *b*, we shall expect four types of offspring in approximately equal numbers—those that received both dominant alleles, *A* and *B*, from the heterozygous parent; those that received only one dominant, either *A* or *B*; and those that received neither. Table 6.1 gives an example.

Thus, two mutants in different pairs of chromosomes can provide for four different types of gamete. If there were three chromosome pairs with each pair differing by one mutant gene, the number of gametic types would be doubled to eight. This number is given by the formula 2^n, where n is the number of chromosome pairs in which the two mem-

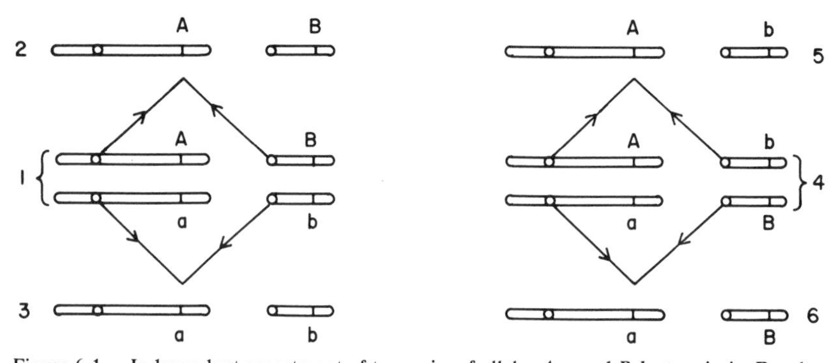

Figure 6.1. Independent assortment of two pairs of alleles *A, a* and *B, b* at meiosis. Results of the first meiotic division only are shown, and only one chromatid is shown for each chromosome, which would really have two similar chromatids. 1,4 -paired chromosomes; 2,3,5,6-chromosomes separated into daughter cells. Centromeres are indicated by open circles. If *A* and *b* are alleles inherited from the mother and *a* and *B* are inherited from the father, separation of chromosomes as on the left (2,3) will give new combinations of alleles *AB* and *ab*, while that shown on the right will give parental combinations (5,6). (After Srb, Owen, and Edgar, 1965).

Table 6.1. An example of independent assortment of genes in the tomato: Result of cross between T;C and t;c plants and test cross of T/t;C/c offspring with t/t;c/c plants

Phenotype of offspring	Genotype of offspring	Number of offspring
Tall, cut leaf	T/t;C/c	77
Tall, potato leaf	T/t;c/c	62
Short, cut leaf	t/t;C/c	72
Short, potato leaf	t/t;c/c	73
Total		284

The semicolon is used to separate genes on different pairs of chromosomes. The slash separates genes on the two chromosomes of a pair. T, tall; t, short; C, cut leaf; c, potato-like leaf. Adapted from Srb, Owen, and Edgar (1965).

bers differ genetically. As we have 23 pairs of chromosomes, if each pair was heterozygous for one mutant gene we could produce 2^{23}, or about 8,375,000 genetically different gametes.

LINKAGE AND RECOMBINATION

Linkage

However, there are a number of genes in each chromosome, so each chromosome that enters into the nucleus of a developing sperm or egg brings in a whole group of genes. If all the genes on one chromosome of a pair are normal, but the other chromosome carries two mutant genes *c* and *d*, one would expect half the gametes formed to get a chromosome with normal genes *C* and *D*, and half to get a chromosome carrying both mutants, *c* and *d* (Figure 6.2). In this case the genes are not assorted independently but are said to be *linked*.

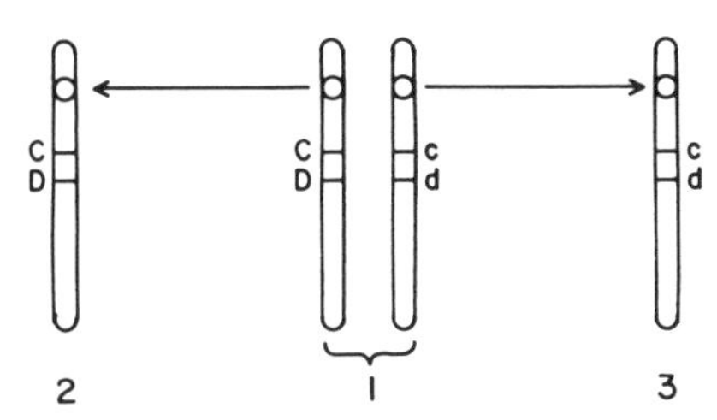

Figure 6.2. Behavior at meiosis of mutant genes closely linked in the same chromosome. Only the first meiotic division is shown, and only one chromatid per chromosome. 1) paired chromosomes; 2,3) chromosomes separated into daughter cells. Centromeres are indicated by open circles. Only the parental combinations of alleles are obtained because of linkage.

Recombination

Linkage is not always complete, however. Even genes on the same chromosome may sometimes be separated and inherited independently. This separation occurs at the first meiotic division by an exchange of segments between two chromatids (half-chromosomes) belonging to opposite members of a pair of chromosomes (Figure 6.3). The exchange process is referred to as *crossing over* or *recombination*, and the resulting chromatids are called *crossover* or *recombinant* chromatids.

There may be more than one exchange between chromatids of a pair of chromosomes. At each exchange, any two of the four chromatids may be involved. Thus, double crossovers may involve the same two chromatids; or three chromatids, one of which is involved in both exchanges; or all four chromatids (Figure 6.4).

Crossing over may occur between any two genes on a chromosome. If one chromosome pair were heterozygous for ten different pairs of alleles, it would be possible to get 2^{10} different chromosomes (including the two parental chromosome types) each differing in the linear sequence of alleles. This gives us a figure of 1024 different combinations—just for one chromosome pair.

After mutation has provided a few hundred mutant genes, millions of new genomes can be obtained through the mechanisms of independent assortment and recombination.

INHERITANCE OF QUANTITATIVE CHARACTERS

Many inherited traits, like hemophilia or sickle cell anemia in the human species, represent clear-cut qualitative differences from the "norm" determined by single gene mutations. (Actually, this is a simplification. Marked differences in eye color, for instance, may depend on differences in the amount of one of the eye pigments. This is a case of a clear-cut phenotype based on an underlying quantitative difference). There are also characteristics in which we differ from one another *quan-*

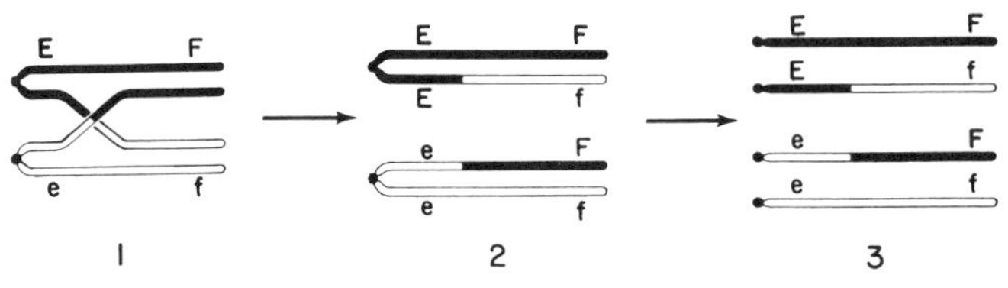

Figure 6.3. Crossing-over between chromatids of paired chromosomes of first meiotic division. 1) Pairing and exchange; 2) The resulting chromatids; 3) Chromosomes separated into gametes. Centromeres are indicated by solid circles. The exchange between two of the four chromatids results in new combinations of alleles (*Ef* and *eF*) while the other two chromatids retain the parental combinations. (After Srb, Owen, and Edgar, 1965).

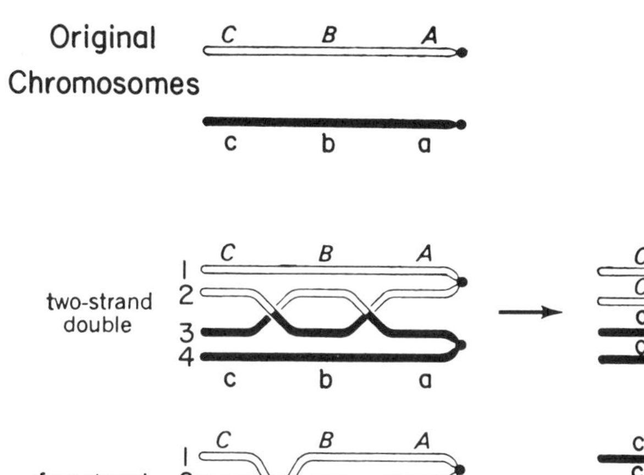

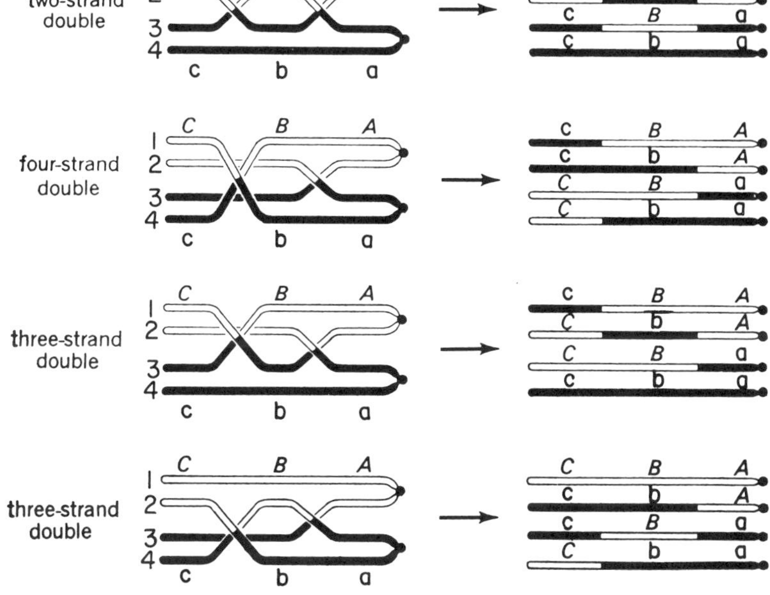

Figure 6.4. Different types of double crossover. First row shows original chromosomes. Centromeres are indicated by solid circles. Second row, the same two chromatids (2 and 3) are involved in both crossovers. The second exchange cancels the effect of the first and there is no recombination of alleles of A and C (right). Third row, the two chromatids not involved in one exchange, are involved in the second (left). All chromatids become recombinant chromatids, 1 and 4 from the left side exchange and 2 and 3 from that on the right (right). Fourth row, three-chromatid exchange (left) in which one *abc* chromatid (3) is involved in both exchanges, while the other *abc* chromatid (4) is a noncrossover (right). Fifth row, three-chromatid exchange (left) in which *one* ABC chromatid (1) is a noncrossover and the other *ABC* chromatid (2) is involved twice (right). (After Srb, Owen and Edgar, 1965).

titatively, such as height, weight, or IQ. Differences in such measurements between males and females are discussed in Chapter 18.

Continuous Variation

If a large number of people are measured for height, their heights represent a virtually continuous range of variation, from the shortest to the tallest. There are very many such continuously varying traits, covering every measurable part of the body and a number of psychologic and emotional characteristics. These traits provide a vast reservoir of variation in the human population.

Normal Frequency Distribution

When the heights are grouped into size classes and the number of people in each class is plotted on a graph, the frequency distribution for different size classes approximates a hump-backed curve, symmetric about the midpoint (Figure 6.5a). This midpoint not only represents the most frequent height, but is also close to the average height for the whole set of measurements. This type of curve is called the *normal curve* of frequency distribution.

Because men tend to be taller than women, separate curves for the two sexes will be somewhat different. This is true for many quantitatively varying traits (see Chapter 18), some of which give higher values for men than women (such as height) and some the opposite (body fat). Also, in some cultures these sex differences are very much less than in others. This is true for the height difference, said to be almost negligible in one of the American Indian tribes (Oakley, 1972).

For some unknown reason, the range of variation for quantitative characters is usually greater for men than for women, so that the male curve is broader in relation to its peak height and the female peak is higher in relation to the spread of the curve (Figure 6.7).

Suppose that we have one pair of alleles, *H* and *h*, determining height in such a way that height increases with increasing number of *H* alleles (*H* being partially dominant). Then homozygous (*HH*) individuals will be taller than heterozygous (*Hh*), who will be taller than the homozygous recessive (*hh*) individuals. If two parents who are *HH* and *hh* have children, all of them will be heterozygous *Hh*. If two *Hh* parents have children, the chances that any child will have two (*HH*), one (*Hh*) or no (*hh*) *H* alleles are 1:2:1 (see Figure 5.1).

Now, suppose that there are *two* genes affecting height, *H* and *H'*, each with a pair of alternative alleles, *h* and *h'*; and suppose that they assort independently at meiosis, and that the *H'* allele also is partially dominant and adds its effect to that of *H*. If two parents are both heterozygous for both pairs of alleles (*HhH'h'*), the chances that each of their children will have 4, 3, 2, 1, or 0 of the semidominant alleles for increased height are 1 : 4 : 6 : 4 : 1 (Figure 6.8). Taking the series one step further by adding a third pair of alleles, *H"* and *h"*, we obtain seven genetic size groups ranging from 6 to 0 semidominant alleles in the proportions 1:6:15:20:15:6:1. If the sets of proportions which we have obtained with one, two and three pairs of alleles are represented as frequency distributions on graph paper, they will look like Figure

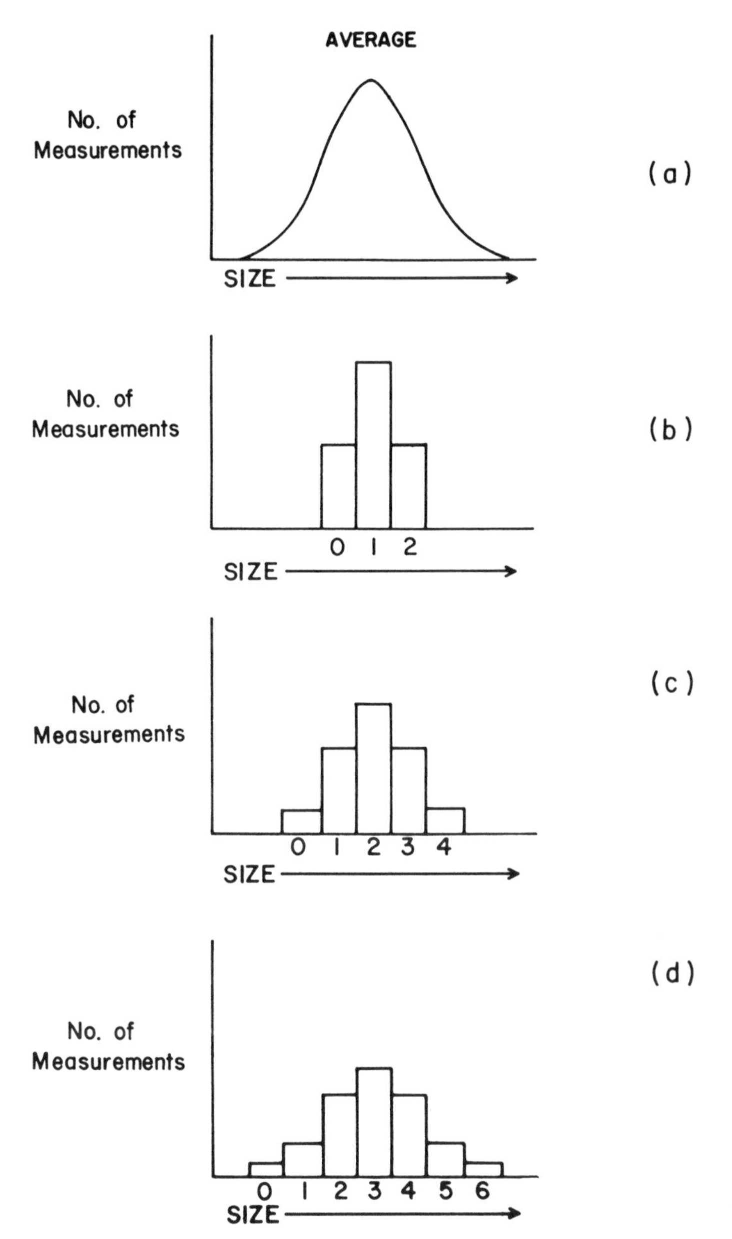

Figure 6.5. A) Normal distribution curve formed by plotting measurements of a character showing continuous variation against the frequency with which they occur. Distribution of individuals in size classes determined by B) one pair of genes; C) two pairs of genes; and D) three pairs of genes. (After Herskowitz, 1965).

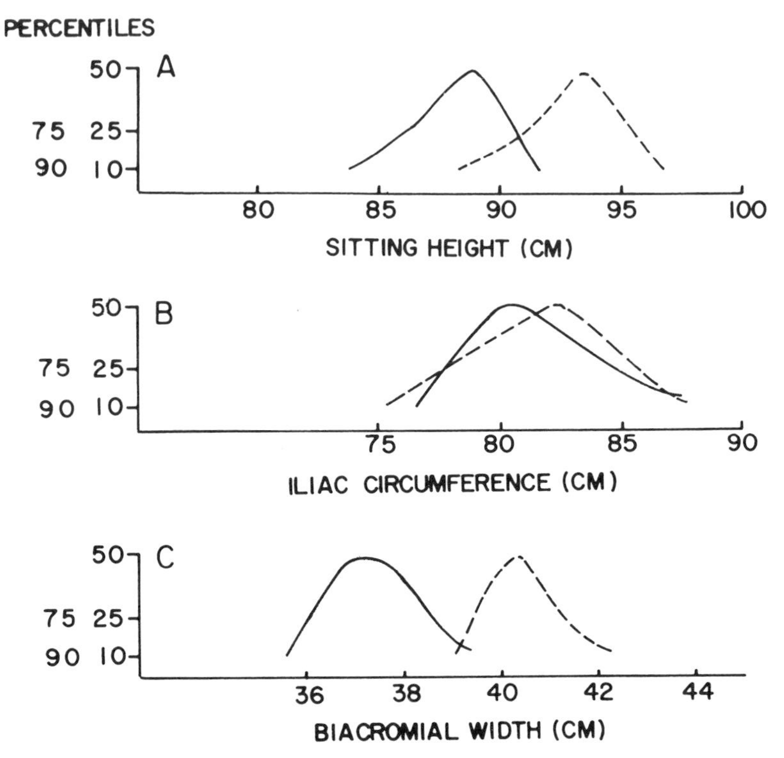

Figure 6.6. Frequency distribution curves for measurements of mature men (broken lines) and women (solid lines) showing different degrees of overlap. A) Sitting height—moderate overlap. B) Iliac circumference (hip measurement)—large overlap. C) Biacromial width (shoulder width)—slight overlap. (Based on percentile figures taken from McCammon, 1970).

6.5b, c, and d. The general formula for finding frequencies of genotypes with 0, 1, 2—$2n$ semidominant alleles (H) acting additively on a character is $(H + h)^{2n}$, where n is the number of gene pairs involved. It is easy to see that if there were many more gene pairs, the frequency distribution would approximate to the normal curve (Figure 6.5a).

GENES AND ENVIRONMENT

How can continuous variation due to a number of genes acting additively be distinguished from a similar range of variation due to a continuous range of environmental conditions?

Obviously, environmental factors such as nutrition affect our growth. Hulse (1968) found that the average height of Swiss immigrants to the United States was not different from that of their brothers who

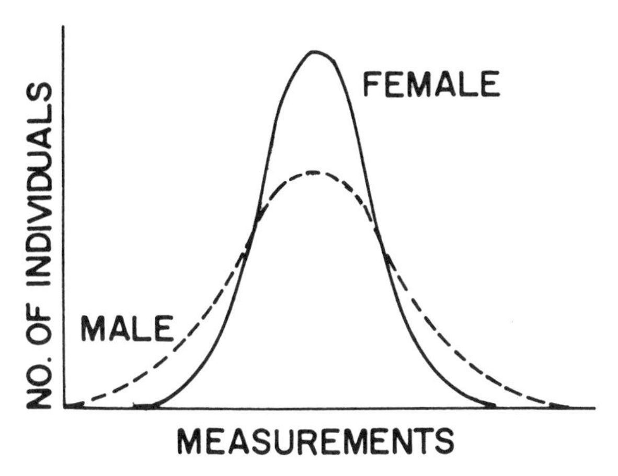

Figure 6.7. Curves showing typical differences between males and females in range of distribution for a quantitatively varying character.

HhH'H ♂ SPERM

HhH'h' ♀ EGGS

EGGS \ SPERM	$\frac{1}{4}$ HH'	$\frac{1}{4}$ Hh'	$\frac{1}{4}$ hH'	$\frac{1}{4}$ hh'
$\frac{1}{4}$ HH'	HHH'H' $\frac{1}{16}$ 4	HHH'h' $\frac{1}{16}$ 3	HhH'H' $\frac{1}{16}$ 3	HhH'h' $\frac{1}{16}$ 2
$\frac{1}{4}$ Hh'	HHh'H' $\frac{1}{16}$ 3	HHh'h' $\frac{1}{16}$ 2	Hhh'H' $\frac{1}{16}$ 2	Hhh'h' $\frac{1}{16}$ 1
$\frac{1}{4}$ hH'	hHH'H' $\frac{1}{16}$ 3	hHH'h' $\frac{1}{16}$ 2	hhH'H' $\frac{1}{16}$ 2	hhH'h' $\frac{1}{16}$ 1
$\frac{1}{4}$ hh'	hHh'H' $\frac{1}{16}$ 2	hHh'h' $\frac{1}{16}$ 1	hhh'H' $\frac{1}{16}$ 1	hhh'h' $\frac{1}{16}$ 0

Figure 6.8. Expected proportions of offspring with different combined numbers of H and H′ alleles, from a mating of two individuals heterozygous for H and h and for H′ and h′. Numbers at bottom of squares indicate number of semi-dominant alleles for increased height (H or H′). (See text for more information.) This result corresponds to the frequency distribution shown in Figure 6.5C.

stayed in Switzerland. There was a difference of about 4 cm, however, between the average height of California-born children of Swiss immigrants and that of Switzerland-born children of nonimmigrants. At the same time, there is a strong genetic component in the determination of height. The continuous range of heights could be due to differences in genetic or environmental factors, or a mixture of the two.

Controlled Experiments

With experimental plant or animal material, it is relatively easy to distinguish, at least in part, between genetic and environmental effects, especially when we are dealing with single gene differences. We can control the environment so that differences between individuals raised in the same environment are then presumably genetically determined. Or, we can take individuals from pure lines, which are genetically homozygous and so breed true for a given trait, and rear them in different environments to see whether specific environmental factors can modify the given trait. Such experiments are possible in certain plants which inbreed naturally, producing homozygous strains (Johannsen, 1903), and also in animals, such as mice, where inbred strains are available.

But, there are other difficulties. We cannot simply subtract environmental effects from genetic effects, or counterpose them to one another, because they are interdependent and interact with one another. For instance, normal red-eyed fruit flies have the same amount of pigment regardless of the temperature at which they are raised. So do mutant white-eyed flies, which lack eye pigments altogether, at any temperature. One would say that the difference between the red- and white-eyed strains is determined entirely by genetic difference. On the other hand, the amount of pigment and hence the eye color in flies with the mutant gene *blood* shows a continuous range of variation, depending on the temperature. Knowing this, one would say that the environment plays a major part in determining eye-color differences, but the gene difference between red and *blood* plays a part, too. If one confined experiments to higher temperatures, however, where the difference between *blood* and normal eyes is large, one would say that the genetic difference is relatively more important than the environmentally produced differences.

Thus, the importance of genes in determining the phenotype depends on environmental conditions, and vice versa. Some genes respond more than others, and, as a matter of fact, even in opposite ways, to the same environmental change. Given a genotype, we cannot always predict what the phenotype will be in an untested environment. The effects of some mutations can virtually be nullified by the proper treatment. For instance, a diet low in the amino acid phenylalanine can mitigate the effects of the gene for phenylketonuria, an enzyme deficiency that results in brain damage (Cavalli-Sforza and Bodmer, 1971).

THE HUMAN CONDITION

When dealing with human beings, the difficulties described above are compounded. It is much more difficult to distinguish between genetic and environmental causes of continuous variation. If, for instance, one wants to determine to what extent different subgroups in the population (such as economic, ethnic, or racial) differ in height, and the relative importance of nature and nurture in determining such differences, we can neither make experimental matings nor control cultural differences (such as diet) so as to provide the same developmental conditions for samples from all groups.

Intelligence

Concerning characteristics such as intelligence, the difficulties are virtually insuperable, including the question of whether IQ is a valid measure of intelligence. Nevertheless, various claims have been made about intelligence (Herrnstein, 1971; Jensen, 1969; Shockley, 1967). Jensen, for instance, claims that 80% of differences in intelligence are genetically determined and only 20% are due to environment. In fact, such statements are not meaningful (Feldman and Lewontin, 1975; Layzer, 1974). As we have just considered, the value for the genetic component is valid at best only in the limited range of environments available. The value for the environmental component depends on the nature of the genes involved and of their alternative alleles and their relative frequencies in the population being studied.

Incidentally, much of the data used by Jensen and other hereditarians was derived from the papers of Cyril Burt, whose work has recently been seriously called in question (Hearnshaw, 1979).

Twin Studies

Some of the statements about heredity and IQ are based on twin studies. There are two types of twins, monozygotic and dizygotic, sometimes called *identical* and *nonidentical* (or *fraternal*). Monozygotic twins develop from a single fertilized egg or zygote. After the zygote has started to divide and form an embryo, the cells separate into two (or more) clusters which continue to develop as separate embryos, instead of remaining together to form a single individual. The twins (triplets, quadruplets, etc.) formed in this way are genetically identical to one another and of the same sex, as they develop from a single cell. Dizygotic twins, as indicated by the term, develop from two separate zygotes. They may therefore be of the same or different sex, and are no more alike genetically than siblings born at different times.

Any differences between a pair of identical twins must obviously be due to environmental influences or to newly arisen mutations. Even though environmental differences may be minimal when twins are brought up together, these differences could be greater in cases where they are adopted separately and reared apart. For nonidentical twins of the same sex, reared together, differences in upbringing also tend to be minimal, and differences between the two members of a pair might be expected to be largely genetic. It is, however, very hard to determine the extent to which the environment differs for monozygotic and dizygotic twins reared together. We only know that one member of a pair of monozygotic twins may be right-handed and one left-handed, that they may differ in weight at birth, and that one of a pair may generally show more initiative than the other. As the genes are the same, this implies that they have been subjected to different environmental influences, or have newly arisen mutations.

Table 6.2 gives some average differences between pairs of twins. The figures come from three different sets of data. In each set, some of the monozygotic twins were brought up together and some separately. The dizygotic twins were pairs of the same sex.

It appears that the environmental range in these studies was limited. The point has been made, moreover, that the number of cases collected to date falls far short of the number required to give a statistically significant result.

CONCLUSIONS

A vast store of genetic variation is available to the human species. It arises through independent assortment and recombination of genes at meiosis, as well as by the occurrence of new mutations.

The continuous range of variation shown by characters such as height, weight, and IQ is influenced by a large number of genes as well as by environmental factors. With experimental material it is possible to demonstrate fairly clearly the effects of multiple genes, on the one hand, and of environment on the other. Studies of twins give some information on the separate effects of nature and nurture in humans, but they are limited in two ways. First, not enough twins have been available for study. Second, and even more important, only a very limited range of environments has been covered in the cases studied.

Some points can be made about quantitative variations, however.

Table 6.2. Mean differences and correlations for height, weight, and IQ for sets of monozygous twins reared together and reared apart, and of dizygous twins.

Trait measured	Monozygous, together	Monozygous, apart	Dizygous[a]	Authors
Height in cm	1.7 (+0.932)	1.8 (+0.969)	4.4 (+0.645)	Newman et al., 1937
	1.3 (+0.98 ♂)	2.1 (+0.82 ♂)	4.5	Shields, 1962
	(+0.94 ♀)	(+0.82 ♀)	(+0.44 ♀)	Shields, 1962
	(+0.962)	(+0.943)	(+0.472)	Burt, 1966
Weight in lb	4.1 (+0.917)	9.9 (+0.886)	10.0 (+0.631)	Newman et al., 1937
	10.41 (+0.79 ♂)	10.5 (+0.87 ♂)	17.3	Shields, 1962
	(+0.81 ♀)	(+0.37 ♀)		Shields, 1962
	(+0.929)	(+0.884)	(+0.586)	Burt, 1966
IQ	5.9 (+0.881)	8.2 (+0.670)	9.9 (+0.631)	Newman et al., 1937
	7.4	9.5	13.4	Shields, 1962
	(+0.76 ♀)	(+0.77 ♀)	(+0.51 ♀)	Shields, 1962
	(+0.925)	(+0.874)	(+0.453)	Burt, 1966

Correlations in parentheses.
[a] Some reared apart.
From *The Genetics of Human Populations*, L. L. Cavalli-Sforza and W. F. Bodmer. W. H. Freeman and Company. Copyright © 1971.

First, there are no human pure lines. Therefore, when the extreme phenotypes at each end of a continuous range are due in part to particular constellations of alleles of a large number of genes, our understanding of recombination and independent assortment tells us that such constellations are very unlikely to be inherited intact. The children of exceptional individuals (for height, IQ, or whatever) will not necessarily be exceptional in the same respect. Conversely, an exceptional constellation can arise fortuitously through the combination of alleles inherited from two parents whose genotypes for the trait in question were both in the middle of the range.

As far as sex differences in quantitatively varying traits are concerned, even if the range of variation differs for men and women, we can be fairly sure that the ranges overlap (Figure 6.6). Within this area of overlap are women and men who are equal in the trait being measured. If we set up male and female stereotypes or ideals for which the norms fall in nonoverlapping regions of the two curves, we tend to push all individuals in the overlapping region toward their respective (female or male) norms, increasing conformity and restricting the acceptable range of variation for both men and women. If, on the other hand, we are open-minded about what is "masculine" and "feminine," and regard the overlap area as acceptable for both sexes, we permit men and women alike a greater range of variation and individuality. We would certainly take this attitude toward the range of variation in height: why not also for other traits such as assertiveness, intuitiveness, or the expression of feelings? On this basis, the fear of losing important differences between males and females, suggested by Margaret Mead, seems to be illusory. In addition to this, apart from the characteristics of the reproductive systems, the same qualitative differences can appear in both sexes. It would seem, then, that the way to combat homogeneity and maximize individuality in our society would be to recognize and utilize the whole spectrum of variation found in men and women, rather than selecting masculine and feminine norms,—and obliterating by socialization the overlapping parts of the male and female ranges between these norms.

REFERENCES

Bleier, R. 1979. Social and political bias in science: an examination of animal studies and their generalizations to human behavior and evolution. *In* R. Hubbard and M. Lowe (eds.) Genes and Gender II. Gordian Press, Staten Island, N.Y., pp. 49-69.

Burt, C. 1966. The genetic determination of differences in intelligence: A study of monozygotic twins reared together and apart. Brit. J. Psychol. 57:137-153.

Cavalli-Sforza, L. L., and W. F. Bodmer. 1971. The Genetics of Human Populations. W. H. Freeman & Co., San Francisco.
Feldman, M. W., and R. C. Lewontin. 1975. The heritability hang-up. Science 190:1163-1168.
Hearnshaw, L. S. 1979. Cyril Burt, Psychologist. Cornell University Press, Ithaca, N.Y.
Herrnstein, R. J. 1973. I.Q. in the meritocracy. Atlantic Monthly Press 228: 43-64.
Herskowitz, I. H. 1965. Genetics, 2nd Ed. Little, Brown, & Co., Boston.
Hulse, F. S. 1968. The breakdown of isolates and hybrid vigor among the Italian Swiss. Proceedings of the 12th International Congress Genetics 2:177. (Science Council of Japan, Tokyo.)
Jensen, A. R. 1969. How much can we boost IQ and scholastic achievement? Harvard Educational Review 39:1-123.
Johannsen, W. 1903. Uber Erblichkeit in Populationen und reiner Linien. [On heritability in populations and pure lines.]. Fischer, Jena.
Layzer, D. 1978. Heritability analysis of I.Q. scores: Science or numerology? Science 183:1259-1266.
McCammon, R. W. 1970. Human Growth and Development. Charles C Thomas, Springfield, Ill.
Mead, M. 1949. Male and Female: A Study of the Sexes in a Changing World. William Morrow & Co., Inc., N.Y.
Newman, H. H., F. N. Freeman, and K. J. Holzinger. 1937. Twins: A Study of Heredity and Environment. University of Chicago Press, Chicago.
Oakley, A. 1972. Sex, Gender and Society. Harper, San Francisco.
Raymond, J. G. 1979. Transsexuals: An issue of sex-role stereotyping. *In* R. Hubbard and M. Lowe (eds.), Genes and Gender II: Pitfalls in Research on Sex and Gender. Gordian Press, New York.
Shields, J. 1962. Monozygotic Twins Brought Up Apart and Brought Up Together. Oxford University Press, London.
Shockley, W. 1967. A "Try Simplest Cases" approach to the heredity-poverty-crime problem. Proc. Nat. Acad. Sci. 57:1767-1774.
Skeels, H. M. 1966. Adult Status of Children with Contrasting Early Life Experiences: A follow-up Study. The Society for Research in Child Development, Monograph Series, Vol. 31, no. 3.
Srb, A. M., R. D. Owen, and R. S. Edgar. 1965. General Genetics. 2nd Ed. W. H. Freeman, San Francisco.

7 Endocrine Glands

GENERAL PROPERTIES

The endocrine glands produce substances called hormones that are secreted directly into the blood vessels, and are transported by the blood throughout the body. Hormones affect the metabolism of cells in other parts of the body. In this chapter, all major endocrine glands are mentioned, and the chemical nature as well as the functions of the hormones related to the female genital system are also given briefly. The next chapter is devoted to a description of the microscopic structure of the glands most involved in the biology of women.

Relation to Blood Vessels

The primary relation of endocrine glands is to blood vessels. Regardless of their origin they secrete their products internally, into blood vessels, in contrast with exocrine glands, which secrete onto a surface.

Generally speaking, the endocrine gland cells are in more intimate contact with blood vessels than exocrine gland cells, as there may be less intervening material between the cell surface of the endocrine gland cell and the lining cells of the minutest blood vessels called *endothelial cells.* In addition, the endothelial cells may be thinner-walled in regions adjacent to endocrine cells, than in regions where exocrine cells are present. Also, the endothelial cells adjacent to endocrine cells may be more active in fluid transport (pinocytosis) and may be characterized by having many fenestrae—that is, places where the cytoplasm is so thin, if it exists at all, as to be invisible even with the electron microscope (Figure 7.1). These structural features are presumed to be adaptations for more efficient operation of endocrine functions.

Relation to Ducts and Surfaces

In some instances, an organ may have mixed functions, exocrine and endocrine. For example, in the pancreas, the islets of Langerhans constitute the endocrine portion. These may vary in size from 15 μm (one or two

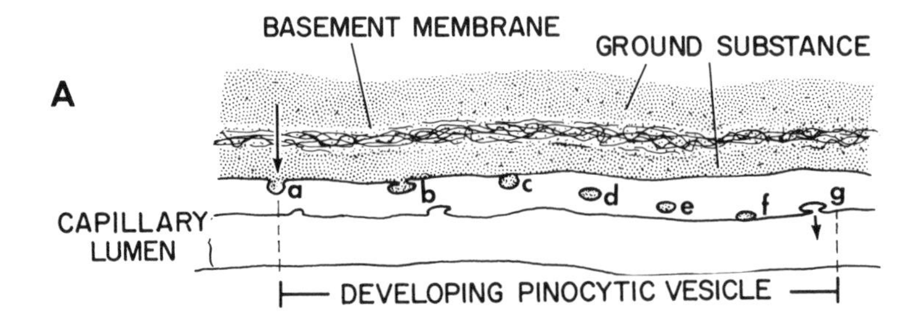

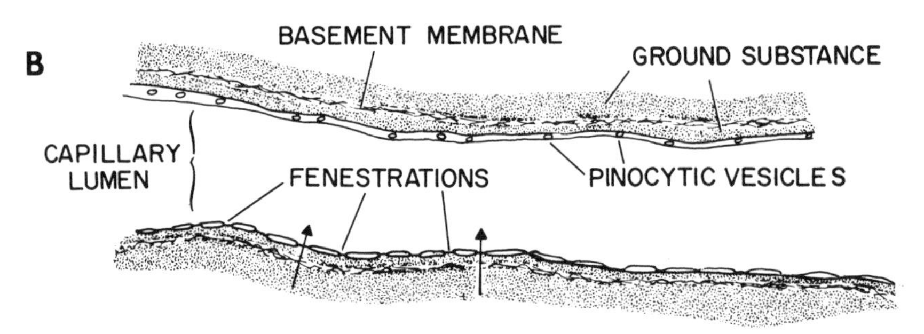

Figure 7.1. A) Passage of fluid from outside a capillary to blood plasma in the lumen by the process of pinocytosis in the cytoplasm of the endothelial cell. This could be how some hormones enter the circulation, or leave when the process is reversed. B) Passage of fluid from outside a capillary to the blood plasma through thin membranous windows (fenestrae) interposed between thicker cytoplasmic portions of the endothelial cell. The basement membrane may also be thinner or even absent in such capillaries. Basement membranes are composed of collagen and protein-polysaccharide complexes of the ground substance. (After Junqueira, Carneiro, and Contopoulos, 1975).

cells) to 1 mm or so. All of them are attached to columns of cells, which are related to ducts, or arose embryologically from ducts and later separated from them. But they secrete into the blood, not into the ducts. Most (90% or more) of the gland is exocrine, and secretes its product into a duct that empties into the upper part of the intestine. Similarly, many endocrine glands arose embryologically from some surface, from which they later became separated, the final contact with the rest of the body being through the vascular system. This is true of the anterior lobe of the hypophysis and the thyroid gland.

Size and Compactness

The size of the endocrine structure varies considerably from one or a small number of cells (islets of pancreas) to a structure visible with the naked eye (parathyroid, thyroid, pituitary gland). The endocrine structure can be diffuse (interstitial cells of the testis) or clearly demarcated (corpus luteum). Some are permanent (adrenal cortex or medulla), some temporary (corpus luteum, egg follicle of the ovary). Or they can be limited both in time and to only a fraction of the population, and marked by sudden death or atrophy, like the placenta and the hormones it secretes until it separates from the uterine wall before its expulsion at the end of pregnancy.

Variation in Endocrine Chemistry

The hormonal products that are manufactured (synthesized) by endocrine gland cells vary greatly and include: modified amino acids (thyroxine, adrenalin); polymers of amino acids called polypeptides, some of which are small (three amino acids), some fairly large (51 amino acids); glycoproteins, which are large polypeptides to which are attached a sugar polymer or polysaccharide; and steroids. These terms are defined in the next section.

Receptor Sites and Cell Activation

The products synthesized by endocrine gland cells always involve the action of enzymes in the gland cells. Endocrine gland cells have receptor sites on their surface. When these receive and combine with a stimulus, which arises in most instances from the blood, a chain of events occurs that influences the hereditary or genetic molecules (DNA) in the nucleus. The appropriate parts of the DNA make a corresponding messenger, and ribosomal and transfer RNA. These pass into the cytoplasm and there result in the synthesis of the hormones or of the appropriate enzymes necessary for the synthesis of the hormone elsewhere in the cytoplasm. Some of the hormones are stored in large quantities (in the thyroid gland and adrenal medulla), while others are stored in minimal amounts (in the stroma cells of the ovary).

Target Organs and Systems

Some hormones have very general actions on all cells, for example, thyroglobulin. It affects the basal metabolic level of all cells in the body. Some hormones have general, but more limited actions; e.g., parathormone and calcitonin produced in the parathyroid and thyroid glands respectively affect calcium metabolism; aldosterone produced in the adrenal cortex affects salt metabolism. Some hormones have more restricted actions, the effects being limited primarily to certain sites called *target organs.* For example, estrogen produced in the ovary affects chiefly the female genital tract and the mammary glands, which have receptors for the hormone.

Integration

Hormones collectively are one of the two major integrative systems in the body; the other is the nervous system. The endocrine system acts more slowly than the nervous system, and is more general in its action; it lacks the precision and delicacy of the nervous system (or, at least, the greater part of it). The two integrative systems are in turn integrated with each other.

HYPOPHYSIS (PITUITARY GLAND)

The hypophysis is suspended from the hypothalamus at the base of the brain, and lies in a recess of the sphenoid bone above the roof of the mouth, about two-thirds of the way back from the lips. It is further protected by its enclosure in the meninges, which are membranes enclosing the central nervous system for protection and for support of the blood vessels to the regions of the brain (Figure 7.2). At its point of attachment are many blood vessels, some of which constitute the arterial supply and venous drainage of the hypophysis and hypothalamus. The gland itself is rounded, and approximately 14 mm × 9 mm × 6 mm; and the stalk attaching it to the hypothalamus is about 2 mm in diameter.

When observed in a midsagittal section, the pituitary gland is made up of two major subdivisions: the anterior lobe and the posterior lobe (Figure 7.3), usually separated in mammals by a cleft. In people, the anterior lobe is about five times the size of the posterior lobe. The posterior lobe is made up of two parts that interdigitate in many places: a neural part, the pars nervosa or infundibular process, connected to the

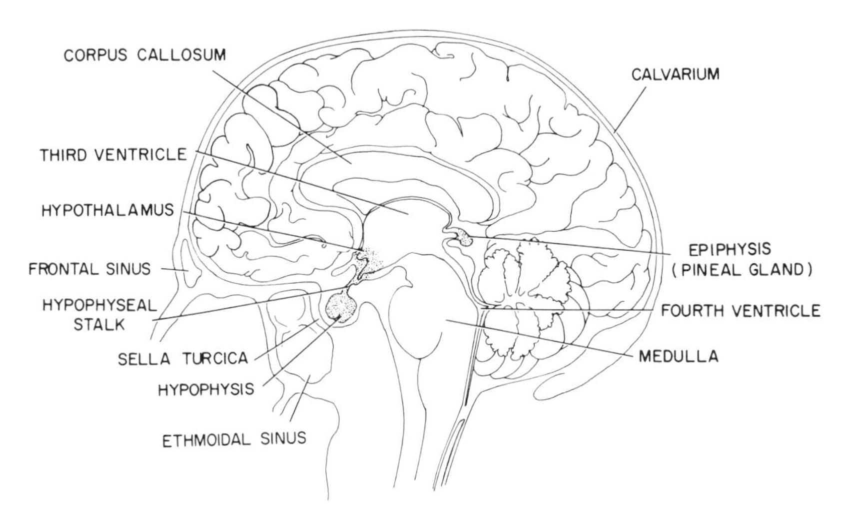

Figure 7.2. The hypophysis is a single small gland that lies at the base of the brain attached by a short stalk to the hypothalamus. It consists of three parts: an anterior lobe, derived embryologically from the epithelium of the roof of the mouth, an intermediate part derived from the same region, and a posterior lobe, derived from the base of the brain. Regardless of their embryologic origins, the various parts are connected with the rest of the body and to each other by the circulating blood. The pineal gland, like the neural portion of the hypophysis, is also neural in origin.

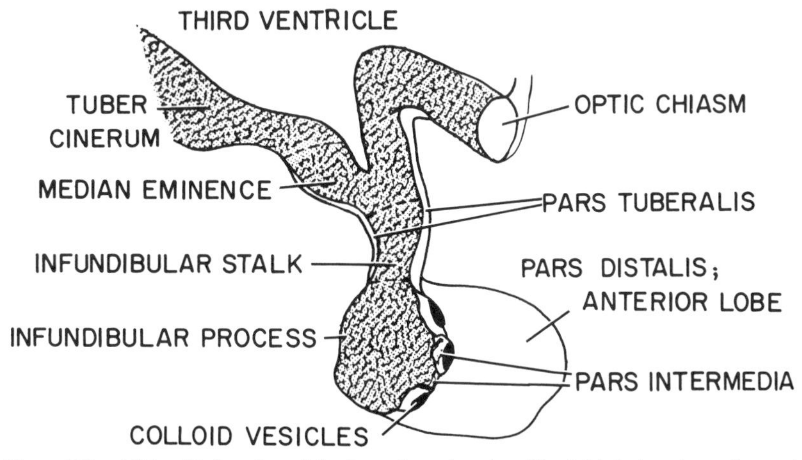

Figure 7.3. Midsagittal section of the hypophyseal region. The table below shows the parts arranged according to their embryologic origin. (After Bloom and Fawcett, 1975).

Adenohypophysis (Rathke's pouch)	Pars distalis (anterior lobe)	
	Pars tuberalis	
	Pars intermedia	Posterior lobe
Neurohypophysis	Pars nervosa (infundibular process)	
	Infundibulum	Infundibular stem Median eminence of tuber cinereum

hypothalamus by the stalk, and an epithelial part (the intermediate part or pars intermedia) that is derived in embryos from the same part of the epithelium of the roof of the mouth that gives rise to the anterior lobe. Some epithelial cells, known as the *tuberal part*, or *pars tuberalis*, of the pituitary gland, partly enclose the neural stalk, and may reach the base of the brain.

The pituitary gland secretes a number of very important hormones, all polypeptides. The amino acids comprise the essential building blocks of polypeptides. There are 20 or more amino acids, all of them characterized by the presence of a carboxyl (COOH) group and an amine (NH_2) group attached to a carbon atom. Below are shown the structural formulae of two common amino acids (alanine and glycine). When a molecule of water is removed from them under the proper conditions, they combine to form a dipeptide, glycylalanine:

$$\begin{array}{c} \mathrm{H{-}N(H){-}CH(H){-}C({=}O){-}OH} \\ \text{GLYCINE} \end{array} + \begin{array}{c} \mathrm{HN(H){-}C(H)(CH_3)(H){-}COOH} \\ \text{ALANINE} \end{array} \longrightarrow \begin{array}{c} \mathrm{H{-}N(H){-}C(H)(H){-}C({=}O){-}N(H){-}C(CH_3)(H){-}C({=}O){-}OH} \\ \text{GLYCYLALANINE} \\ \text{(a dipeptide)} \end{array}$$

In this way, tri-, tetra-, penta-, and polypeptides may be built, each with its backbone,

$$-N-C-\overset{O}{\overset{\|}{C}}-N-C-\overset{O}{\overset{\|}{C}}-N-$$

and which may be several hundreds of amino acids long. Not all of the polypeptide is necessarily essential for the hormonal activity, since certain parts of the polypeptide may be removed without affecting the activity of the residue. The active part of the polypeptide is called the *essential* or *active site* and without it, activity is essentially nil.

The Anterior Lobe

The following hormones are synthesized and secreted by the cells of the anterior lobe of the hypophysis. The hormones are carried by the blood to the various target sites.

1. Adrenocorticotropic hormone (ACTH) is a polypeptide containing 39 amino acids, of which 23 are essential. It stimulates the synthesis and secretion of steroid hormones from the adrenal cortex. It also affects the general metabolism of the body by stimulating the uptake of glucose and amino acids by muscle, and lipolysis by fat cells.
2. Thyroid stimulating hormone (TSH) is a glycoprotein—a large protein that contains a sugar polymer as an integral part of the molecule. It consists of two polypeptides joined by many disulfide bonds. The α chain contains 96 amino acids, and the β chain 113 amino acids. It stimulates the thyroid gland to form and secrete thyroxine and triiodothyronine. These hormones influence the basal metabolism of the cells of the body and stimulate fat tissue to release lipid.
3. Somatotropic hormone (STH) or growth hormone (GH) is a fairly large polypeptide (or protein) consisting of a single chain of 191 amino acids. It enhances protein metabolism, DNA and RNA synthesis, and lipolysis generally throughout the body, but more especially in the epiphyseal (or growth) regions of cartilage, whose growth results in increased length of long bones.
4. Luteinizing hormone (LH) (in women) or interstitial cell-stimulating hormone (ICSH) (in men) is a glycoprotein that consists of three parts, an α chain of 96 amino acids, and a β chain of 119 amino acids (joined by many disulfide bonds) and a carbohydrate polymer. In women, under suitable conditions, it causes the ripened follicle to rupture, thus releasing the egg. The residual cells together with surrounding thecal cells are converted to a corpus luteum. The hormone also controls the synthesis and secretion of progesterone (also, to a lesser degree, that of estrogen). In men, it helps maintain the interstitial cells of the testis and stimulates them to secrete androgens.

5. Follicle stimulating hormone (FSH) is glycoprotein also comprising two chains, an α chain of 89 amino acids and a β chain of 115 amino acids. In women, it causes Graafian follicles to mature; acting with LH, it causes ovulation to occur. It also controls the synthesis and secretion of estrogen by the follicular cells.
6. Prolactin (PRL), also called *lactogenic hormone*, is a large polypeptide, or protein, consisting of a single chain of 198 amino acids. Under suitable conditions (see Chapter 16) prolactin promotes the growth of gland cells of the mammary gland and is involved in milk production and expulsion.

The Hypothalamus and the Anterior Lobe

In recent years, tremendous advances have been made in analyzing the controlling effects of hormones of the lower levels of the hypothalamus on the cells of the anterior lobe of the hypophysis. The cells that secrete the hormones are nerve cells, and the process is an example of neurosecretion. The hormones are secreted into microscopic blood vessels of the hypothalamus. These are gathered together to form larger vessels. These course down the infundibular stalk and break up again in the anterior lobe into capillaries (Figure 7.4). By this mechanism, the neurosecretory products of the hypothalamus are delivered in concentrated form into the circulation of the anterior lobe, without the dilution and loss of effectiveness that would have followed secretion into blood vessels leading directly to the heart. The regulatory hormones seem to be small polypeptides and those that are known range from three to fourteen amino acids. The convention is to refer to them as hormones when their chemical composition and structure are known. Until they can be characterized chemically, they are referred to as *factors*.

Releasing Factors These factors include the following:

1. Corticotropin releasing factor (CRF) is a part of a feedback system concerned with the release of hormones by the adrenocorticotropic cells of the anterior lobe of the hypophysis.
2. Thyrotropin releasing hormone (TRH), a tripeptide, is a part of a feedback system concerned with the release of thyrotropin or thyroid stimulating hormone (TSH) by appropriate cells of the anterior lobe of the hypophysis.
3. Luteinizing releasing hormone is a decapeptide that acts to release LH *and* FSH by appropriate cells of the anterior lobe of the hypophysis. The evidence indicates that the releasing hormone (which has these two actions) is a single decapeptide, and for this reason it is referred to as *LH/FSH-RH* or *gonadotropic* releasing hormone (GN-RH).

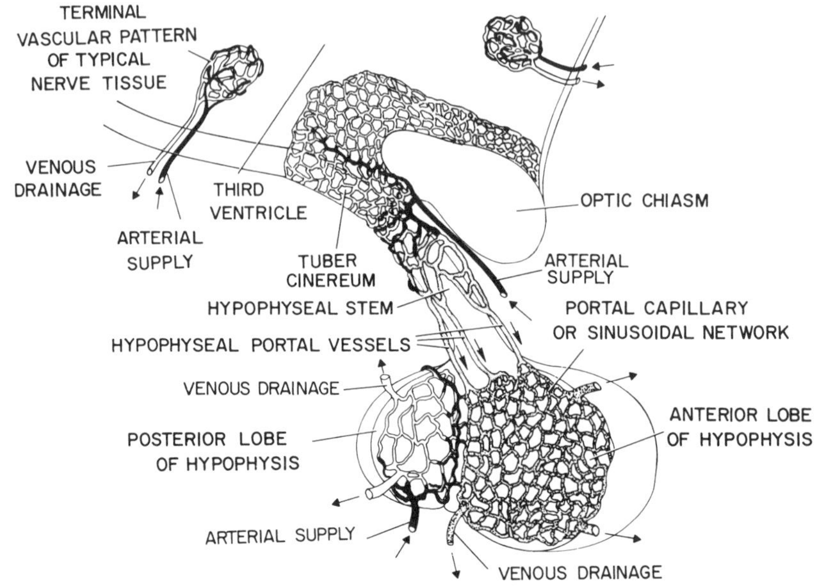

Figure 7.4. The vascular interrelations of hypothalamus and hypophysis as currently conceived by most neuroendocrinologists. The capillaries of the central nervous system are very largely terminal. But in the hypothalamus, the capillary net or plexus is not drained directly into the venous system to return the blood to the heart. Instead, the capillaries form "portal veins" that course along the hypophyseal stalk, only to break up again into capillaries, known as *sinusoids*, in the anterior lobe of the hypophysis. The blood only then enters the venous drainage pattern of the anterior lobe. The capillary bed of the neurohypophysis resembles that of most of the central nervous system in that it is largely terminal and drains directly into the venous system. There are some anastamoses between these two circulatory systems across the intermediate part of the hypophysis. (After Junqueira, Carneiro, and Contopoulos, 1975).

4. Growth stimulating-releasing factor (GHRF) or somatotropin releasing factor (SRF) is concerned with the release of somatotropin by the cells synthesizing it in the anterior lobe of the hypophysis.

Release-Inhibiting Substances Also secreted by the neuroendocrine cells of the hypothalamus are two release-inhibiting substances. These cause the cells of the anterior lobe of the hypophysis that are specifically affected to *inhibit* the release of their secretory products.

1. Growth hormone release-inhibiting hormone (GHIH) or somatotropin release-inhibiting hormone (SRIH) or somatostatin is a tetradecapeptide that probably is part of a complex feedback system controlling metabolic aspects of growth.
2. Prolactin release-inhibiting factor (PIF) may be concerned with the

onset of lactation—its secretion must be inhibited before lactation may take place.

The Hypothalamus and the Posterior Lobe of the Hypophysis

Two hormones, oxytocin and antidiuretic hormone, are secreted by the neural portion of the hypophysis and certain cells of the hypothalamus. These hormones arise in certain nerve cells, especially in the supraoptic and paraventricular nuclei of the hypothalamus, and stream down the nerve fibers into the neural portion where they are stored for later release (Figure 7.5). Both are small polypeptides, each of nine amino acids, each bound to a specific protein called *neurophysin.*

1. Oxytocin is thought to be involved in the contraction of smooth muscle, especially in the uterus of pregnant women near term, and in that way to aid in the expulsion of the fetus. It is also thought to be one of a series of factors important in lactation, aiding in the release of milk through contraction of myoepithelial cells around secretory parts and ducts of the mammary gland.
2. Antidiuretic hormone (ADH) or vasopressin, affects large arteries, causing them to contract and raise blood pressure, and is involved also in the conservation of water by the kidney.

Pars Intermedia

The intermediate part of the posterior lobe produces melanocyte-stimulating hormone (MSH). Its function in people is unknown. In amphibians it acts on pigment cells, causing the pigment granules in them to spread out, resulting, for example, in a darkening of the skin when a frog is kept in the dark. There is also a melanocyte-stimulating (hormone) releasing hormone (MRF).

ADRENAL GLANDS

Medulla

Each adrenal gland sits anteromedially (or suprarenally) on each kidney. This surface is concave, adapting its shape to the surface of the kidney. The other surfaces are somewhat conical or suggestively pyramidal (Figure 7.6A). It is roughly 3-5 cm × 2.5-3.0 cm × 1 cm. The central portion is called the medulla, and is surrounded by the cortex (Figure 7.6B). The medulla is dark in color because of its contained blood, and frequently so is the inner part of the cortex. The medullary cells are derived from sympathetic ganglion cells in embryos, while the cortex arises in the embryo from the adrenogenital ridge. The extremely rich vascularity of both parts is intimately interrelated.

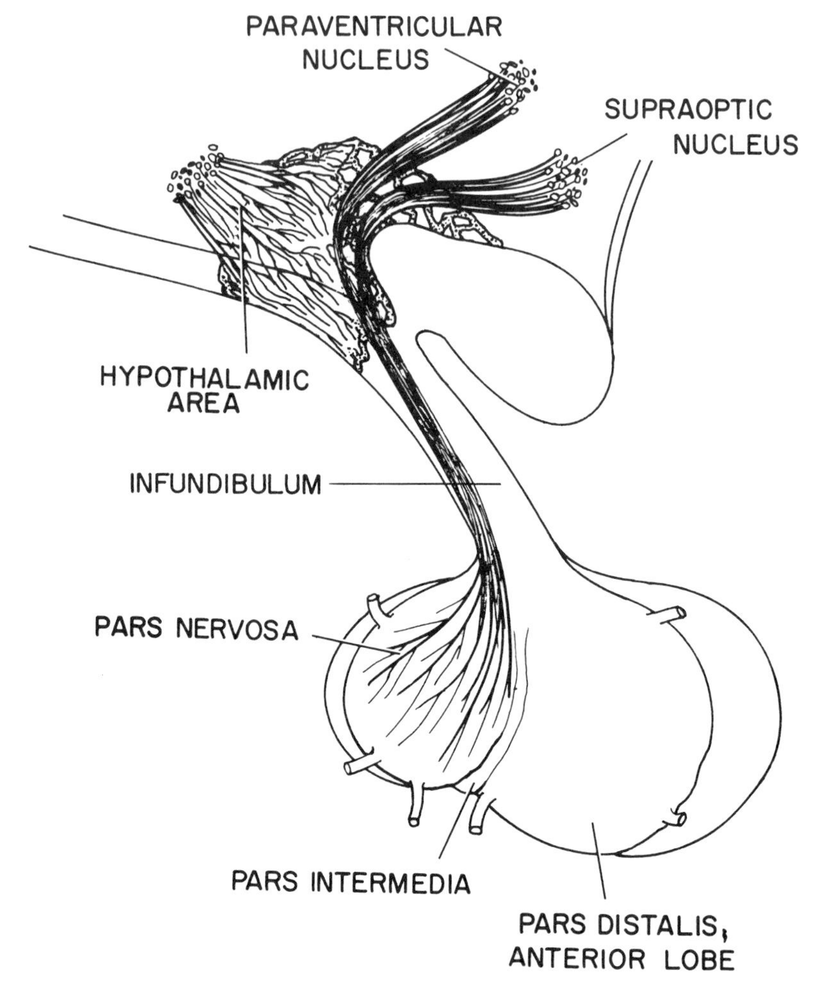

Figure 7.5. Superposition of the cell secretory pattern on the vascular pattern to show the secretory pathways of the neurohypophysis and hypothalamus. (Refer to Figure 7.4 for relations of nerve fibers to the capillary bed, tuber cinereum, and hypophyseal stalk, and also for the relations to the portal vessels and capillaries of the neurohypophysis.) There are several secretory pathways: 1) The secretory products, the various hormones and release-inhibiting factors, arise in the nerve cell bodies of the hypothalamus and pass along the nerve fiber to the first capillary bed in the tuber cinereum and hypophyseal stalk. The secretory products then enter the portal vessels in the stalk and pass in them to the anterior lobe of the hypophysis. There the secretory products leave the vascular system and become attached to the target cells. 2) The target cells in the anterior lobe of the hypophysis secrete their hormones directly into the capillaries or sinusoids and only then enter the general circulation of the body. 3) The secretory products of the supraoptic and other cell groups in the hypothalamus pass along the nerve fibers, where they are stored and then secreted extracellularly. They then enter the capillaries of the neurohypophysis to be carried into the general circulation. (After Junqueira, Carneiro, and Contopoulos, 1975; Noback and Demarest, 1975; and Carpenter, 1975).

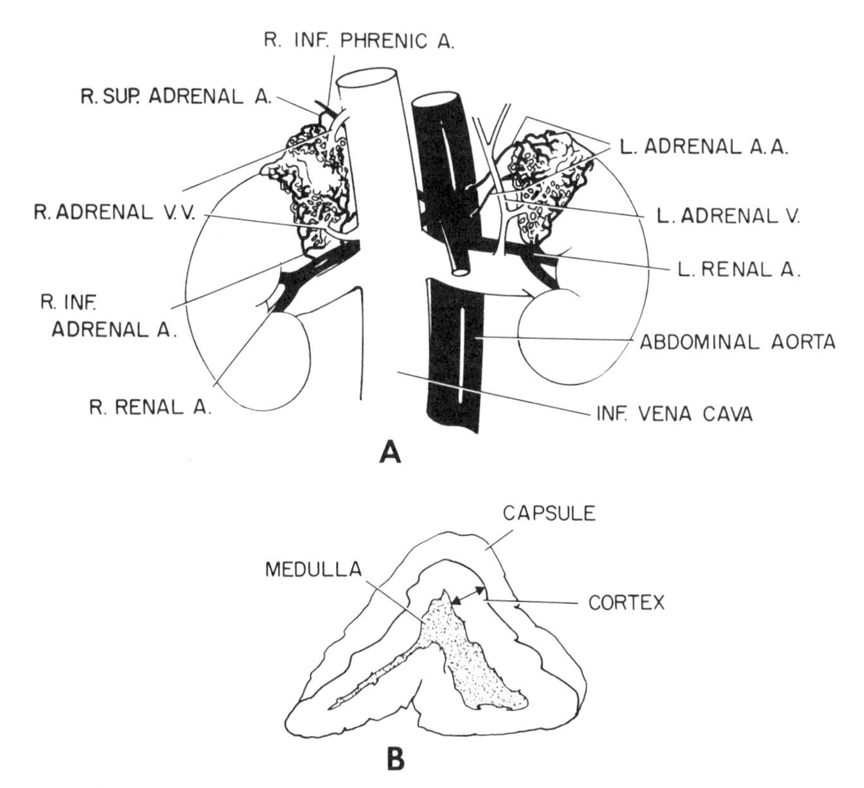

Figure 7.6. A) The adrenal glands are located on the anterior pole of each kidney. B) Their shape is highly irregular, and in section they appear tricornuate.

The medulla produces two hormones, both derivatives of the amino acid phenylalanine, epinephrine and norepinephrine.

Phenylalanine

$$C_6H_5—CH_2—CH(NH_2)—COOH$$

Epinephrine

$$(HO)_2C_6H_3—CH(OH)—CH(OH)—NH—CH_3$$

Norepinephrine

$$(HO)_2C_6H_3—CH(OH)—CH_2—NH_2$$

Epinephrine affects skeletal, cardiac, and smooth muscle, especially that in arterioles. Thus it is important for cardiac action and blood pressure. Through promoting glycogenolysis and lipolysis it affects the level of blood sugar. Norepinephrine has similar actions.

Cortex

The adrenal cortex is mostly yellowish, but the inner part of the cortex may be dark because of the blood it contains. The outer portion of the cortex just beneath the capsule is called the *glomerular zone*; the inner portion is called the *reticular zone*; the intermediate portion is called the *fascicular zone*. The zones are named because of the arrangement of the gland cells, and this arrangement is in relation to the organization of the microscopic vessels that course through the cortex from the outside to the medulla (Figure 7.7).

All three zones of the cortex synthesize many steroid substances, all of which are related chemically and form a series in which one can be converted to another through the action of cellular enzymes. All of them are derived from cholesterol, which is synthesized by the cells from very small and simple building blocks; i.e., acetate. Some simple relations of various steroids are shown in the diagrammatic structural formulae in Figure 7.8.

Aldosterone is synthesized and secreted into the blood by the cells of the outer, or glomerular zone. Progesterone, cortisol, corticosterone and probably other steroids are synthesized and secreted by the intermediate, or fascicular, zone, and probably also elsewhere. Testosterone is synthesized and secreted by the cells of the inner, or reticular zone.

1. Aldosterone acts on various parts of the nephron, the basic unit of renal structure, and regulates the elimination of Na^+ in the urine. This influences markedly the water retained in the plasma, and thus regulates the plasma volume of the body, and through this blood pressure.
2. Deoxycorticosterone has a similar, but weaker, action.
3. Corticosterone also acts similarly, and is intermediate in potency.
4. Cortisone acts on most cells of the body, and in general is antagonistic to insulin. For example, it decreases the synthesis of nucleotides and protein, as well as of triglycerides and other lipids. It also causes a reduction in the synthesis of sulfated mucopolysaccharides.
5. Androgenic substances (including testosterone) are thought to be secreted by the cells of the zona reticularis, the innermost cortical zone.

OVARIES

The ovaries are paired, flattened, and about 2 cm × 3 cm × 1 cm (Figure 7.9). Each is more or less fixed to the lateral wall of the pelvis, in relation

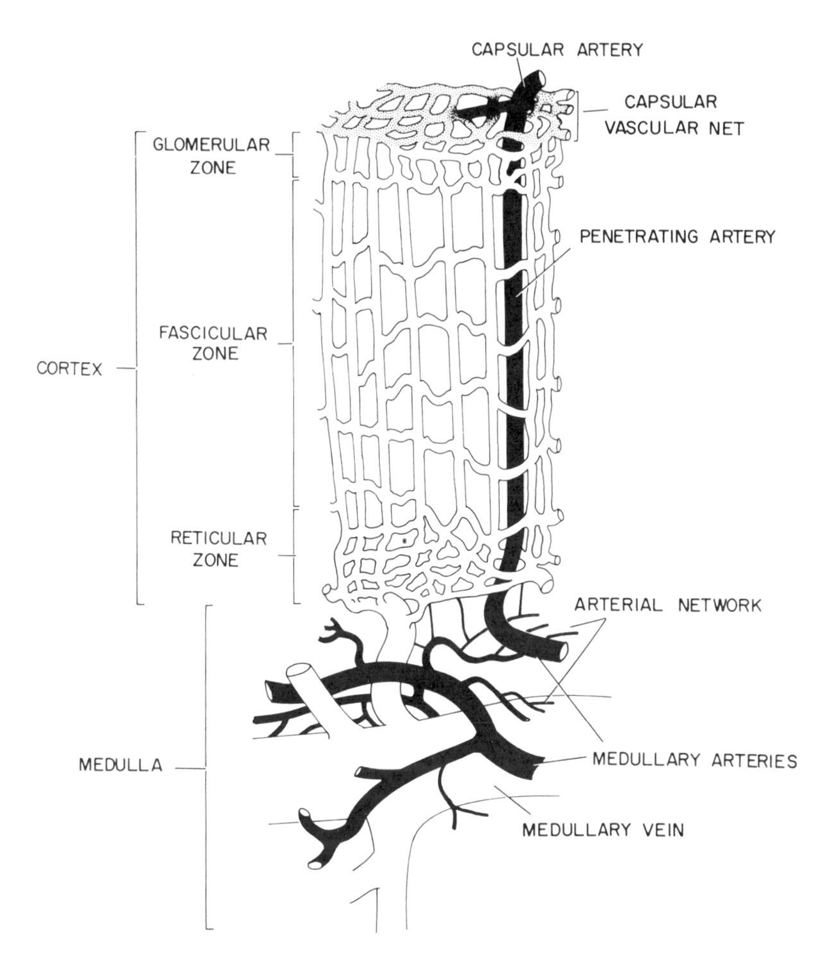

Figure 7.7. Circulatory pattern in the adrenal gland of the mouse, which closely resembles that of people. In the capsule, the arteries break up into capillaries forming a close net. Some of these penetrate the cortex in the outer or glomerular zone. Then the capillaries plunge deeper, forming a wide-meshed net in the fascicular zone. They are gathered together in the reticular zone before ending in branches of the medullary vein. The arterial system of the medulla is almost independent of this capillary circulation, and consists of blood vessels that penetrate the cortex before branching in the medulla. (After Gersh, 1941).

to the Fallopian tube. Their structure and appearance vary with age and with the time of the menstrual cycle. The surface may bulge as the egg follicle develops, and it may be slightly deformed by the yellow corpus luteum. The follicle might have recently ruptured and the site be marked by a fresh clot, or the corpus luteum might have become the corpus albicans, reduced in size and whitish in appearance. The remainder of the ovary, except for the surface cells, consists largely of stroma and blood

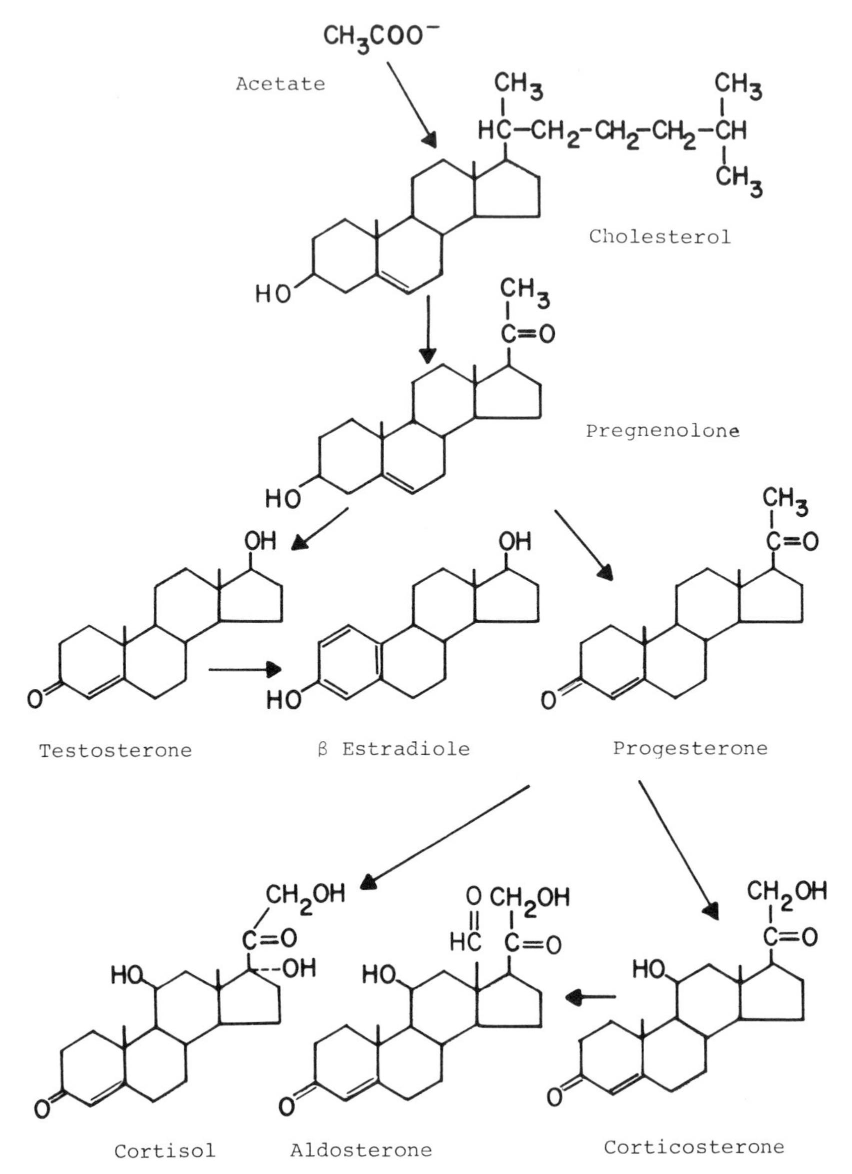

Figure 7.8. Structural formulae of some steroids, presented to show their interrelations and origin from acetate and cholesterol (arrows). At least one enzyme is involved in each metabolic step.

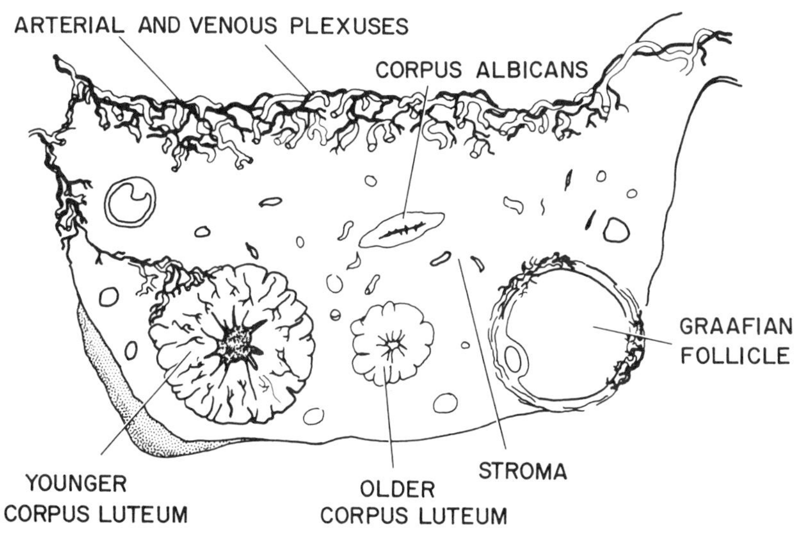

Figure 7.9. The ovaries lie in the pelvis, one on each side, below the oviducts and medial to their fimbriated ends. Vascular relations of the ovary are shown together with ripening Graafian follicles and stages in development of corpora lutea.

vessels. The ovary produces and secretes three hormonal substances or groups of substances:

1. Estrogens are members of the steroid family (see Figure 7.8) and are probably synthesized by cells of the ripening egg follicle and also by the syncytioblast cells of the placenta, and to a very slight extent by cells of the corpus luteum and the adrenal cortex. The levels reached in the blood of menstruating women are cyclic. These hormones act on the female genital tract, stimulate the development of the secondary sex characters, and make an important contribution to the growth and development of the duct system of the mammary gland. They are in general antagonistic to androgenic hormones. The chief estrogen is estradiole, but estrone and estriole are also important. Some estrogen is derived from unknown extraovarian sites via androstenedione, which itself may originate in the stromal cells of the ovary or adrenal cortex. Estrogens form a part of the negative and positive feedback system with the neuroendocrine cells of the hypothalamus and with certain cells of the anterior lobe of the hypophysis (see Chapter 8). This system controls to a large extent the changes that take place during puberty, menstruation, and pregnancy.
2. Progesterones are also members of the steroid family (see Figure 7.8). Progesterone is produced and secreted by the corpus luteum, the adrenal cortex, and the placenta; the concentration it achieves in the

blood is cyclic in menstruating women. In men, some progesterone may arise in the testis as well as in the adrenal cortex. The hormone plays a role in ovulation, the reconstruction of the endometrium after menstruation, the preparation of the uterine wall for implantation of the fertilized egg, and the continuation of pregnancy in the early months. The hormone is part of the feedback system mentioned above.

3. Relaxin is a medium-sized polypeptide produced and secreted by the corpus luteum. It is not known how important this substance is for parturition or for the birth of the infant. In some other mammals (guinea pigs, some bats) it causes enlargement of the ventral parts of the birth canal by causing dissolution of bone, cartilage, and dense connective tissue in the region of the symphysis pubis. This enables passage of the large head of the newborns through the birth canal without excessive trauma.

TESTES

The testes are paired, oval, slightly flattened whitish bodies about 4 cm × 2.5 cm × 2 cm (Figure 7.10). The surface is smooth. Most of the testis is made of seminiferous tubules; crowded between them are clusters of interstitial cells. These produce and secrete into the blood vessels the androgenic substances, chief of which is the steroid testosterone (see Figure 7.8). The hormone affects all the sex organs and secondary sex characters of men. It strongly influences protein synthesis. The level of testosterone in women is low. More than half is derived from some unknown extraovarian source as androstenedione, while the remainder probably originates in the ovary and adrenal cortex (probably the zona reticularis).

PLACENTA

Several hormones are synthesized and secreted by epithelial cells of the placenta:

1. Several steroid hormones (estrogen and progesterone) identical with those originating in the ovary.
2. Human chorionic gonadotropin (HCG) is a large glycoprotein with α and β chains of amino acids. It maintains the corpus luteum and its secretion of progesterone during the first third of pregnancy.
3. Human placental lactogen is a large polypeptide of 191 amino acids. With prolactin, the hormone stimulates growth of the mammary gland during pregnancy.

Other endocrine organs not directly concerned with reproduction are

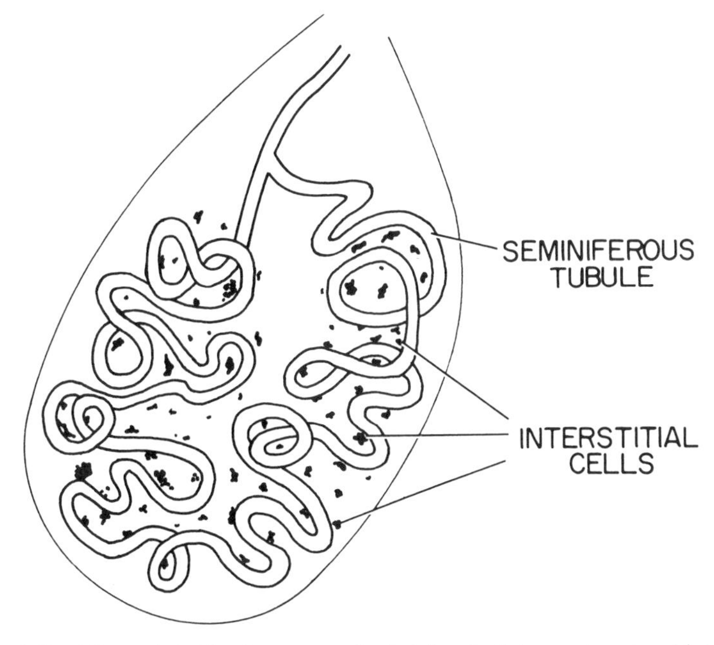

Figure 7.10. The testis is like the pancreas in that they both have external and internal secretions. The external secretion of the testis (the sperm cells) are the product of the seminiferous tubules, which are contained within the many lobules of the testis. The internal secretion of the testis is produced by the interstitial cells, which occupy the interstices between the tubular convolutions. They occur in microscopic islets of irregular size and shape in the connective tissue between the coils. One lobule is shown.

listed in Table 7.1, together with some of their hormonal actions and the composition of the hormones. In the table also are references to figures in this chapter that pertain to some general properties of endocrine glands as described in the first part of this chapter.

INTERRELATIONS OF MAJOR ENDOCRINE GLANDS AND THE CENTRAL NERVOUS SYSTEM IN WOMEN

Both of these control systems contribute to the maintenance of the relative constancy of the state of activity of the body, as well as to the specialized activity of certain target organs and systems. For example, the thyroid gland usually maintains normal levels of metabolism of body cells. The adrenal cortex, the parathyroid glands, and the islets of Langerhans in the pancreas, are each concerned with some general or specific metabolic aspects (such as calcium or sugar metabolism). Some endocrine organs have more limited functions; their hormones influence primarily certain target organs, for example, follicle stimulating hormone and the ovarian

Table 7.1. Some endocrine glands not directly involved in reproduction.

Source	Chemical Composition	Actions
Thyroid glands (Figure 7.5)		
Thyroglobulin, specifically thyroxin and diiodotyrosine	Glycoprotein	Oxygen consumption of cells (basal metabolism)
Calcitonin	Polypeptide	Fall in Ca and PO_4 metabolism; inhibition of bone formation
Parathyroid glands (Figure 7.5)		
Parathormone	Polypeptide	Rise in Ca and PO_4 metabolism, accelerates bone destruction
Calcitonin	Polypeptide	See above
Pancreas (islets of Langerhans) (Figure 7.6)		
Insulin	Polypeptide	Affects metabolism of sugar utilization, glycogen metabolism, synthesis of fatty acids and protein of most cells, especially fat, liver, and muscle.
Glucagon	Polypeptide	Affects glycogenolysis and gluconeogenesis in liver, stimulates lipolysis in fat
Gastrointestinal tract		
Gastrin	Small polypeptide	Stimulates HCl secretion in stomach, blood flow in gastrointestinal tract, and contraction of smooth muscle
Secretin	Polypeptide	Stimulates pancreas to secrete bicarbonate
Cholecystokinin	Polypeptide	Stimulates emptying of gall bladder and secretion by pancreas
Kidney		
Erythropoietin		Accelerates formation of red blood cells in bone marrow
Renin	Tetradecapeptide	Gives rise to angiotensins, affects sodium excretion, water balance, blood vessel contraction, and blood pressure
Pineal gland		
Melatonin	Derivative of serotonin	

follicle, the luteal hormone and the corpus luteum of the ovary, estrogen of the ovary and the uterus, cervix, vagina, and mammary glands. To a greater or lesser extent, then, all the endocrine glands affect the state of activity of the body cells, keeping this state within certain limits, called "normal," in most people. This endocrinologic control of the internal environment of the body does not operate independently—it operates in conjunction with the nervous system. Nerve stimuli arising autonomously within the person (as thoughts, anxieties, or pleasure), or originating as a result of external stimuli (visual, auditory, or tactile) are funneled to the hypothalamus, where they impinge on the endocrine glands primarily through the neuroendocrine system (the various hormone-releasing and release-inhibiting hormones).

Thyroid Gland Feedback Mechanism

A great part of the control of the activity of endocrine glands consists in feedback mechanisms that are not well understood. Perhaps the simplest feedback system is that controlling the relatively constant level of secretion of triiodothyronine (T_3) and thyroxine (T_4) by the thyroid gland. Certain basophil cells of the anterior lobe of the hypophysis usually secrete enough

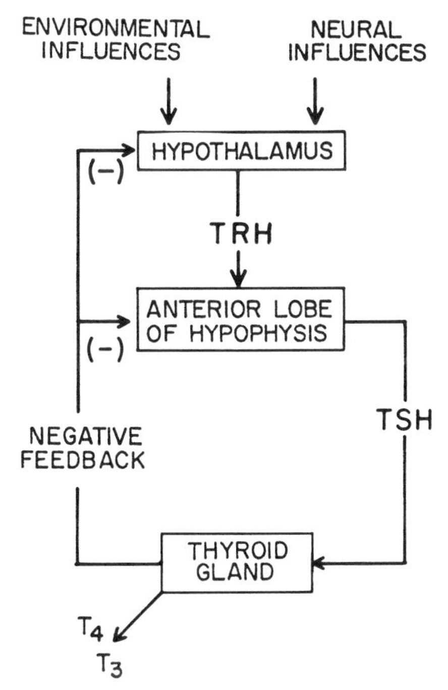

Figure 7.11. Diagram of a simple type of feedback system involving the production of thyroid-releasing hormone.

thyrotropic hormone to maintain a relatively constant level of secretion by the thyroid gland. If too much is secreted, the basophil cells reduce their activity by negative feedback. This is reinforced by negative feedback also, which results in reduced secretion of the thyrotropic-releasing hormone by the neuroendocrine cells of the hypothalamus, shown in Figure 7.11.

Ovulatory Feedback Mechanism

Another more complicated example is that of the feedback mechanisms operating during the ovulation and the menstrual cycle. The following sequence of events is supposed to occur, but will have to be reconciled with recent work indicating that FSHRH and LHRH are identical. The ovarian follicles are induced to grow and secrete estrogen by follicle-stimulating hormone secreted by the anterior lobe of the hypophysis. The rising plasma level of both of these hormones peaks abruptly just before ovulation. At the same time, the plasma level of the corpus luteum-stimulating hormone (LH) rises, ovulation takes place and a corpus luteum develops

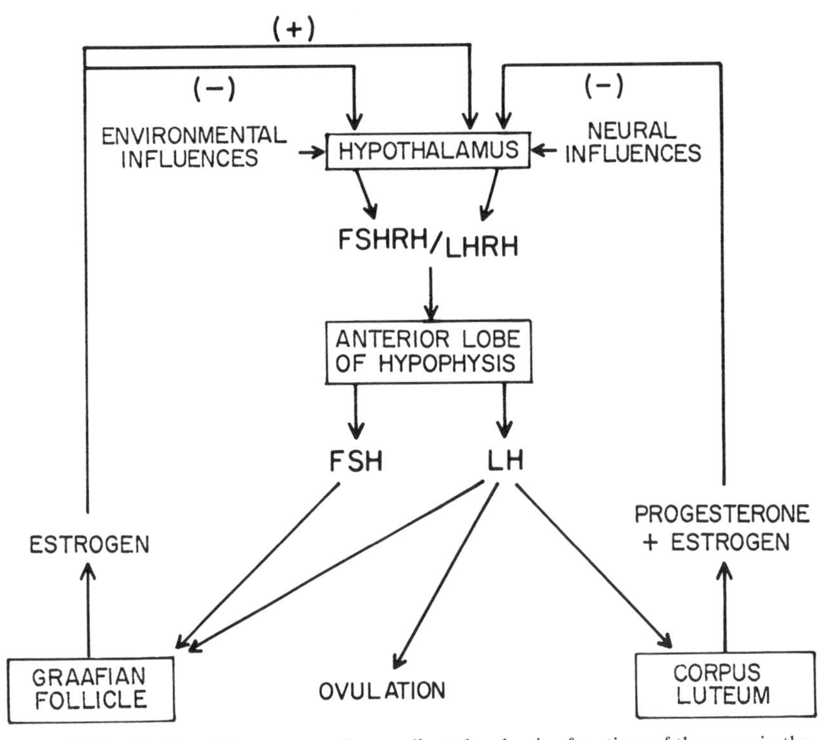

Figure 7.12. In view of the enormously complicated endocrine functions of the ovary in the course of a single menstrual cycle, it is no surprise that the feedback processes include both positive and negative reactions, as shown in this diagram.

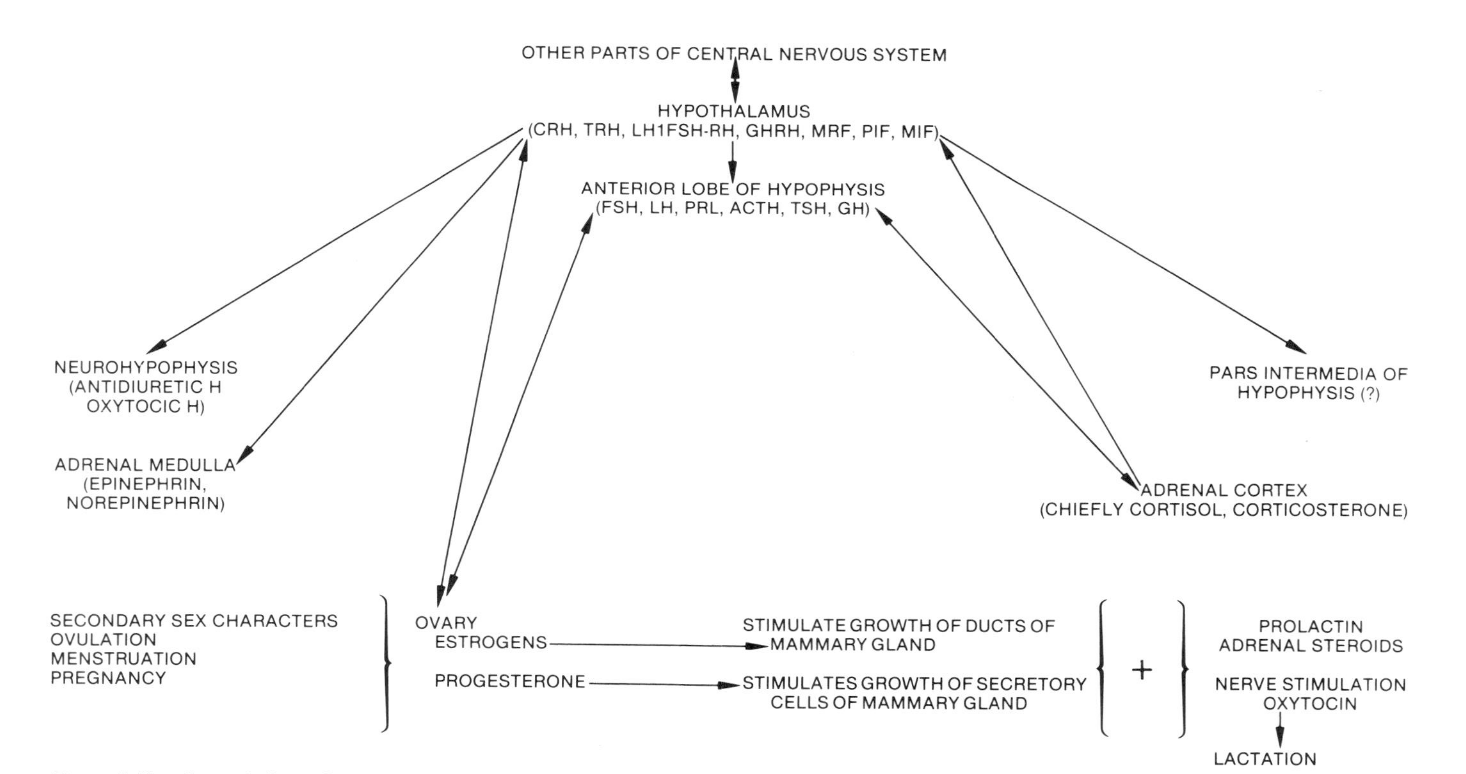

Figure 7.13. Interrelations of some endocrine glands and the hypothalamus in women.

and secretes its hormone, progesterone, together with some estrogen. The secretion of LH is promoted by the positive feedback of the high level of estrogen on the neuroendocrine cells of the hypothalamus which secrete luteal hormone-releasing hormone, as shown in Figure 7.12. The high plasma level of progesterone reduces the secretory activity of the neuroendocrine cells containing LHRH by negative feedback, and the secretion of LH is reduced. As the estrogen is now low, the cycle begins again with the increased secretion of follicle stimulating hormone.

Up to this point, the control of the activity of each endocrine gland has been treated separately. But obviously these endocrine activities are all integrated in most people so as to permit only narrow deviations of all hormones from relatively constant levels. The chief role as integrator is the hypothalamus. These overall interrelations of endocrine, neuroendocrine, and neural factors are shown in simplified form in Figure 7.13.

In addition, more strictly localized neural activity is involved in certain activities, for example, the secretory activity of the adrenal medulla, the suckling effect on lactation, the erection of the erectile tissues, and the secretion of oxytocic and antidiuretic hormones. The more strictly neurologic aspects are treated in Chapter 9.

Some of these hormonal interrelationships can become very complicated. For example, the action of the antidiuretic hormone is related to renin, and the angiotensins involve the liver and kidney; the angiotensins in turn have effects on the aldosterone thought to be secreted by the cells of the glomerular zone of the adrenal cortex. The hormonal interrelations of Figure 7.13, complex as they are, indicate only the main features of hormonal homeostatic mechanisms that are normally at work in the body at all times.

REFERENCES

Beer, A. E., and R. E. Billingham. 1976. The Immunobiology of Mammalian Reproduction. Prentice-Hall, Englewood Cliffs, N.J.

Bensley, R. R. 1911. Studies on the pancreas of the guinea pig. Am. J. Anat. 12:297-388.

Bloom, W., and D. W. Fawcett. 1975. A Textbook of Histology. 10th Ed. W. B. Saunders, Co., Philadelphia.

Bumpus, F. M. 1977. Mechanisms and sites of action of newer angiotensin agonists and antagonists in terms of activity and receptor. Fed. Proc. 36:2128-2132.

delPozo, E., J. Hiba, I. Lancranjan, and H. J. Künzig. 1977. Prolactin measurements throughout the life cycle: endocrine correlations. *In* Proceedings of the Serono Symposium, Vol. II, P. G. Crosignani and C. Robyn (eds.), Prolactin and Human Reproduction. Academic Press, New York.

Gersh, I. 1941. The vascular pattern of the adrenal gland of the mouse and rat and its physiological response to changes in glandular activity. Contributions to Embryology, No. 183, Carnegie Institution of Washington, Washington, D.C.

Greep, R. O. (ed.). 1974. International Review of Science. Physiology Series One. Vol. VIII. Reproductive Physiology. University Park Press, Baltimore.
Guillemin, R. 1978. Peptides in the brain: The new endocrinology of the neuron. Science 202:390-402.
Jensen, E. V., and E. R. DeSombre. 1973. Estrogen-receptor interaction. Science 182:126-134.
Johnson, L. R. 1977. Gastrointestinal hormones: physiological implications. (Symposium of the American Physiological Society) Fed. Proc. 36:1929-1951.
Junqueira, L. C., J. Carneiro, and A. Contopoulos. 1975. Basic Histology. Lange Medical Publications, Los Altos, Cal.
Karim, S. M. M., and B. Rao. 1975. General introduction and comments. *In* S. M. M. Karim (ed.), Advances in Prostaglandin Research. Prostaglandins and Reproduction. University Park Press, Baltimore.
Kolata, G. B. 1977. Hormone receptors: How are they regulated? Science 196:747, 748, 800.
Larner, J. 1977. Polypeptide hormone receptors. Fed. Proc. 36:2110-2132.
McCann, S. M. 1977. Localization of hypophysiotropic hormones. (Symposium of the American Physiological Society) Fed. Proc. 36:1952-1983.
Mountcastle, V. B. 1974. Medical Physiology. 13th Ed. Vol. I and II. C. V. Mosby Co., St. Louis.
O'Malley, B. W., and A. R. Means. 1974. Female steroid hormones and target cell nuclei. Science 183:610-620.
Pastan, I. 1972. Cyclic AMP. Sci. Am. 227(2):97-105.
Raiti, S., and N. K. Maclaren, 1974. Advances in human growth hormone research. Fed. Proc. 33:1682-1685.
Ruch, T. C., and H. D. Patton (eds.). 1973. Physiology and Biophysics. Vol. III, Digestion, Metabolism, Endocrine Function and Reproduction. 20th Ed. Howell's Textbook of Physiology. W. B. Saunders, Co., Philadelphia.
Schally, A. V. 1978. Aspects of hypothalamic regulation of the pituitary gland. Science 202:18-28.
Schreiber, V. 1974. Adenohypophyseal hormones: regulation of their secretion and mechanism of their action. *In* H. V. Rickenberg (ed.). Biochemistry of Hormones. University Park Press, Baltimore, pp. 61-100.
Schwabe, C., and J. K. McDonald. 1977. Relaxin: A disulfide homolog of insulin. Science 197:914-916.
Stumpf, W. E., and L. D. Grant (eds.). 1975. Anatomical Neuroendocrinology. S. Karger, New York.
Tagatz, G. E., and E. Gurpide. 1973. Hormone secretion by the normal human ovary. *In* R. O. Greep (ed.), Handbook of Physiology, Section 7, Endocrinology. Vol. II, Female Reproductive System, Part I. American Physiological Society, Washington, D.C., pp. 603-612.
Yalow, R. S. 1978. Radioimmunoassay: A probe for the fine structure of biologic systems. Science 200:1236-1245.

8 Cytology and Histology of the Endocrine Glands

GENERAL CYTOLOGY

In order to understand the action of cells which synthesize and secrete hormones, it is essential to discuss cells in general, with particular emphasis on those features of cellular activity that take part in the synthesis and release of hormones by cells. A brief description of how cells are arranged to form tissues, and how the tissues are organized as organs follows. The rest of this chapter discusses the specific cells and tissues of the major endocrine organs concerned with synthesis and secretion of hormones, especially those involving sex hormones.

Interphase Nucleus

The cell consists of two major parts, the nucleus and the cytoplasm (Figure 8.1). The nucleus contains the prime genetic material, which originates many aspects of cell activity, and perpetuates its orderly or programmed activity for the life of the cell.

The total amount of DNA per nucleus is, with few exceptions, the same in all nuclei in the body. In interphase nuclei, the amount of DNA per nucleus is the same as the amount of DNA in the chromosomes of dividing cells. In the latter, the DNA is nearly (or entirely) genetically inactive or heterochromatic. It is tightly packed or condensed during the process of mitosis. In interphase nuclei, most of the DNA is condensed, but a small proportion (about 10% or so) is loosely dispersed in the nucleus. This is the active or euchromatic portion of the DNA. This fraction contains two main kinds of genes: 1) those concerned with all the cellular activities common to all cells, and 2) those particular to certain cell types. The second group of genes is largely different in each cell type from those in other cell types.

In some way that is not understood, each pattern of active genes is associated with a certain pattern of organization of all the remaining

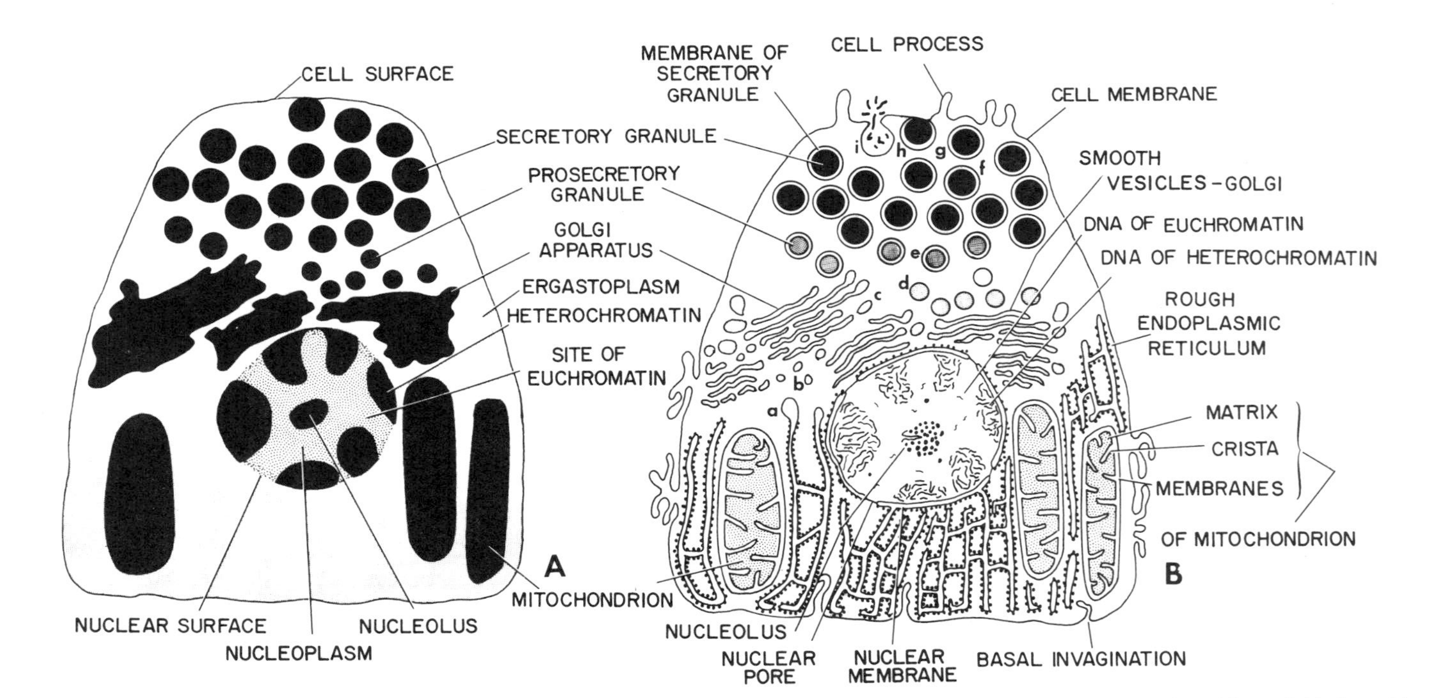

Figure 8.1. A) Externally secreting cell as seen with the light microscope. B) Externally secreting cell drawn to the same scale as Figure 8.1A, but with the improved resolution of the electron microscope.

DNA into *chromatin*, the genetic material of the interphase nucleus—so named because it is readily stained by certain dyes. Of the 46 recognizable chromosomes of most people, only two can be identified as recognizable chromosomes in the interphase nucleus. These are the presumably inactive X chromosome of women and the Y chromosome of men.

Chromatin The chromatin consists essentially of DNA double helices associated with proteins, primarily histones. Histones are basic proteins with many amino acids capable of combining with the phosphate groups of nucleic acid to form nucleoprotein. Histones are organized in definite proportions (two molecules of each of four different histones) to form submicroscopic particles around which the double helix of DNA (see Figure 3.3) is wound twice. These particles are called *nucleosomes*. Thus the DNA helix is arranged in the form of beads on a string, the helix joining the nucleosomes to each other. The exact manner in which the positive and negative charges are arranged in nucleosomes is not known. Nor is it known how the nucleosomes of actively transcribing DNA differ from those of less active or inert DNA. Other proteins in the nucleus are more acidic than the histones and thus known as *acidic proteins*. These may combine with histones even more strongly than the latter are combined with DNA. In the process of binding strongly with histones, the acidic proteins may free the DNA to transcribe messenger and/or other RNA.

The nucleoprotein of cell nuclei is not distributed uniformly throughout the nucleus of cells in the interphase nucleus. It is aggregated into lumps or granules in a manner that is probably characteristic of each cell type when the cell is engaged in ordinary activities. These clumps, or *heterochromatin*, may be large enough to be visible with the light microscope, or they may be too small to be visible except with the electron microscope. Part of the nucleoprotein may be even more finely dispersed, the *euchromatin*, even as individual double helices.

The pattern of distribution of DNA as chromatin is not static. It varies in two major directions: 1) When the cell is stimulated to activity in excess of the normal, the chromatin becomes more finely distributed; that is, the heterochromatin is reduced and the euchromatin is increased, the total amount of DNA per nucleus remaining the same. 2) When the cellular activity is reduced markedly, the chromatin becomes coarser; in extreme conditions the nucleus is shrunken and uniformly dense, a state known as *pyknosis*. In cells with reduced activity, the heterochromatin is increased, while the euchromatin is reduced.

Nucleolus Another feature of chromatin is that in many cells most of it is associated with the nuclear membrane. But a small part of the chromatin is always attached to the nucleolus. All cells have a nucleolus

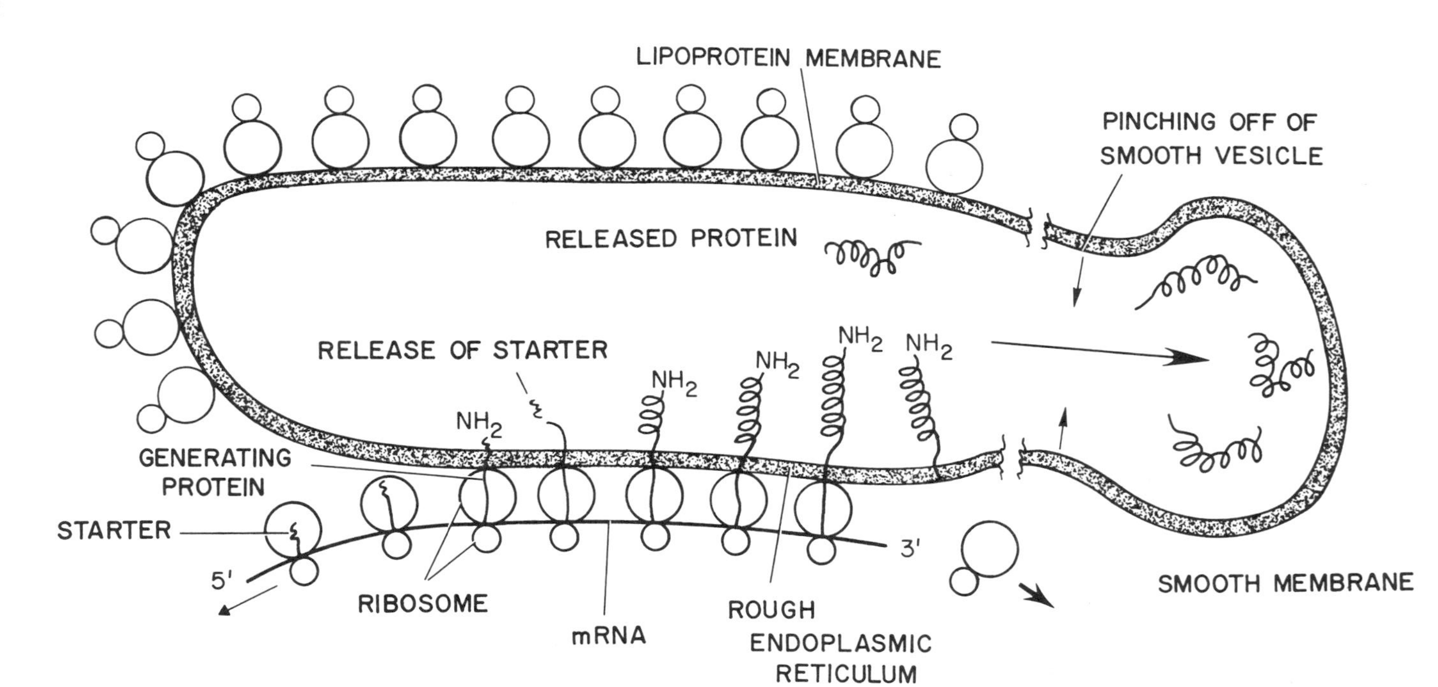

Figure 8.2. Early stages of the process of protein synthesis and secretion—a close-up view of rough endoplasmic reticulum (After Lodish and Rothman, 1979).

at some time in their life history. The nucleolus is a dense body that contains submicroscopic particles composed primarily of ribosomal RNA and associated proteins. The RNA is formed by transcription from DNA, which is visible in the electron microscope and frequently in the light microscope as perinucleolar or sometimes intranucleolar chromatin. It is called *nucleolar-associated DNA*. From one or many sites, depending on the type of cell, the ribosomal RNA genes transcribe RNA. This RNA peels off as long molecular filaments which, with associated proteins, form ribosomal subunits. These may accumulate to form a nucleolus. In times of excessive activity, more ribosomal RNA is formed from the DNA and the nucleolus is larger, and in times of reduced activity the nucleolus may be barely detectable or may be lacking altogether. Thus the appearance of the nucleolus is not static but varies in size and number per cell depending on the state of activity of the cell.

The RNA destined to be in ribosomes passes into the nucleoplasm, which fills the interphase nucleus not occupied by DNA and RNA structures, and then through the nuclear membrane. The nuclear membrane is a submicroscopic structure composed primarily of proteins and lipids whose arrangement as thin sheets is not fully understood.

Cytoplasm

Ribosomes and Cytoplasmic Activity Ribosomes may occur singly in the cytoplasm, or may be aggregated with messenger RNA as polysomes (Figures 8.1B and 8.2). In both cases, presumably, each ribosome consists primarily of three different RNA molecules associated with some 50 proteins to form a granule about 150–200 Å in diameter. The polysomes are frequently organized on discontinuous, sheetlike lipoprotein membranes, which are irregularly interconnected. These are known for historical reasons as *rough endoplasmic reticulum* (RER). These structures are nearly always submicroscopic. When the ribosomes are stained, they appear as a homogeneously stained or colored material, visible with the light microscope (if it large enough and dense enough) and known as *ergastoplasm*, or *chromophile substance* of the cytoplasm, or as *chromidial bodies* (Figure 8.1A). With a few exceptions, all proteins of cells are made in association with the polysomes by translation of the message or code carried by the messenger RNA.

Secretory Granules In some way not clearly understood, the newly synthesized protein can be condensed to form a granule that can be stored and released at a later time during a period of heightened activity (Figure 8.1B). Such a granule is called a *secretory granule*. Or the newly synthesized protein may be stored, without being condensed, for later release, or it may be released directly without storage. The condensation of secretory granules (and other structures) involves in a way not yet clear, another structure in the cytoplasm known as the *Golgi apparatus*.

The Golgi Apparatus The Golgi apparatus consists of membranous sacs that are connected with each other to form a system of interrelated channels (Figure 8.1B). These function in relation to ribosome-studded (rough) endoplasmic membranes or sacs. The relationship is thought to be the following: Protein molecules synthesized at polysomes are extruded into the rough endoplasmic sacs or cisternae. They are passed along in some way to a portion of the cell not far from the Golgi structure where the ribosomes become sparse. The membrane with its enclosed and newly synthesized molecules is pinched off as a vacuole (Figure 8.2). The contents enter into the Golgi channel system, where they are concentrated ("packaged") and fused. All these membranes are free of ribosomes and are part of the smooth endoplasmic reticulum. Numerous enzymic activities must take place there, for it is the site where sugars are added to protein to form glycoproteins, for example, mucins. In some cells, where, for example, certain steroid hormones are synthesized, the smooth endoplasmic reticulum is enormously enlarged.

Despite the general acceptance of the theory on the supramolecular and molecular cytoplasmic events during synthesis and secretion of proteins, some aspects require investigation. While it is true that after extreme stimulation of gland cells secretory granules are extruded from such cells, it is also true that within physiologic or even stronger ranges of stimulation, enzymes may be secreted without any reduction or movement of secretory granules, even under direct observation of the living cells. But even within the present view of secretion, it stretches the imagination to account for the synthesis and arrangement in the Golgi lamellae of a suitable pattern of all the enzymes required to synthesize the sugar component of mucin, for example, and its attachment to the protein component, as well as sulfation, water transport to achieve concentration, the packing of molecules in each secretory granule as it enlarges and the polarized movement of the granules during their maturation and secretion. Whenever a membrane is pinched off from the rough endoplasmic reticulum and is transferred to the Golgi structure, it is presumably reconstructed. Then as the membrane eventually moves toward and joins the surface membrane, the molecular changes in its chemical composition as well as in its response to changes in its internal and external environment must be very great indeed.

Lysosomes Another segregated structure related to the Golgi apparatus is the lysosome. This is a granule or vacuole generally visible in the light microscope, which is simply packed with degradative enzymes (proteinases, lipidases, amylases, phosphatases, etc.). The enzymes may operate in the vacuole (when, for example, the vacuole fuses with another containing some degradable substrate); or they may operate extracellularly if the granule or vacuole is extruded from the cell (Figure 8.3). Lysosomes may play an important role in menstruation, in reconstruction of pelvic viscera after parturition, and in atrophy of the postlactating breast.

Ground Substance of Protoplasm Many of the proteins synthesized by cells are released by the cell and serve elsewhere in some capacity (for example, hormones); but many proteins are retained by the cell. Some may act there as structural proteins that can serve as a cytoskeleton, while others are specialized further as a contractile mechanism. The former includes the system of microfilaments and microtubules. The filaments vary in frequency from cell type to cell type, probably occurring in all cells at some time in their life history, and may be contractile in nature. The microtubules are especially prominent as spindle fibers during mitosis. Both microfilaments and microtubules seem in a state of constant change in cells, but the factors controlling their self-assembly and disassembly in cells are not known. But most varieties of proteins

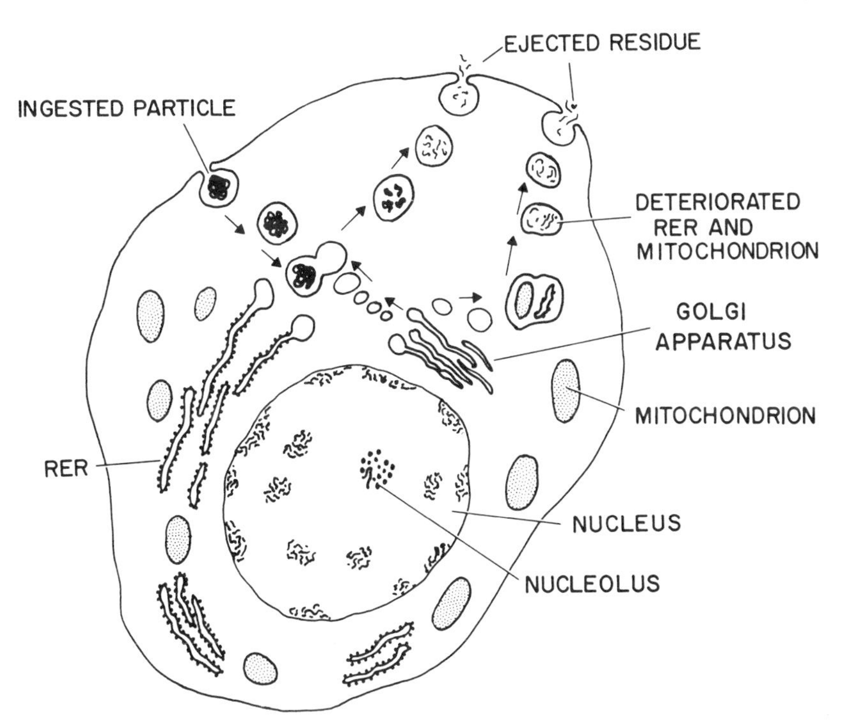

Figure 8.3. The current concept of the formation and fate of lysosomes. RER, rough endoplasmic reticulum. (After Junqueira, Carneiro, and Contopoulos, 1975).

synthesized by the cell are enzymes concerned with some aspect of cell metabolism. The enzymes may be concerned with the synthesis or destruction of cell proteins, lipids, or carbohydrates, or with the generation of energy required for cellular activity. All of these multitudinous molecules (except the contractile apparatus of voluntary and cardiac muscle) are not visible with the light microscope, and together comprise the *ground substance* of the cytoplasm. *Ground substance* is a literal English translation of the original term in German, *Grundsubstanz*. It means something like background substance, without visible structure, nearly transparent or translucent. There is no doubt whatever that *all* of the components of the ground substance are organized into functional units (like, for example, the rough endoplasmic reticulum) and are not homogeneously distributed, as if in a mixture or solution in a flask.

Mitochondria Mitochondria are specialized structures in the cytoplasm of cells, and are concerned with lipid synthesis, oxidation, and the generation of energy for metabolic purposes of cells. Most of the proteins of mitochondria are synthesized in the rough endoplasmic reticulum of the ground substance of the cytoplasm, and are somehow organized to form the very highly complex and particular structure of mitochondria. The remaining small proportion of mitochondrial proteins is synthesized by a circular form of DNA, which is the only cytoplasmic or non-nuclear DNA of all cells. (Other cytoplasmic sites of DNA are the basal bodies of cilia of protozoa, and chloroplasts of green plants.) It is very small in amount, but essential. As commonly prepared, two mitochondrial membranes are prominent: an outer membrane and an inner one. The inner one has many folds, called *cristae.* Lining the inner membrane, including all its folds, are submicroscopic granules, which may well be artifacts of the method of preparation. Between the cristae are numerous granules that together constitute the matrix. While several enzyme preparations may be obtained from mitochondria, their localization in the mitochondrial matrix is subject to debate.

A plausible view supported by some morphologic studies is that mitochondria are comprised of many unit compartments, each enclosed in lipid-rich membranes. It is conceivable that when such uniformly compartmented structures are fixed, the vagaries of fixation are such that the compartments may be preserved to varying degrees, resulting in the appearance of cristae or tubules, depending on the nature of the cellular environment.

Cell Membrane The cytoplasm is in its entirety enclosed in a membrane too thin to be visible in the light microscope, hence the label *cell surface* in Figure 8.1A. It consists chiefly of protein, lipid, and some form of sugar complex associated with the protein. All the substances entering or leaving cells must pass through the cell membrane. The adhesiveness of cells to form organized tissues and organs ultimately is related to the properties of the cell membrane. The cell membrane also contains receptors that respond to changes in the environment of the cell and to stimuli from other cells nearby or remote, such as hormones. These specialized functions are associated with structural modifications of the cell membrane, some of which are visible with the electron microscope as cell processes or junctions of several kinds. In epithelial tissues, where the cells are arranged in a single layer, the free (unattached) surface and basally attached surface frequently show specializations, such as microfilamentary structures, cell processes, and basal infoldings. Such cells are considered *polarized*. The polarization process extends into the cytoplasm, where the rough endoplasmic apparatus, Golgi structure, secretory granules, mitochondria, and even the nucleus may show preferential localization in accordance with the functional polarization.

The basic plan of organization of the cell membrane seems very much the same in all cells. The membrane is thought to consist of two layers of lipid molecules, with the lipophilic ends of the molecules engaging each other, and the lipophobic (hydrophilic) ends facing the inner (cytoplasmic) and outer surfaces. The double molecular membrane varies, however, in different cells in the number and nature of intercellular and cellular processes, in the number and position of basal ingrowths, in the occurrence and frequency of phagocytic or pinocytic sites, and in the nature and number of proteins, or protein-polysaccharide complexes, and lipids. (Phagocytosis is the ingestion of solid particles, plus some water, by the cytoplasm; pinocytosis is the ingestion of watery material that sometimes contains some solid particles.) The protein and protein-polysacharide complexes may extend through the thickness of the double lipid membrane and function in cell transport, as receptors for hormones, or other chemical substances, or as surface enzymes. The regularity of the lipid membrane is also broken up by various steroids. And finally, at various kinds of cell junctions the membrane may appear to be thickened by a special junctional substance and by a special kind of intercellular substance. The membrane in its totality is regarded as in a state of change, depending on the state of cellular activity, and to be replenished by membranes from secretory products, smooth membranes of various sorts, such as, pinocytic and phagocytic vesicles, as well as presumably by protein molecules originally synthesized by polysomes. Cell membranes may thus be polarized in accordance with the functional polarization of the cell.

It is important to realize that the genes (DNA of chromatin) are active not only in the synthesis of a new (the first) protein molecule of a kind; they are also active in the continuous activity of the cell in the interphase period, in periods of excessive activity, and in cell maintenance and repair. After all, proteins, like other wet substances, deteriorate and must be replaced if cellular activity is to continue in a normal way. The same processes go on at a heightened level during cell growth and during regeneration.

Some cells, like muscle or nerve cells, show little or no tendency to regenerate. The total number of nerve cells after some early time in childhood is fixed. If any nerve cells die, they cannot be replaced. Virtually the same is true of skeletal muscle cells. The lifetime of at least some of the cells is probably the same as the lifetime of the person. This is not true of other kinds of cells in the body, that can live for a few days or months, and then be replaced by other cells of the same kind. Reservoirs of each of these replacement cells exist for each cell type, and such cells divide, yielding some cells that mature as rapidly as the aged cells fall away and replace them. This is also considered to be a form of regeneration, without which no person could survive.

GENERAL HISTOLOGY

A tissue is a group of cells that have similar functions, together with the extracellular material surrounding them. Histologists differentiate among four tissues: nerve, muscle, epithelial, and connective tissue.

Nerve Tissue

This type of tissue is highly specialized primarily for rapid conduction of nerve impulses. Highly organized cells effect precise integration of the various parts of the body within the person and of the person as a whole and the environment. The structure, organization, and functions of nerve cells and the nervous system are discussed in Chapter 9. In addition to nerve cells, nerve tissue also includes supporting cells (*neuroglia*), which have the same origin as nerve cells. Finally, although all nerve cells secrete some substance important in conduction at their endings, some are particularly specialized to secrete hormones; these, called *neuroendocrine cells*, are discussed in Chapter 7.

Muscle Tissue

Cells of this type of tissue are highly specialized primarily for contraction and relaxation. They are characterized by the occurrence in them of a contractile substance called *actinomyosin* together with other associated substances, all of which are very highly structured. Voluntary muscle is particularly specialized for rapid, precise movement. Heart muscle is specialized for prolonged, rhythmic contraction. Smooth muscle is concerned with slower, rhythmic, or persistent activities. The first two are very highly organized at the light microscopic and molecular levels. The organization of the third has not yet been generally elucidated.

Epithelial Tissue

These specialized tissues are very diverse in origin, structure, and function, but share one feature in common. The tissues are primarily cellular (up to about 80%), with relatively little extracellular material. They may originate from ectoderm, entoderm, or mesoderm. They usually line a surface, or are derived from cells which lined a surface in the embryo. The following are examples: The anterior lobe of the hypophysis in embryonic life is connected to the epithelium of the roof of the mouth. The epithelium of the skin and all its numerous appendages (hair, fingernails, sweat glands, oil glands), are derived from the outermost layer of embryos, the ectoderm. The whole lining of the gastrointestinal tract including the liver and pancreas is derived from the entoderm or inner layer of embryos. On the other hand, the adrenal cortex and testis are derived from mesoderm or middle layer of embryos, and were not in contact with the outside of the body in embryonic life. It can be deduced that epithelial tissues: 1) protect the body to prevent dehydration; 2) are a means of communication with the environment; 3) absorb nutrients (through the gastrointestinal tract) and take up oxygen (through the lungs); 4) secrete mucins, digestive enzymes, hormones, sperm, and other substances; 5) lubricate (e.g., the thoracic and peritoneal lining around the lungs, intestines, etc.); 6) line the inner layers of blood vessels, heart, lymphatic vessels, etc.; and 7) excrete watse products of the lungs, the intestines, and the kidneys.

Connective Tissues

These specialized tissues are very diverse in appearance, function, and consistency. Their common feature is that, by contrast with epithelium, the extracellular substance is generally very prominent and may constitute up to 80% or more of the tissue. The density may vary from that of bone, cartilage, tendon, and ligament to that of blood, lymph, and fat. Connective tissues hold all parts of the body together as functioning units. For example, they enclose muscle cells and join them together to form muscles. They join muscles to bone, and bone to bone, thus making movement possible. The circulating blood, together with the lymph and tissue fluid, distributes water, ions, nutrients and metabolites, besides removing waste products. Connective tissues serve as the main storehouse and reserve for the ions of the body, and are in equilibrium with the components of the circulating blood and lymph. In this capacity, the connective tissues are essential for the maintenance of the homeostasis of the body. They serve, essentially as great buffers, maintaining the composition of blood, lymph, and tissue fluid relatively stable and constant despite great, potentially serious shifts. The connective tissues also serve in the repair of wounds and in inflammation. Some cells of the connective tissues constitute the basis of immunity, and protect people against bacterial and parasitic diseases.

TISSUES AND ORGANS

Normally, tissues exist always in relation to the structure and function of organs or of the organism as a whole. For example, in organs such as the liver, submaxillary gland, or prostate gland, all tissues coexist in each organ, and contribute to the orderly functioning of the gland. The same is true of all endocrine glands, and of all the organs of the male and female genital tract.

Anterior Lobe of the Hypophysis

The cells that synthesize and secrete the hormones are packed to form thin cords of cells one to three cells thick. These cords connect with adjacent ones to form fine networks that are separated from each other by thin strands of connective tissue. The numerous delicate small blood vessels run in the connective tissue strands. Some of the cells lining the blood capillaries are capable of *phagocytosis*; that is, they take up particles in the blood plasma. The other cells lining the blood capillaries are extremely thin—they may be submicroscopic (0.2 μm thin or less, below the limit of resolution with the light microscope). The connective tissue space between the gland cells, and between them and the blood capillaries may be even thinner. The blood capillaries are believed to arise from portal veins, which in turn arise from a capillary network in the hypothalamus. The rate of blood circulation in the anterior lobe must be slow, and most researchers think the blood moves from the hypothalamus to the anterior lobe. Some reports disagree, suggesting that blood flow is in the opposite direction, at least part of the time.

Gland Cells At least six varieties of gland cells and possibly seven, can be identified by special stains with the light microscope, by histochemical tests that identify the polypeptide produced, and by electron microscopic differences. About half the cells are called *chromophobes*. Their nuclei are surrounded by scant cytoplasm that is not well stained by dyes (hence the name). The remainder are alpha and beta cells. The alpha cells are characterized usually by large granules that are stained by dyes such as eosin or orange G. They are of two sorts:

1. *Somatotrophs*, which are large and ovoid, tend to have smaller granules and are thought to synthesize and secrete growth hormone.

2. *Prolactin cells*, which are irregular and have coarser granules, are thought to synthesize and secrete prolactin.

The beta cells are characterized by their granules, which stain with aniline blue and other basic dyes. The granules are usually smaller than those of alpha cells. Beta cells are of three or possibly four types:

1. The cells that secrete thyrotropic hormone show characteristic changes and increased size after surgical removal of the thyroid gland and are irregular in shape.
2. The cells that synthesize and secrete follicle stimulating hormone are more rounded than the thyrotropic hormone-secreting cells, and in rats show cyclic behavior: they become more prominent just preceding estrus, less prominent during anestrus.
3. These cells synthesize and secrete luteinizing hormone (in females) or interstitial cell-stimulating hormone (in males).
4. The cells that synthesize and secrete adrenocorticotropic hormone are not very well characterized, but probably form a separate class of beta cell. The submicroscopic structure of these kinds of cells is shown diagrammatically in Figure 8.4.

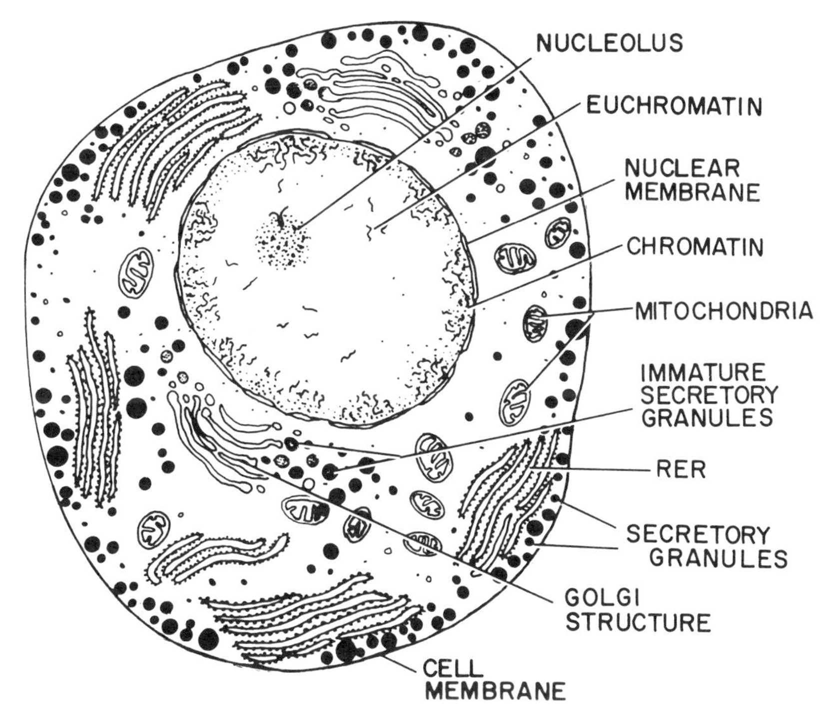

Figure 8.4. Endocrine cell synthesizing and secreting a protein hormone (growth hormone). RER, rough endoplasmic reticulum. (After Bloom and Fawcet, 1975).

These six (or seven) varieties of cells are scattered irregularly throughout the anterior lobe of the hypophysis, sometimes singly or in small groups. It seems likely that for each hormone produced by the gland there is a cell type. When they have released their granules, they revert to chromophobes, which can be regarded as reserve cells. At some later time, this cell type may synthesize the same hormone or a different hormone, which would then be stored as granules of the alpha or beta cells. It is important to recall that, during each cycle of hormone synthesis, changes probably take place in the nuclear DNA associated with gene action and with protein synthesis in the cytoplasm, and that the receptors on the cell membrane in some way are responding to some stimulus arising from the circulating blood in the blood vessels.

The synthesis of hormones by cells of the anterior lobe of the hypophysis is probably directly related to gene action. That is, the messenger RNAs concerned with the synthesis of the protein hormones are derived from the DNA of the specific genes for these hormones. This is in contrast with the synthesis of steroid hormones. These are synthesized as a result of the action of enzymes in mitochondria and probably the adjacent ground substance of the cytoplasm (excluding the rough endoplasmic reticulum). It seems probable, however, that the enzymes concerned with steroid synthesis are themselves synthesized by polysomes whose messenger and ribosomal RNAs are derived from DNA of the appropriate genes. Thus, the synthesis of steroid hormones would seem to be one step removed from direct genic action.

Hormonal Substances of Uncertain Origin Discordant notes on the question of localization are sounded in connection with a series of apparently related hormonal substances of unknown function. The sequence of the 13 terminal amino acids of ACTH very closely resembles the 13 amino acids of α-melanocyte-stimulating hormone (αMSH). Also, positions 4-10 of ACTH are identical with amino acids in positions 7-13 of β-melanocyte-stimulating hormone (βMSH). In animals with dissectable pars intermedia, both MSH hormones are localized strictly to this part of the hypophysis. In amphibia, the hormone causes melanophore granules to be dispersed more widely in the cell processes of the melanocytes, causing the skin to darken. But the function of the hormones in mammals is unknown. ACTH also occurs in the pars intermedia, and it has been suggested that in the pars intermedia, both αMSH and βMSH are synthesized from it and then degraded to form still smaller bioactive peptides that are secreted to perform some function not presently known.

Most of the hypophyseal ACTH is found in the anterior lobe in corticotropic cells. Also present in the same cells, perhaps even in the same secretory granule is β-lipotropin, a small protein having 91 amino acids. This molecule can be cleaved to yield βMSH as well as β-endomorphin and methionine enkephalin. These are small polypeptides associated with receptors at certain synapses in the central nervous system. The complete function of β-lipotropin in the body is unknown, although it is known to influence the metabolism of fatty acids. Clearly there are many uncertainties concerning the interrelationships of a series of closely related substances of a hormonal nature associated with the anterior lobe and the pars intermedia of the hypophysis. Also uncertain is the origin and function of several of these compounds localized in and presumably confined to discrete regions of the brain.

Ovaries

Each ovary is partially enclosed by a cuboidal epithelium that is continuous with the flat cells on the surface of the peritoneal lining. Immediately beneath this layer of cells is a thin layer of connective tissue. This layer overlies the cortex of the ovary, which in turn encloses the poorly demarcated medulla. The medulla consists primarily of connective tissues in which are distributed blood vessels, nerves, and lymphatic vessels. The cortex contains the germ cells, associated follicular cells and their derivatives, connective tissues, nerves, and blood and lymphatic vessels.

Germ Cells The germ cells arise in the yolk sac in very early embryos (see Figure 2.3); they migrate to the genital ridge and multiply so rapidly that they number about 600,000 at 2 months. In the 5-month old fetus, they number about 7,000,000. At birth the number has gone down to about 2,000,000, and at puberty, the number is further reduced about ten-fold. This loss of egg cells continues throughout the reproductive period of life until menopause when very few egg cells are visible in sections of the ovary. Ovarian loss of egg cells through ovulation (about 500 during the whole reproductive life) is a very small fraction of the total number lost. The stimulus for the tremendous cell multiplication of egg cells during fetal life, and for the death of so very many of them during fetal life and thereafter is not known. Nor is it known what is involved when a single one of thousands of egg follicles goes on to maturity while hundreds nearby may die and disappear.

The Egg Follicle: Development Each egg is enclosed in a cellular envelope, and the whole is known as a follicle (Figure 8.5). The least developed stage is called the *primordial follicle*. It consists of an egg surrounded by only a few flattened follicular cells. In the next stage, known as *primary follicle*, the egg cell is larger, and the follicular cells are taller (cuboidal) and more numerous. In later stages, the cuboidal follicular cells have increased in number, and they may be disposed in several layers. Then a noncellular, fluid-filled space appears between some layers, known as the *follicular cavity*. From this point on, the egg cell is

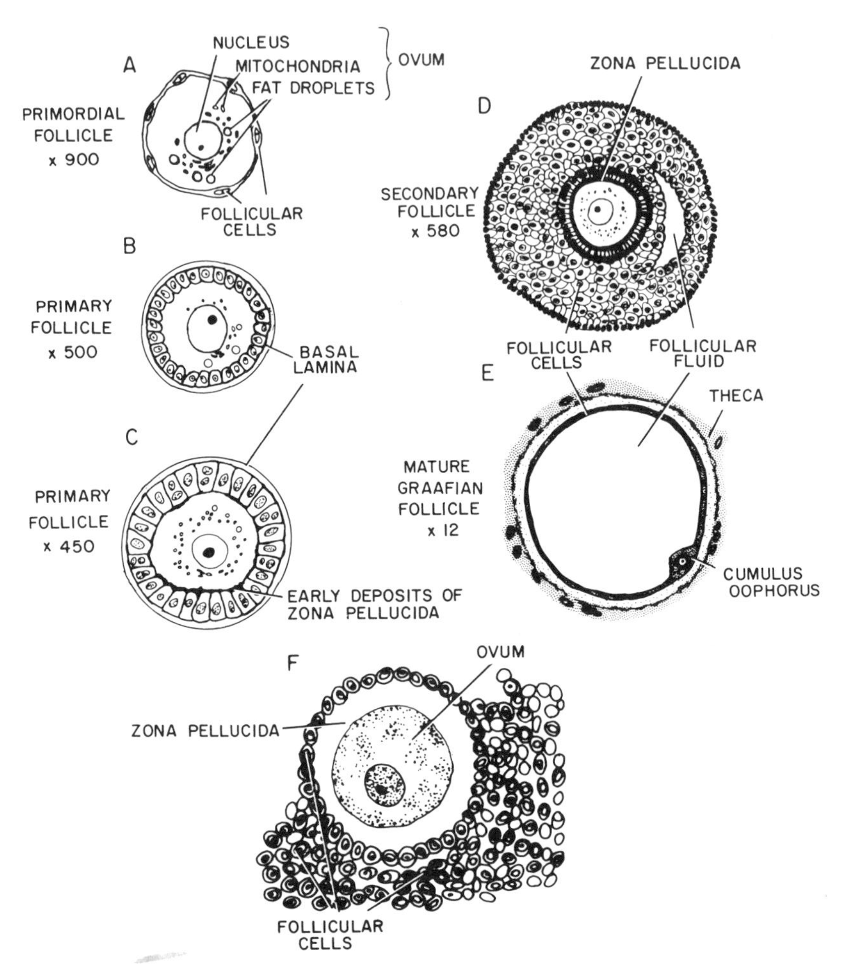

Figure 8.5. Stages in the growth and development of the ovarian follicle. Approximate magnifications are given for each stage. Part of the cumulus oophorus, zona pellucida, is unstained in F, but stained in D. Whether stained or unstained, it appears homogeneous with the light microscope. (After Hamilton, Boyd, and Mossman, 1972; Bloom and Fawcett, 1975).

progressively more eccentrically placed in the follicle. As the follicle expands, it is surrounded by a delicate connective tissue capsule. The fully mature follicle is known as the *Graafian follicle*.

Maturation During the process of maturation, the egg cell has enlarged. The nucleus is clear, with little microscopically visible chromatin, and contains a very prominent nucleolus. In other words, the nucleus of the egg cell has all the signs of marked genetic activity, as if it were busily engaged making messenger and ribosomal RNAs at a great rate. The products of cellular activity are stored in the cytoplasm. After fertilization, there does not seem to be any further genic activity until the late, blastula or gastrula, stage. It can be said that the unfertilized egg in the mature follicle has made enough messenger and ribosomal RNA of the right kinds to last through blastula to the gastrula stage.

The egg is enveloped in a membrane known as the *zona pellucida*. The membrane is penetrated by submicroscopic processes of follicular cells attached to it. These cells radiate from the membrane and form a small hillock known as the *cumulus oophorus*. The whole process of maturation takes about 2 weeks, when the follicle is 10–12 mm in diameter. The connective tissue of the capsule of the follicle also matures. It becomes thicker, more highly vascularized, and more cellular. The part closer to the follicle is known as the *theca interna* and becomes largely cellular. The follicular cells are believed to synthesize and secrete the steroid hormone, estrogen. The cells are ovoid, and fine lipid droplets accumulate in the cytoplasm. The mitochondria, as viewed with the electron microscope, are peculiar, differing from those of most other cells of the body in that they have spaces that have been interpreted as tubular in nature. These characteristics (fine lipid droplets in the cytoplasm and tubular mitochondria) seem to be shared by most other cells that secrete steroid hormones, i.e., the corpus luteum cells, cells of the adrenal cortex, and the interstitial cells of the testis (Figure 8.6). Outside of this part of the capsule the connective tissue forms a capsule resembling the adjacent connective tissue and containing numerous small blood vessels.

Rupture (Ovulation) As the follicle expands to its maximum size, the amount of follicular fluid increases. The whole region is well vascularized. Then at one point near the surface of the ovary, the blood flow stops, and the region appears pale. The follicle projects beyond the surface, and then ruptures. As the viscous follicular fluid oozes out, follicular cells float out into the peritoneal cavity, and with them the cumulus oophorus. This sticky mass enters the open end of the oviduct, picked up by the sweeping action of the beating cilia of the fimbria, which is itself in constant motion over the ovary. At the same time, some vessels rupture, releasing some blood into the follicular cavity, some of which may also enter the oviduct. Most of the follicular wall persists, but loses its spherical shape as the walls fold and occupy partly the space formerly filled by the follicular cavity. This is the beginning of the corpus luteum. Before describing the origin and structure of this organ, it is necessary to review the process of mitosis and meiosis in the egg cell.

As discussed earlier, the primary oocytes multiply rapidly, reaching a peak of about 7,000,000 at about 5 months of fetal life. This process of cell multiplication is achieved through mitosis, with its cyclic process

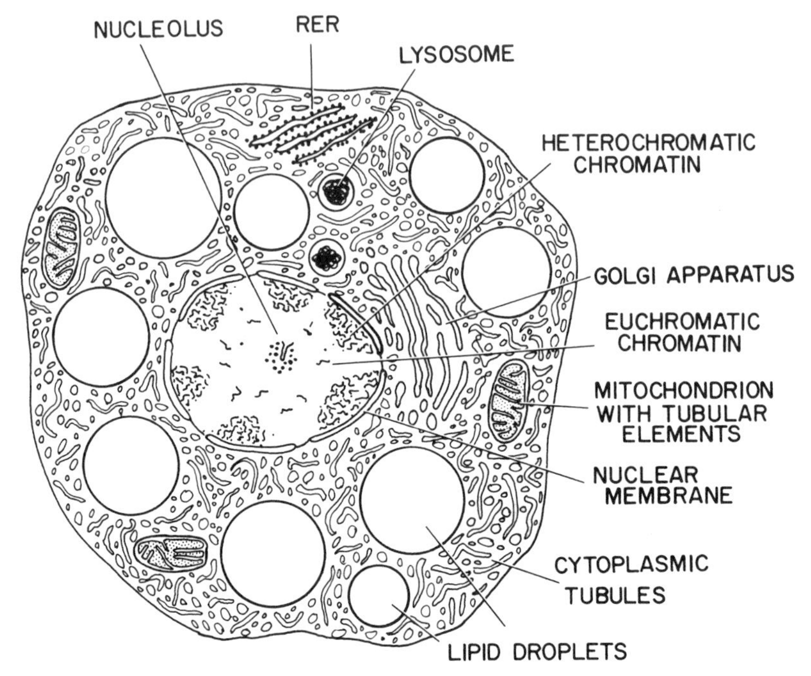

Figure 8.6. Endocrine cell synthesizing and secreting a steroid hormone, such as a cell of the corpus luteum, an interstitial cell of the testis, or a cell of the adrenal cortex. RER, rough endoplasmic reticulum. (After Junqueira, Carneiro, and Contopoulos, 1975).

of replication of DNA as in all other cells of the body (see Chapter 3). The oocytes have the diploid number of chromosomes. This process is completed in fetal life, and the prophase of the first meiotic division begins. The egg cells remain in this condition until puberty, when the first meiotic division is completed. The number of chromosomes is now haploid, or half of the diploid number. The second meiotic division takes place just before fertilization. The first and second polar bodies resulting from these two meiotic divisions degenerate, leaving one mature ovum with the haploid number of chromosomes (see Chapter 4).

Corpus Luteum The formation of the corpus luteum begins with ovulation. When the follicle collapses, the follicular cells increase in size, as do the cells of the theca interna. In both, fine lipid droplets and fine pigment granules appear. The two types of cells soon become indistinguishable. The connective tissue of the capsule penetrates between the epithelial gland cells, and carries with it small blood vessels, so that the corpus luteum is well supplied with nutrients. The corpus luteum is now a "solid" endocrine organ synthesizing and secreting steroid hormones, especially progesterone. The gland cells resemble other steroid-secreting cells because they have numerous, fine fat droplets in the cytoplasm and tubular mitochondria (Figure 8.6).

The connective tissue is sparse, and there is little space between gland cells and the capillaries, whose walls are very thin. If the egg released by the mature follicle is not fertilized, the corpus luteum degenerates; the dead cells are replaced by a scar of whitish connective tissue called the *corpus albicans*. The essential processes—growth and maturation of the ovarian follicle, ovulation, formation of the corpus luteum and its subsequent degeneration—take place normally during menstrual cycles. If the egg passes into the oviduct and is fertilized, the corpus luteum is enlarged to 20-30 mm or more and may persist until late in pregnancy.

Testes

These are nearly completely enclosed by a smooth layer of flat mesothelial cells originally derived from the peritoneum of the abdomen. This surmounts a thick connective tissue capsule (about 0.5 mm thick). From this, strands and sheets of connective tissue penetrate the testis, separating it into compartments or lobules (see Figure 1.16). These are occupied chiefly by the seminiferous tubules. These will not be described, except to note that there are two major cell types—the *supporting* or *Sertoli cells* and the *spermatogenic cells*. After the final mitotic division, the latter go through two meiotic divisions as a result of which four cells are produced from each diploid cell, each with the haploid number of chromosomes. Between the seminiferous tubules are some connective tissue cells and fibers and numerous fine capillaries. Associated with the latter are the interstitial cells of the testis. These cells, scattered in small, microscopic clusters between the tubules (see Figure 7.11), comprise a diffuse endocrine organ, from which testosterone is synthesized and secreted. They are large cells, ovoid or polygonal in shape, with large nuclei like other steroid-secreting cells. The cytoplasm contains numerous fine lipid droplets and the mitochondria are tubular. They also contain pigment granules and rodlike crystalloids.

Adrenal Glands

Cortex The adrenal cortex surrounds and encloses the centrally placed adrenal medulla. The cortex is covered by a connective tissue capsule. From it, strands penetrate the cortex, and bear with them the whole arterial circulation of the cortex and much of the medulla. Connected with these strands are fine connective tissue layers that enclose cell cords and blood vessels. As in most endocrine glands, these connective tissue strands are extremely thin, as are the smaller blood vessels.

The finer blood vessels of the cortex are called *sinusoids*, and the cells lining them are of two sorts: 1) those which ingest particles from the plasma as they impinge on the cell surface, as in the anterior lobe of the hypophysis, and 2) those which resemble capillary lining cells of most parts of the body, except for their extreme attenuation. (The circulatory pattern is described more fully in Chapter 7.)

The three layers of the adrenal cortex are not clearly demarcated (Figure 7.7). The outermost layer, the *zona glomerulosa*, is the thinnest. Here the gland cells are arranged in cell cords which may be aggregated to form ovoid or columnar groupings. These cells have few lipid droplets in their cytoplasm and the mitochondria are not tubular; instead, the internal architecture of the mitochondria is marked by numerous partitions (cristae), much as are most other cells of the body when prepared in the same way. The gland cells are thought to synthesize and secrete aldosterone.

Beneath this layer is the *zona fasciculata*. Here the gland cells form cell cords one or two cells thick. The cell cords are somewhat parallel to each other, and are surrounded by the blood sinusoids. The gland cells are polyhedral in shape, and the cytoplasm is filled with numerous fine lipid droplets. The mitochondria are tubular in nature. The gland cells are thought to produce the glucocorticoid hormones (cortisone, for example).

The innermost layer of the cortex is the *zona reticularis*. The cell cords are highly irregular and form a dense network associated with the equally irregular blood sinusoids. This layer is somewhat thinner than the zona fasciculata. The gland cells are smaller than those of the zona fasciculata, and contain fewer lipid droplets. The gland cells are thought to synthesize and secrete androgenic hormones.

Medulla The adrenal medulla differs from the cortex in its embryologic origin and in the hormones that are synthesized and secreted. While the cortex elaborates steroid hormones, the medulla secretes adrenaline and noradrenaline, two hormones also produced elsewhere in the body by nerve cells. Although the cytoplasm of cortical cells contains lipid droplets, the cytoplasm of the medullary cells is frequently filled with minute secretory granules. The common bond between cortex and medulla is that they share very largely the same blood vessels, which is the means of transport of their secretory products to the rest of the body.

REFERENCES

Bergman, R. A., and A. K. Afifi. 1974. Atlas of Microscopic Anatomy. W. B. Saunders Co., Philadelphia.

Bloom, W. and D. W. Fawcett. 1975. A Textbook of Histology. 10th Ed. W. B. Saunders, Co., Philadelphia.

Dodd, E. E. 1979. Atlas of Histology. McGraw-Hill Book Co., New York.

Gersh, I. 1973. Submicroscopic Cytochemistry. Vol. I. Proteins and Nucleic Acids. Vol. 2. Membranes, Mitochondria, and Nucleic Acids. Academic Press, New York.

Ham, A. W. 1974. Histology. 7th Ed. J. B. Lippincott Co., Philadelphia.

Hamilton, W. J., J. D. Boyd, and H. W. Mossman. 1972. Human Embryology. Williams & Wilkins Co., Baltimore.

Isenman, L. D., and S. S. Rothman. 1979. Diffusion-like processes can account for protein secretion by the pancreas. Science 204:1212–1215.

Junqueira, L. C., J. Carneiro, and A. Contopoulos. 1975. Basic Histology. Lange Medical Publications, Los Altos, Cal.

Krieger, D. T., and A. S. Liotta. 1979. Pituitary hormones in the brain: Where, how, and why? Science 205:366–372.

Lodish, H. F., and J. E. Rothman. 1979. The assembly of cell membranes. Sci. Am. 240(1):48–63.

Mercer, E. H. 1967. Cells: Their Structure and Function. 2nd Ed. Anchor Books, Doubleday & Co., Inc., Garden City, N.Y.

Rennels, E. G., and D. C. Herbert. 1979. The anterior pituitary gland—its cells and hormones. BioScience 29:408–414.

Rhodin, J. A. G. 1974. Histology: A Text and Atlas. Oxford University Press, New York.

Thomas, L. 1975. The Lives of a Cell. Bantam Books, N.Y.

Weiss, I., and R. O. Greep. 1977. Histology. 4th Ed. McGraw-Hill Book Co., New York.

9 Neurologic Aspects of Integration of Sexual Functions

The nervous system is involved in the integration of sexual activities and supplements the slower, more prolonged integrative functions of endocrine glands. Examples of its involvement are lactation, fat storage and mobilization, gland secretion, erection, orgasm, and associated activities.

NERVE CELL TYPES

Nerve cells are highly specialized for irritability (or sensitivity to stimuli) and conductivity of the nerve impulse. Synapses make possible the appropriate and precise responses of the body, and the rapid integrative action of billions of nerve cells in one's body.

The chief kinds of nerve cells are: 1) afferent (sensory) cells, which connect with some kind of sensory receptor and conduct nerve impulses to the central or autonomic nervous system; 2) efferent (motor or effector) cells, which carry impulses to muscles and glands; 3) integrative cells, which are involved in correlation of incoming and outgoing nerve impulses; and 4) neuroendocrine cells, which have the properties of most nerve cells and in addition secrete certain hormones.

NERVE CELL CYTOLOGY AND GENE ACTION

There are billions of nerve cells in every person. They vary enormously in size (0.1 mm or less to 1 m or more), in shape (from spherical to highly angular, or pyramidal), in the number of cell processes (from one to hundreds), in the pattern of organization of their cell processes (from tandem to layer on layer to fine mossy branches) and in their interconnectedness and hence function. Despite these diversities, their submicroscopic structure is like that of non-neural cells with respect to the nucleus and its DNA and RNAs, the nuclear membrane, the cytoplasmic ground substance, the smooth and rough endoplasmic reticulum, the Golgi structures, mitochondria, and cell membrane. While not distinctive, some of these constituents are rather characteristically organized at the submicroscopic and microscopic level (Figure 9.1).

The nucleus of large nerve cells is spherical, large and pale-staining except for the nucleolus and a few small chromatin particles, one of which in women is the Barr body. The rough endoplasmic reticulum (RER) of many nerve cells is aggregated into granules known as *Nissl bodies*, which, after staining with basic dyes, are visible with the light microscope. Large numbers of filaments (100 Å in diameter) and microtubules (240 Å in diameter) are present between Nissl bodies, and when suitably prepared they line up side-by-side to form neurofibrils visible with the light microscope.

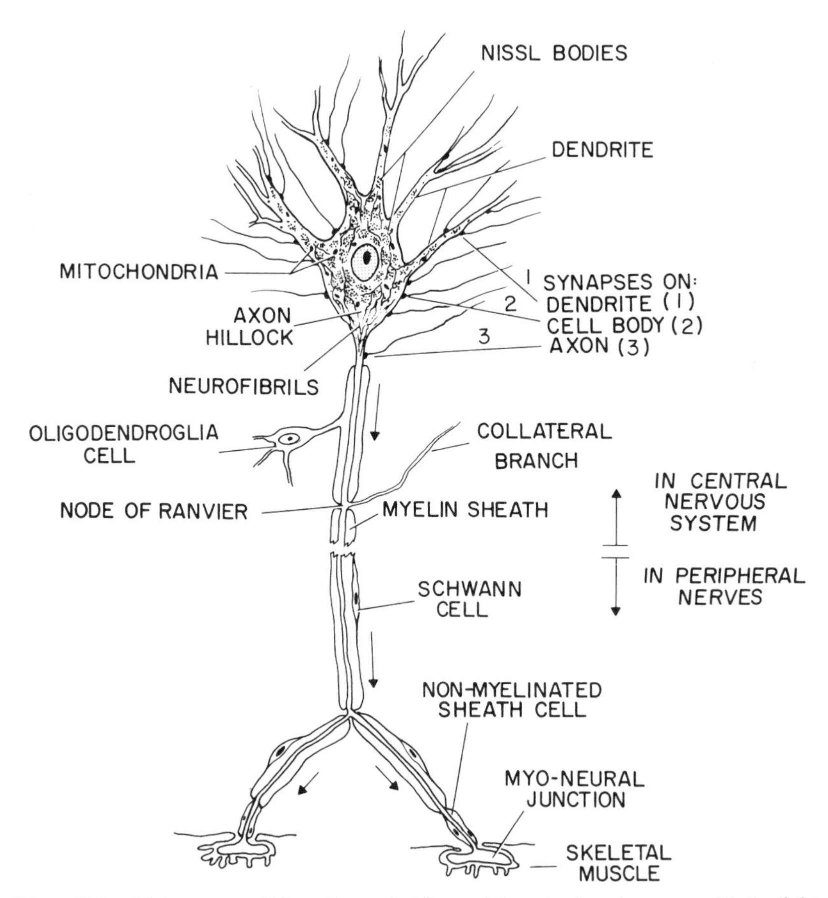

Figure 9.1. Motor nerve cell from the ventral horn of the spinal cord, as seen with the light microscope.

The axon is a nerve cell process specialized for conduction and may be more than a meter long. It is insulated by the myelin sheath, composed of alternating layers of protein and lipid. In some axons, the myelin is so thin as to be not visible with the light microscope: these are known as *unmyelinated nerve fibers.* In others the sheath has so many layers that it is visible with the light microscope: such axons are known as *myelinated nerve fibers.* The myelin sheath of both kinds of nerve fibers is really a modified cell membrane, wrapped around the axon, which arises not from the nerve cell but from neighboring nonconducting or glial cells called *oligodendroglia* (when they occur in the central nervous system) and *Schwann cells* (when they occur in peripheral nerves).

NERVE CONDUCTION

Synapses

All nerve cells have certain characteristic structures called *synapses*, and some efferent cells are characterized by their motor end plates and other myoneural junctions. The synapse is a junction between two nerve cells. While conduction along nerve fibers may take place toward the cell body or away from it, it is polarized by the synapse and can proceed in one direction only—from the presynaptic region of one cell to the postsynaptic region of another. The number of synapses per cell varies enormously in different kinds of nerve cells, ranging from a few to several hundred thousand, depending on the cell type. Synapses are specialized gaps between nerve cells, differ in size or shape, may be inhibitory or excitatory, and function through chemical transmitters. The varied properties of the synapse make possible many more different and controlled responses than is indicated by the obsolete one-to-one, straight line relations of diagrams of nerve cells.

Neurotransmitters

The nerve cell is still the basis of nerve systems, although it is no longer regarded simply as a trunk-line operation. The basic model is for the presynaptic region to be marked by numerous vesicles containing a neurotransmitter substance (Figure 9.2). When an appropriate nerve impulse reaches this region, the contents of one or more of these vesicles are liberated in the synaptic gap. Involved in this process is the receptor in the cell membrane of the postsynaptic cell. This makes possible the transmission across the gap of the nerve impulse, in the sense that the postsynaptic fiber initiates a second nerve impulse. Also, this process establishes the polarity or direction of transmission of the nerve impulse. The transmitter substance is quickly inactivated by an appropriate enzyme and some of

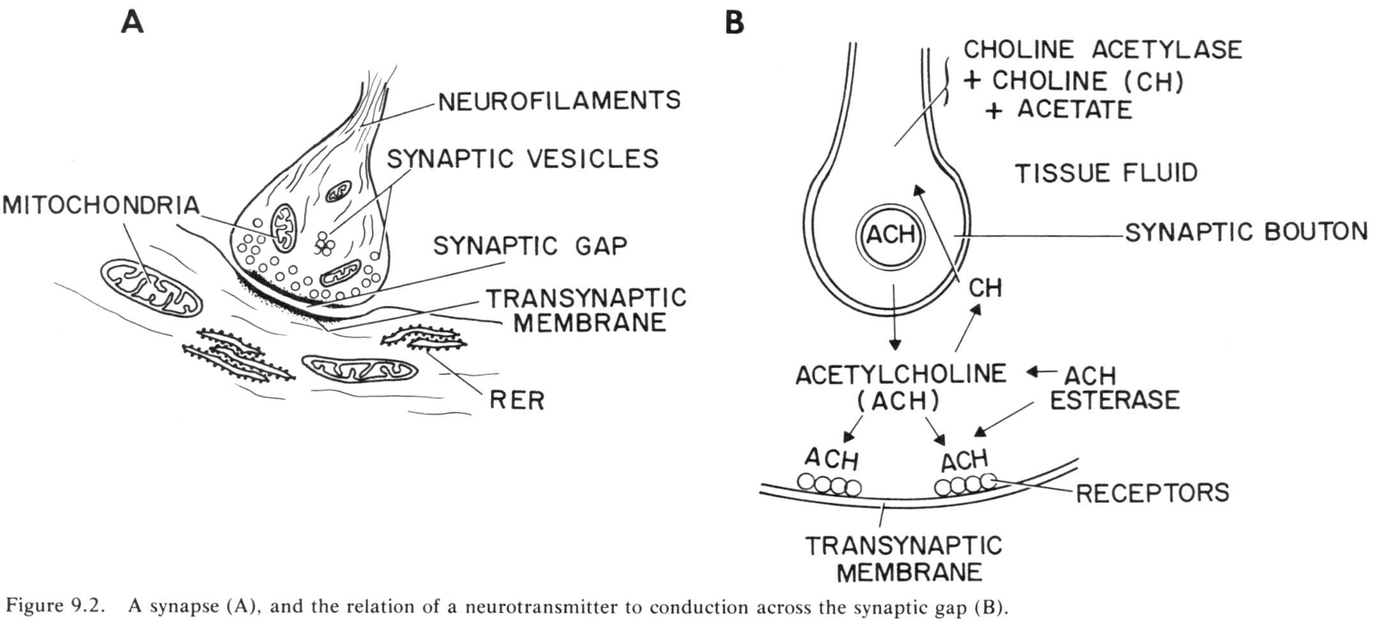

Figure 9.2. A synapse (A), and the relation of a neurotransmitter to conduction across the synaptic gap (B).

the products are reabsorbed, to be degraded, or resynthesized locally to the neurotransmitter, which is stored in vesicles, ready to be released by a new nerve impulse. This group of neurotransmitters includes a wide variety of compounds, of which the following are the classic ones and have been studied for the longest times: acetylcholine, norepinephrine, and epinephrine. Other potential members include serotonin (5-hydroxytryptamine), melatonin, histamine, dopamine, and γ-aminobutyric acid. Presumably these neurotransmitters are synthesized mainly in the nerve cell body and pass down the axon to the synaptic region. Groups of nerve fibers with common sites of origin, course, and termination and with special and characteristic neurotransmitters, may form tracts and interrelated systems of tracts that traverse large parts of the central nervous system. The term *tract* is used for a bundle of nerve fibers originating in one or more related nuclei. The term *nucleus* as used here refers to a group of nerve cells whose distribution may be diffuse or circumscribed, but whose functions and synaptic relations are very similar.

Neuromodulators

Also involved in nerve connection, but not as neurotransmitters, are a relatively new and not yet well-clarified group of compounds, the neuromodulators. These are similar to hormones in that they may arise elsewhere and influence the general environment of synapses. This group includes prostaglandins, corticosteroids, estrogens, testosterone, thyroid hormones, adrenocorticotropic hormone, lipotropic hormone, luteinizing hormone releasing hormone and thyroid releasing hormone, enkephalins, and β-endomorphin. Their exact site or mode of action are not known, nor has their general distribution in the central nervous system been worked out yet.

Summary

In considering the nerve cell in general, it becomes clear that the pale-staining, enlarged nuclei are in a high state of activity, synthesizing much ribosomal, messenger, and transfer RNA, which find their way to the cytoplasm. In the cytoplasm, there are the basic requirements for the synthesis of much protein, which passes with other substances down the axon at a rate of 1 mm to several centimeters per day, depending on the substance being transported. For this and for the processes of synthesis of proteins and other substances needed for cell function and maintenance, the energy requirements are furnished by the numerous mitochondria. The conduction of the nerve impulse takes place at the surface of the cell and its processes, which are insulated from adjacent nerve cells and processes by the myelin sheath and the Schwann cell or oligodendroglial cell. The chemical mediators are stored in minute vesicles, and released in

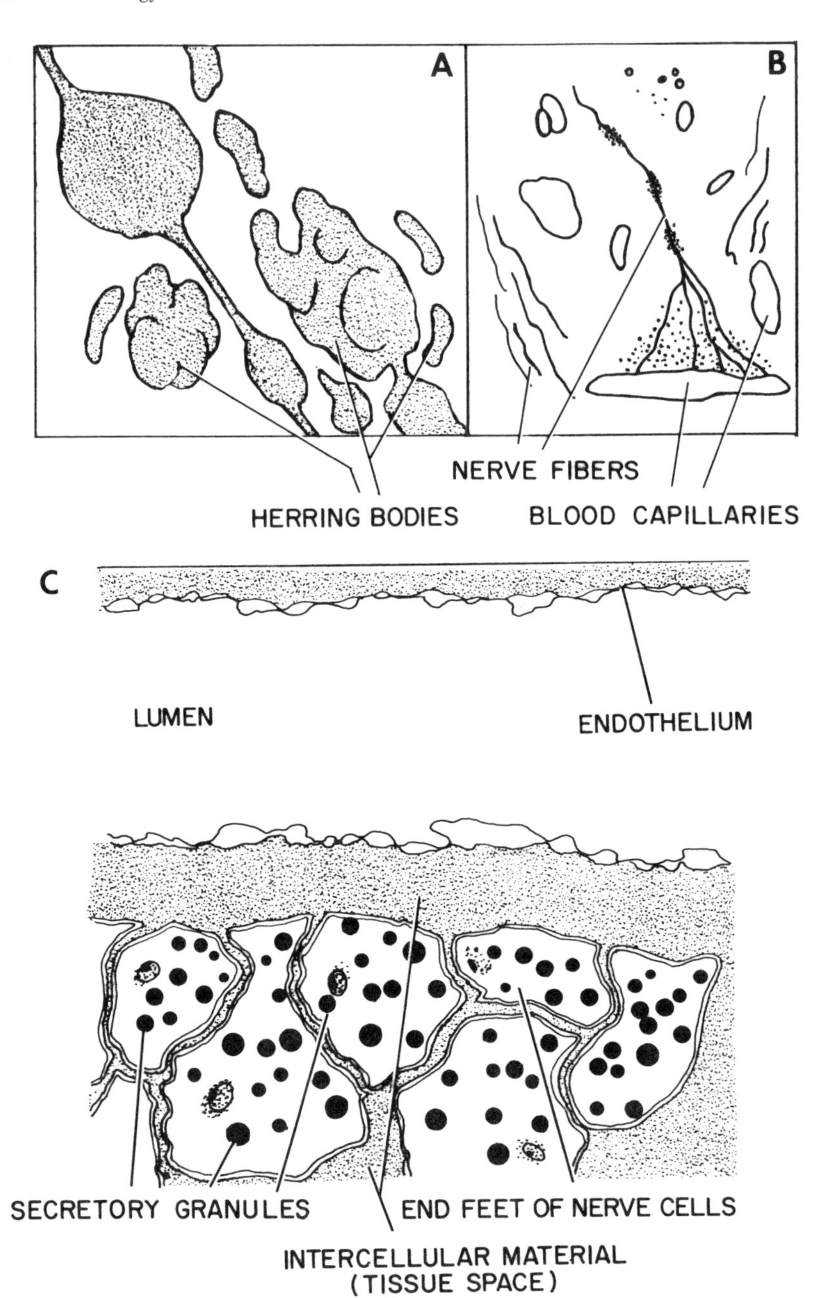

Figure 9.3. Stained bodies, known as Herring bodies, in nerve fibers of the hypophyseal stalk (A). The secretory material is thought to be stored there. When it reaches the terminal branches near capillaries (and perhaps while en route there) it is thought to be extruded and visible with the light microscope (B). The secretory granules appear with the electron microscope in the terminal enlargements (C). (After Ham, 1974 and Palay, 1953).

small discrete packets in a manner that restricts the diffusion of the mediator and limits it to the region of the end organ, the synapse or motor end plate. The synapses are essential parts where the integration of all the input take place, and the nerve ending is where the coordination is transmitted to the effector, or in the case of sensory nerves, where connections are made with specialized sensory receptors.

PROPERTIES OTHER THAN NERVE CONDUCTION

Neuroendocrine cells have certain specialized properties in addition to those associated with nerve conduction:

1. Along the course of their nerve fibers are swellings visible in the light microscope, known as Herring bodies (Figure 9.3A). They can be resolved with the electron microscope as masses of submicroscopic secretory granules either on their way to the axon termination, or stored temporarily, or secreted along the axon.
2. Nerve fibers may terminate close to capillary endothelium (Figure 9.3B). The nerve ending may be enclosed in a sheath of stainable material supposedly on its way to the capillary.
3. The nerve terminations may resemble synaptic junctions, but these junctions contain presecretory material, which is regarded as the neurohormone precursor ready to pass into the intercellular space and thence through the fenestrae into the blood flowing in the lumen of the capillary (Figure 9.3C).

INTEGRATIVE SYSTEMS

Although it is essential to know about nerve cells as such, one must realize that nerve cells function as parts of systems that are integrated into larger and larger systems, and these into still larger hierarchical systems of integration. Underlying these are certain criteria:

1. From the simplest to the most complex hierarchy, the nerve cells and their fibers are organized in an orderly, uniform, regular way.
2. The response to small or large environmental changes is rapid, precise, and punctate.
3. In a massive response to real or imaginary threats (like the fight or flight response), the sympathetic and parasympathetic nervous systems are geared to support the somatic response (increased respiratory, cardiac, and sweating action to match the increased muscle tone).
4. The behavioral response is controlled and homeostatic (i.e., it maintains the overall equilibrium of the body).

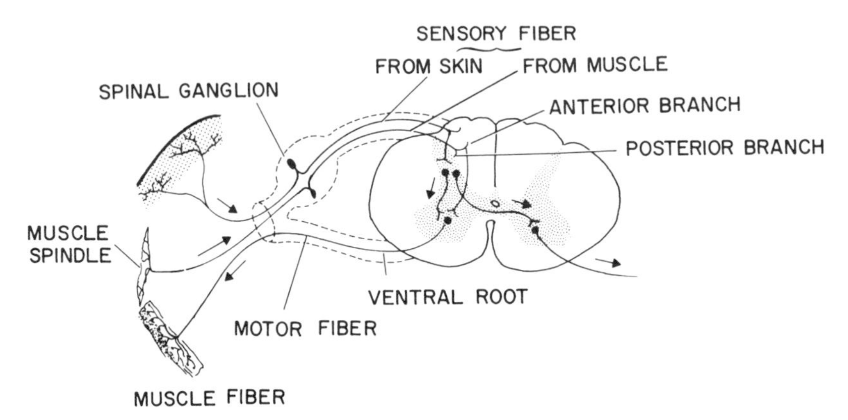

Figure 9.4. The simplest possible reflex connections within a single spinal segment. (After Truex, 1959).

The following examples illustrate different degrees of integration: The reflex level involves two or three neurons (Figure 9.4). A simple controlled movement, even small ones of fingers or arm (such as in flexion and extension) or large ones of the trunk, involve hundreds or thousands of neurons and several tracts. From this level extend a host of higher integrations and hierarchies. The ramifications of each one of them are so numerous and all-extensive that it is virtually impossible to include them all for even one level.

Figure 9.5 attempts to illustrate the ramifications of one level of organization, the hypothalamus, showing the vastness and extensiveness of the neural connections. The same is true of every part of the central nervous system. Figure 9.6 shows the interconnectedness of the hypothalamus with other parts of the central nervous system. In Figure 9.6A, nerve cell fibers that arise in other parts of the central nervous system pass as known tracts to the hypothalamus (stippled area) where they synapse with nerve cells, some of which are neuroendocrine in nature. In Figure 9.6B, fibers originating from nerve cells in the hypothalamus pass as known tracts to other parts of the central nervous system where they synapse with nerve cells from other regions. There is in effect a continuous informational feedback with great parts of the central nervous system.

Hypothalamus

A brief summary of the structure and function of the hypothalamus emphasizes the interrelatedness of its neural and neuroendocrine parts.

The hypothalamus is a small region at the base of the brain whose weight is about 4 g, and whose ventral surface is about 1 cm^2. Continuous with it is the tuber cinereum, which funnels into the underlying infundiblar stalk and lobe of the neurohypophysis. The hypothalamus contains

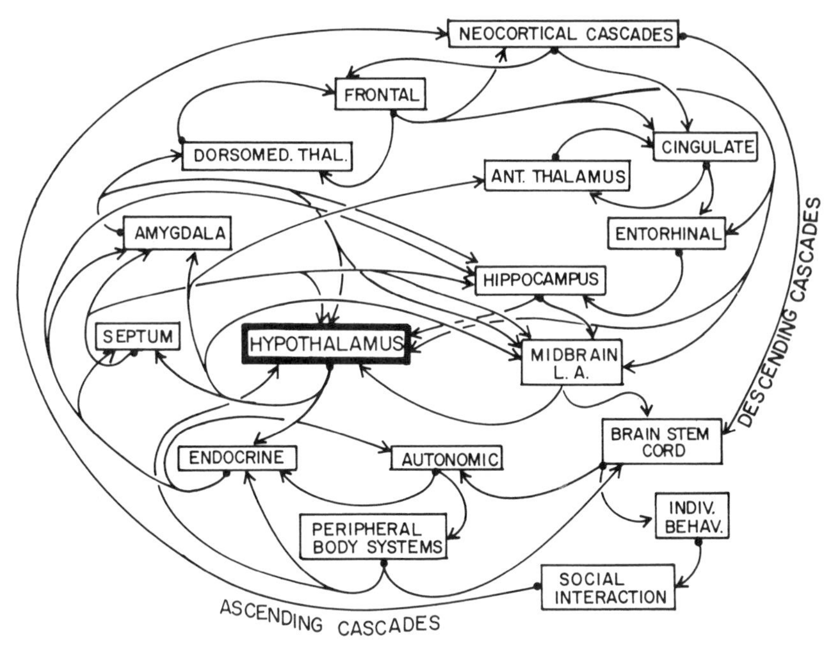

Figure 9.5. Diagram shows minimal ramifications of the hypothalamus with other parts and functions of the central nervous system (After Peter Sterling, personal communication).

a number of important nuclei. The functions of these in humans are only known to a certain extent and are very largely based on experiments with laboratory animals.

The hypothalamus is important for two major reasons. First, it contains the nerve cells where the releasing and release-inhibiting hormones are synthesized. These hormones pass down the axons to the tuber cinereum where they are secreted and enter the blood capillaries. These are gathered together to form the hypothalamo-hypophyseal portal system, whose capillaries are distributed in the anterior lobe of the hypophysis where the hormones act on the cells and affect the rate of release of their hormones (see Chapter 8).

Second, the hypothalamus also contains nuclei named according to their relative positions that mediate certain functions. These include:

1. The control of the water balance of the body through the action of the antidiuretic hormone, a neuroendocrine secretion stored and released by the neurohypophysis
2. The contraction of smooth muscle of the uterus to expel the fetus and placenta through the action of the oxytocic hormone, another neuroendocrine secretion stored and released by the neurohypophysis
3. The contraction of the myoepithelial cells of secretory acini of the

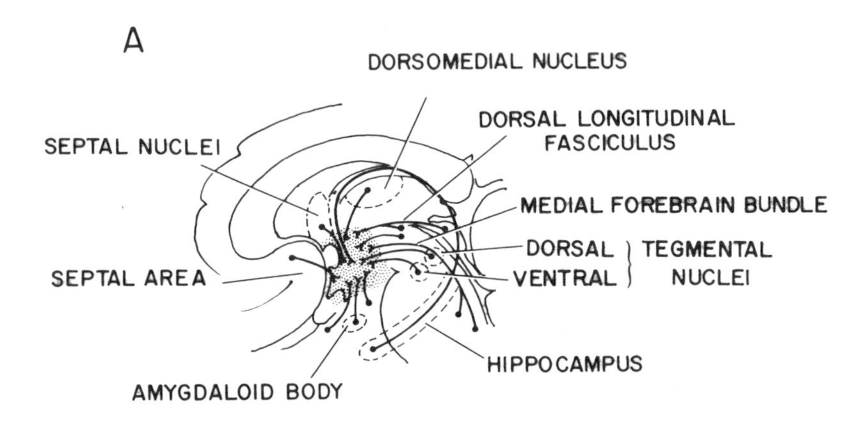

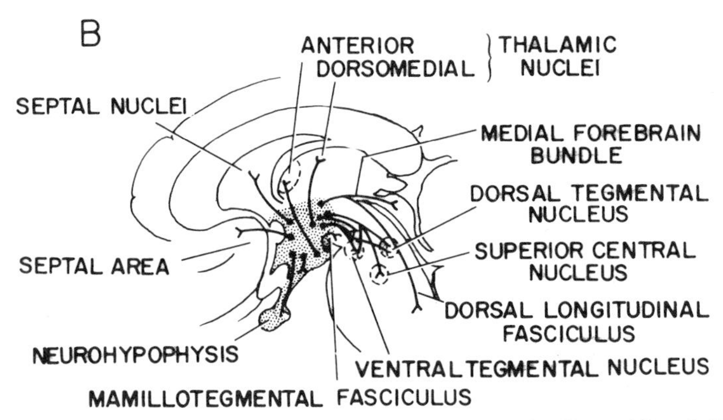

Figure 9.6. Extensive neural connections of the hypothalamic afferent (A) and efferent (B) fibers. (After Noback and Demarest, 1975).

mammary gland and of the small ducts, through the action of the same oxytocic hormone

4. The control of some aspects of sugar and fat metabolism through the "hunger" or "feeding" centers and the "pleasure" and "punishing" centers
5. The control of blood pressure, cardiac rate, and body temperature.

Several of these functions are essential for the maintenance of a relatively constant internal environment in the body. For example, when the body temperature is too low, vasoconstriction takes place in the skin and reduces sweating and loss of heat by convection. There is also increased muscle tone and visceral activity, which generate body heat through appropriate action of the sympathetic nervous system. When the body temperature is too high, vasodilation and increased sweating tend to increase the loss of body heat, all achieved by appropriate stimulation of the sympathetic and parasympathetic nervous system. Again, hypothalamic action may result in an increase in the force and rate of the heartbeat and an increase in blood pressure (via appropriate activation of the sympathetic nervous system), or in a decrease in the force and rate of the heartbeat and a fall in blood pressure (via appropriate action of the parasympathetic nervous system). All of these tremendously important functions are performed by virtue of the synaptic connections made in hypothalamic nuclei by afferent and efferent fibers arising from the neurons in these nuclei. These nuclei are the chief sites of the subcortical centers for the regulation of the sympathetic and parasympathetic nervous systems.

THE LIMBIC SYSTEM

Another level of still higher organization of the nervous system is the limbic system (Figure 9.7). The limbic system includes the fiber paths and great tracts that join one part of it to the others by synaptic fibers and relate them to still other regions of activity of the central nervous system, such as the temporal lobe of the cerebral cortex.

The limbic system is a link between the emotional and cognitive mechanisms of the neocortex and has been implicated in memory for recent events. The limbic system also acts as a link between the emotional centers and the sympathetic and parasympathetic nervous system by way of the hypothalamus. The great range of activities with which the limbic system has been involved in animals includes: respiratory and cardiovascular changes, grooming, erection of erectile tissues, expressive gestures with the shoulders and hands, facial grimaces, activity of the gastrointestinal tract, aggressiveness, and the activity of "pleasure" and "punishing" centers.

In general, the limbic system is critical for activities concerned with individual survival (such as feeding, fighting, and flight) and with survival of the species (such as mating and care of the offspring). It is influenced by sensory input of external or internal origin. The main outlet is the hypothalamus and its central connections are with the sympathetic and parasympathetic nervous system.

THE CEREBRAL CORTEX

The most complex organization of nerve tissue is the cerebral cortex. Its two major parts are interrelated: 1) the archipallium or evolutionarily

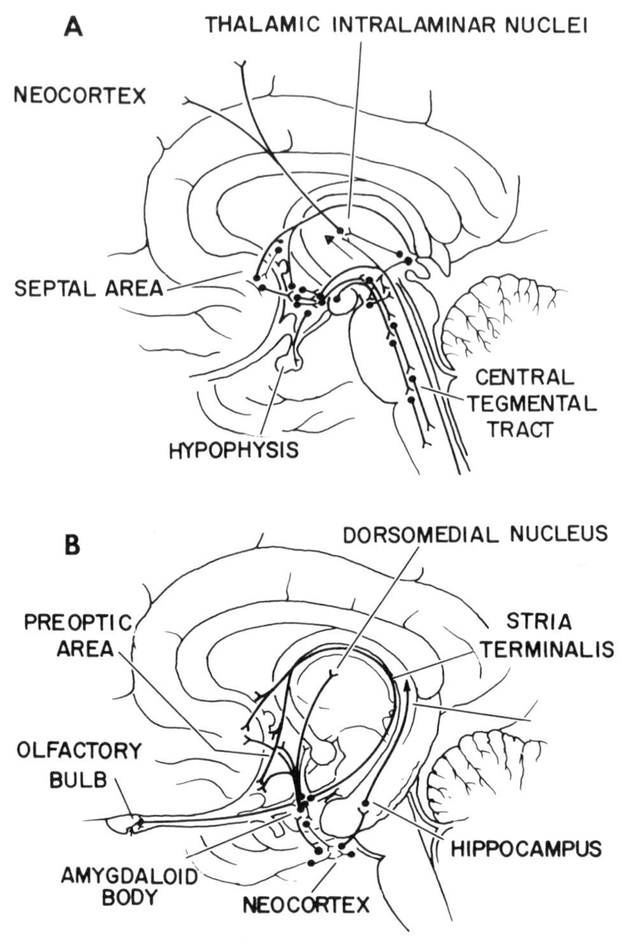

Figure 9.7. Some of the neural ramifications of the hypothalamus through its connections with the limbic system. Afferents to the limbic system are shown in A, efferents in B. (After Noback and Demarest, 1975).

older part of the cortex, recognizable even in fishes and amphibians, and 2) the neopallium or neocortex. The cortex is organized as two halves or hemispheres, joined across the midline by the great *corpus callosum*. It is organized as gray and white matter. The outer layer, or gray matter, is only a few millimeters thick, and is rich in nerve cells. It lies on a core of white matter whose nerve fibers interconnect with all other parts of the cortex as well as with all the lower parts of the central nervous system transmitting "information" from them and sending "instructions" to them after suitable integration in the gray matter. The nerve cells of the gray matter are disposed in several layers, and their organization differs from region to region (Figure 9.8). There are at least 200 areas that can be recognized by the kind and arrangement of their nerve cell bodies and fiber systems (Figure 9.9). The amount of gray matter that can be accommodated in the relatively inexpandable skull is increased by folding to form *gyri* and *sulci*. The term *gyrus* is used for any surface convolution of the cerebral cortex, and *sulcus* for the groove or fissure which bounds the

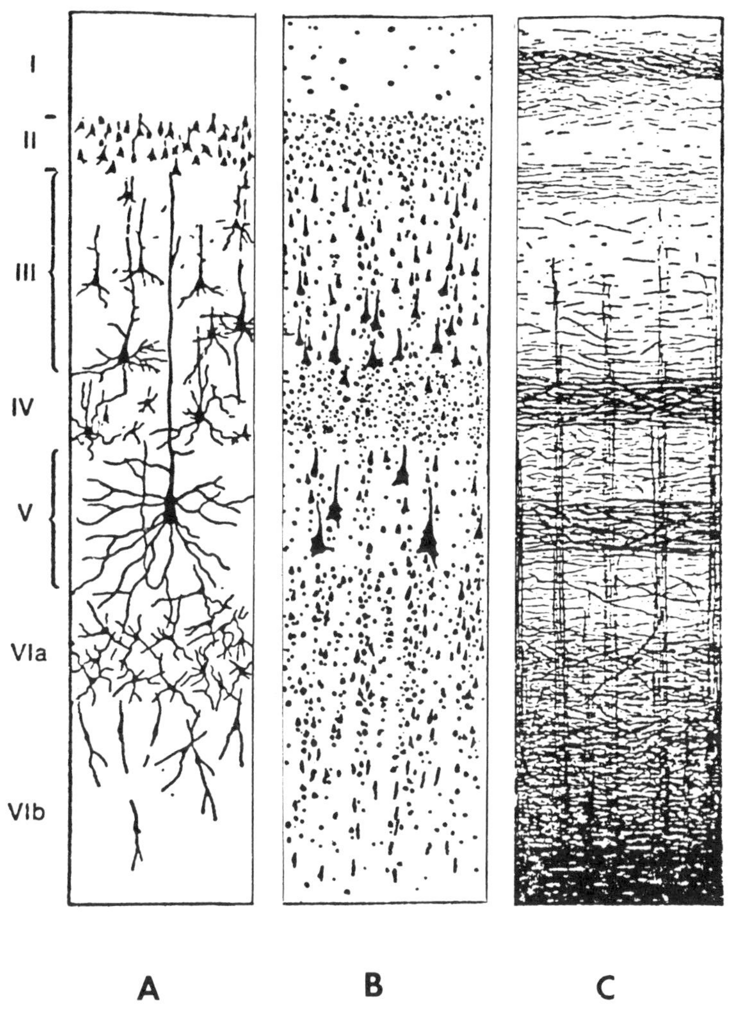

Figure 9.8. Photomicrographs of sections through a single area of the cortex as classified by Brodmann to show some characteristics of the six major layers. Preparation shows cell bodies and their processes (A); cell bodies (B); nerve fibers only (C). Layers are indicated at left. (After Brodal, 1969)

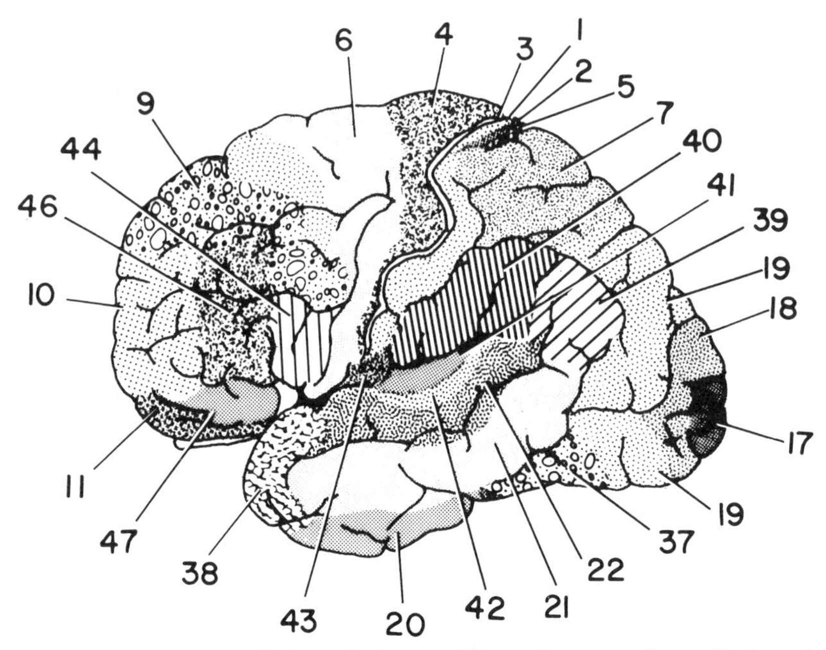

Figure 9.9. Lateral view of the cerebral cortex. The numbers, according to Brodmann's classification, indicate areas in which the layers (like those shown in Figure 9.8) differ cytologically. (After Brodmann, from Brodal, 1969).

gyrus. Only the major gyri and sulci are so constant as to be nameable and identifiable in human brains; the lesser sulci are less constant.

The cortex is divided into several lobes, all of which are visible with the naked eye: frontal (F), parietal (P), temporal (T), and occipital (O), and these lobes each have several gyri and sulci (Figure 9.10). Only four gyri have primary receptor areas that receive information from the environment—1) pain, temperature, touch, and proprioception (or muscle and tendon sensation), 2) the visual receptive center, 3) the auditory receptive center, and 4) the olfactory center (Figure 9.11). These comprise only a small part of the gray matter of the cortex, the major part subserving integrational or coordinating functions through the vast synaptic interconnections mediated by the nerve fibers of the white matter. Within the region of the receptor areas, the localization on the surface of the cortex is extremely precise, orderly, and punctate. Figure 9.12 shows the area localizations of the sensorimotor regions of the cortex that respond to specific stimuli of the various parts of the body. The motor system is similarly precise. Other functions are not so readily localized—i.e., speech, language, handedness, spatial awareness, and nonverbal ideation (Figures 9.11C and 9.12). These involve one or more large regions of in-

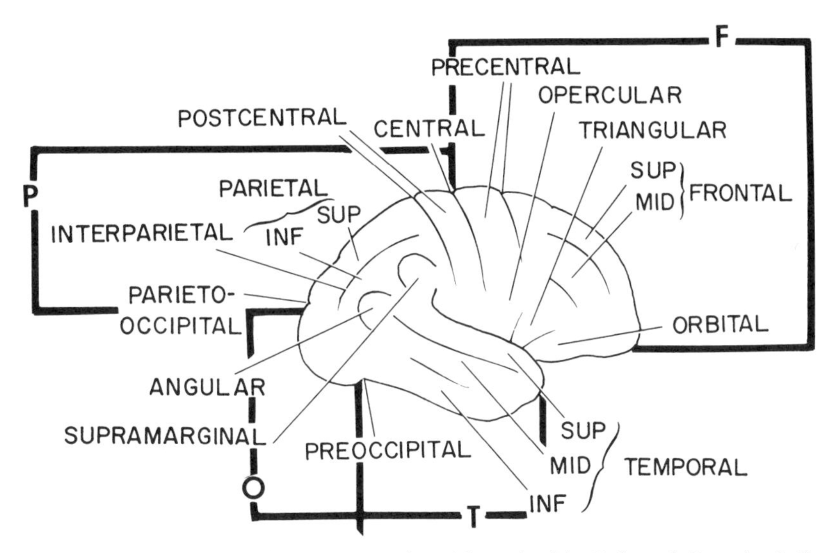

Figure 9.10. Major areas of the cortex as viewed from the side. F, frontal; P, parietal; O, occipital; and T, temporal. (After Willis and Grossman, 1973)

tegrative cortex and are not resolvable into finer parts like the receptive areas. Another interesting feature of the cortex is that although the brain structure may appear to be similar in both halves of the cortex, brain function is not commonly expressed equally on both sides. For example, language is localized to a very large extent in one or two areas of the cerebral cortex. Though apparently anatomically identical on both sides, in most people that on the left side of the brain is actually functional, the other being relatively "silent." This is an example of cerebral dominance. Like language, speech and handedness are expressed in most people in the left hemisphere. The right hemisphere is specialized for spatial appreciation and nonverbal conceptualization.

Neocortex

The neocortex is the great expansion of the most anterior part of the central nervous system. As in all other parts of the nervous system, the major aspects of the organization of the neocortex are predictably orderly, on a structural and functional basis.

The sensory and motor areas are based on the kinds and numbers of nerve cells and fibers in the six layers, which vary from site to site in a single brain, but are quite constant for any one area in many brains. The same constancy of pattern is evident after surgical ablations, and after stimulation during surgery of the brain. This orderly particularity is marked in all primary receptive areas of the neocortex (for vision, audi-

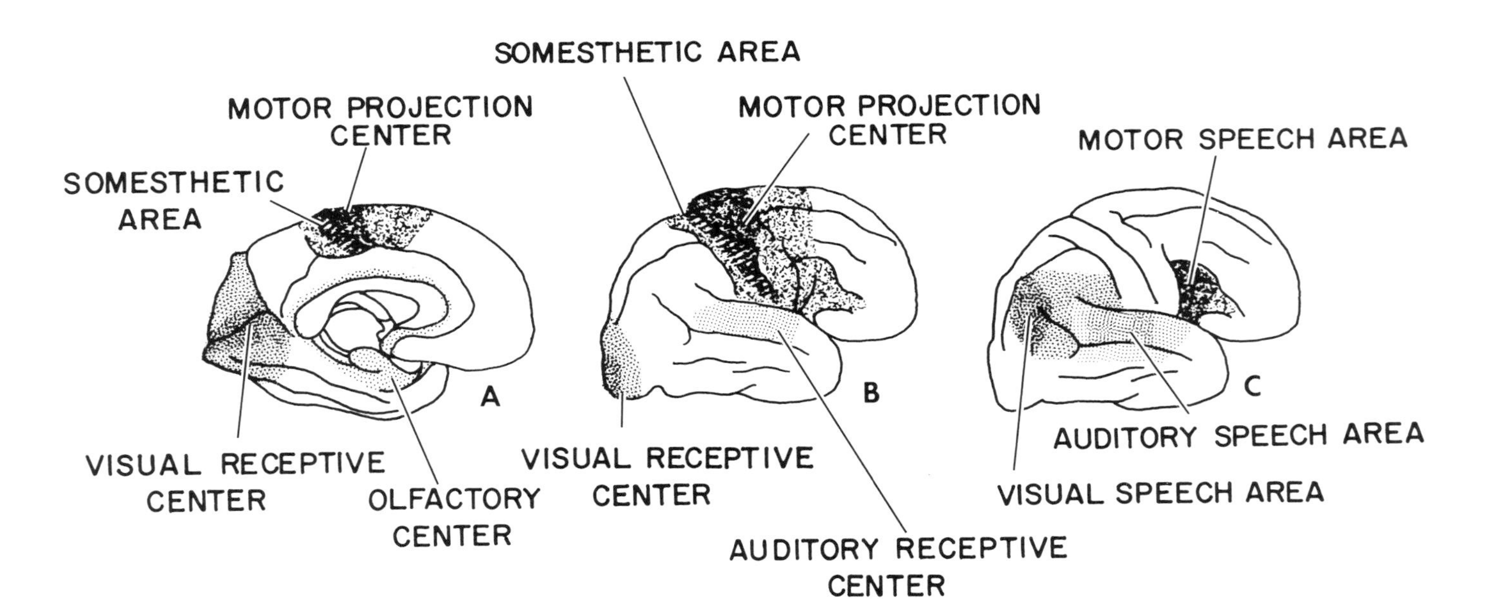

Figure 9.11. Cerebral cortex to indicate some of the surface anatomical and functional features (A and B), as well as some of the parts involved in speech (C). A is as viewed from a medial surface of one-half of the cortex; B and C are lateral surfaces. (After Ranson, 1931)

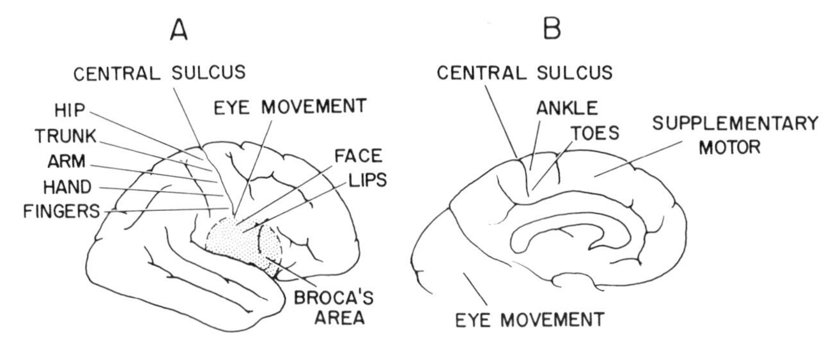

Figure 9.12. Lateral (A) and medial (B) surfaces of the cerebral cortex to show the locations that result in movements specifically of various parts of the body after suitable stimulation. Broca's area is the same area as that in Figure 9.11C labeled "motor speech area." (After Willis and Grossman, 1973)

tion, pain, touch), as well as for motor areas. Such orderly attributes presuppose equally orderly distribution and arrangement of afferent and efferent axons of long and short tracts of each area, even to the very small neurons with processes restricted to a few microns. The tracts are part of the associational and integrative systems which in the end underlie all the neural interactions in the whole body and its behavior in the environment.

Recently, there has been an emphasis on a kind of orderliness in the neocortex that underlies all the preceding properties, i.e., vertical units. These vertical units run through the six cortical planar layers, and are the same in all layers of the neocortex. The basic unit is called a *minicolumn*. The exact number of cells enclosed varies in different regions of the cortex, as does the area involved, and ranges from about 110 cells (in most units) to a maximum of about 260 cells. It forms a cylinder about 30 μm in diameter.

It has been estimated that the human neocortex comprises about 600,000,000 minicolumns. Several hundred of these are organized into processing units which range from 500 to 1000 μm in different areas, and these larger units are called *macrocolumns.* Each area precisely defines the functional parameters of the mini- and macrocolumns, so that a very detailed mapping of the surface areas becomes feasible. Each cortical unit column acts as a complex processing and distributing unit, which links several inputs to several other outputs. Each such area is enclosed by a pericolumnar inhibition, which serves to isolate adjacent minicolumns.

The units are interconnected in a variety of ways, which could be the basis of complex functions that are not necessarily localized in any one unit but are performed by the whole system or great parts of it acting as a unit. Such functions, because of the vast potential in the extensive inflow and outflow patterns, are thought to be accessible to neural influences arising internally or externally from the environment, and hence modifiable.

Most of the research presented in this chapter on the nervous system is based either on special preparations or on assumptions or on dead material. It is all the more exciting to come upon potentially useful methods, whose limits are largely technologic. These are the investigation of active sites in most of the neocortex by the use of radioactive gases and of radioactive metabolites in blood. These methods reveal differences in metabolism that take place after extremely limited and precise functional performances. The application of this method to sex-related functional differences, especially during their periods of origin (as during adolescence), could be very important.

REFERENCES

Barchas, J. D., H. Akil, G. R. Elliot, R. B. Holman, and S. J. Watson. 1978. Behavioral neurochemistry: Neuroregulators and behavioral states. Science 200: 964-973.

Brodal, A. 1969. Neurological Anatomy. 2nd Ed. Oxford University Press, New York.

Dörner, G. 1973. Sex hormone-dependent differentiation of the hypothalamus and sexuality. *In* K. Lissak (ed.), Hormones and Brain Function. Plenum Press, New York, pp. 47-52.

Edelman, G. M., and V. B. Mountcastle. 1978. The Mindful Brain. MIT Press, Cambridge, Mass.

Gorski, R. A. 1973. Mechanisms of androgen induced masculine differentiation of the rat brain. *In* K. Lissak (ed.), Hormones and Brain Function. Plenum Press, New York, pp. 27-46.

Guillemin, R. 1978. Biochemical and physiological correlates of hypothalamic peptides. The new endocrinology of the neuron. *In* S. Reichlin, R. J. Baldessarini and J. B. Martin (eds.), The Hypothalamus. Research Publications: Association for Research in Nervous and Mental Disease, Vol. 56, pp. 155-194.

Ham, A. W. 1974. Histology. 7th Ed. J. B. Lippincott Co., Philadelphia.

Isaacson, R. L. 1974. The Limbic System. Plenum Press, New York.

Joseph, S. A., and K. M. Knigge. 1978. The endocrine hypothalamus: Recent anatomical studies. *In* S. Reichlin, R. J. Baldessarini, and J. B. Martin (eds.), The Hypothalamus. Research Publications: Association for Research in Nervous and Mental Disease, Vol. 56, pp. 15-48.

Krieger, D. T., and A. S. Liotta. 1979. Pituitary hormones in brain: Where, how, and why? Science, 205: 366-372.

Lassen, N. A., D. H. Ingvar, and E. Skinhøj. 1978. Brain function and blood flow. Sci. Am. 239 (4): 62-71.

Meites, J. 1980. Interactions between hypothalamic neurotransmitters and pituitary function. Fed. Proc. 39:288. (See also 2889-2941.)

Noback, C. R., and R. J. Demarest. 1975. The Human Nervous System. 2nd Ed. McGraw Hill Book Co., New York.

Palay, S. L. 1953. Neurosecretory phenomena in the hypothalamo-hypophysial system of man and monkey. Am. J. Anat. 93:107-142.

Raisman, G., and P. M. Field. 1973. Sexual dimorphism in the neuropil of the preoptic area of the rat and its dependence on neonatal androgen. Brain Res. 54:1-29.

Ranson, S. W. 1931. The Anatomy of the Nervous System. 4th Ed. W. B. Saunders Co., Philadelphia.

Truex, R. C. 1959. Strong and Elwyn's Human Neuroanatomy. 4th Ed. Williams and Wilkins Co., Baltimore.

Willis, W. D., Jr., and R. G. Grossman. 1973. Medical Neurobiology. C. V. Mosby Co., St. Louis.

10 Sexual Dimorphism of the Brain

When examined grossly and microscopically the structure of nerve cells and nerve tissues is the same in both sexes at the various levels of organization of the central nervous system. The only known exceptions are the visible presence of an X chromosome (Barr body) in many nerve cells of females and its absence in males; the presence of Y chromosomes in nerve cells of males and their absence in females; and a preponderance of a certain type of synapse in the female hypothalamus, as compared with the male. But as the Barr bodies are considered to be nonfunctional in growing girls and in women, one may conclude that this difference is trivial. One might conclude that the central nervous systems of men and women are thus barely distinguishable cytologically and histologically, and also functionally. This conclusion may be incorrect, as is shown by the following sets of observations.

ANIMAL EXPERIMENTS

Experiments on rats, mice, hamsters, and guinea pigs show that the cyclic behavior of the female reproductive system that develops at puberty is determined at about the time of birth (perinatally). If the testes are removed from a male (castration) at this time, and replaced by one or more grafted ovaries, an ovarian cycle of ovulation and corpus luteum formation is established at maturity. The converse experiment is to implant a testis into a female at birth, or to inject androgen at this time. As a result, the animal's own ovaries will fail to show a cycle at maturity.

These experiments are ineffective unless they are done within a very narrow perinatal time span. Presence of androgen at this time inhibits the cycle and absence of androgen permits the cycle. The effect is determined perinatally, although it is not expressed until puberty.

It has been shown that the perinatal effect of androgen is on the hypothalamus, not on the gonads themselves or the hypophysis. These experiments show that in rats and some other mammals the hypothalamus is permanently modified at birth by the presence of androgen, whether it occurs naturally, as in the male, or is artificially supplied, as in the female. This change results later in functional modification of ovaries if they are present. The perinatal modification takes place normally in males but not in females, so that from birth there is a difference between the brains of male and female animals.

Sexual Behavior

A second set of experiments relates to the effects of hormones on mating behavior in rats. In female rats that have been treated with androgen at birth, the typical mating behavior at estrus tends to be suppressed. If these females are given a further injection of androgen at maturity, not only is the female behavior suppressed, but if they are put in cages with normal females, the treated females behave as males toward the normals, trying to mount them. Conversely, males that have been treated perinatally with an anti-androgen drug may exhibit female mating behavior. Again, this behavior is more marked if the animals are injected with estrogen at maturity (Levine, 1966).

Other Types of Behavior

A third category of observations enlarges the range of perinatal hormone effects. Male and female rats differ in behavior other than mating behavior—for instance, in their activity on an exercise wheel, their aggressiveness, and their exploratory behavior. Females treated with testosterone at birth behave more like males in these respects, and males castrated at birth behave more like females.

This type of behavior modification extends also to primates. Young male monkeys play more aggressively than females, with more threatening facial expressions, and rougher treatment of playmates. Prenatal testosterone treatment of females makes them rougher and more aggressive in their play (Goy, 1968). The time at which this behavior is determined seems to be the second trimester of fetal development (Phoenix, 1974).

It is hard to interpret experiments like these because androgens, estrogens, and cortisone are all steroids and under certain circumstances androgens can be converted to estrogens (as shown in Figure 7.8), and it is therefore doubtful which hormones are responsible for the observed effects.

Perinatal Hormones and Brain Structure

A fourth set of observations is concerned with synaptic counts in various nuclei of the hypothalamus, especially in the preoptic nuclei of the rat. It was reported that in the females, the number of a certain kind of synapse in these nuclei is higher than in males (Raisman and Field, 1971). Castration of the male shortly after birth results in a higher number than in controls, approximating the female level. Also when females four days old are treated with androgen sufficient to abolish cyclicity at maturity, the number of this type of synapse in the preoptic nuclei is reduced, though it is still in the female range. The detailed relationship of synaptic numbers to differences in female and male behavior remains very uncertain.

Extrapolation from Animal to Human Behavior

We are on shaky ground if we try to draw conclusions about human behavior based on the kind of experiments on rats and monkeys just described. As we can see from a comparison of different cultures (Mead, 1967), human behavior is largely determined by culture and thus *learned*, rather than being genetically programed, thus, *instinctive*.

The existence of male domination in our society has sometimes been justified by reference to the social organization of some primate species (Tiger, 1969). Careful study of species such as chimpanzee, gorilla, and baboon shows that the social organization among these species differs. Female chimpanzees may be self-assertive toward males and may travel independently with their offspring (Reynolds and Reynolds, 1965). Gorilla and baboon groups are more cohesive; gorilla groups have one dominant male (Schaller, 1965a), while baboons display a rather rigid dominance hierarchy (Hall and DeVore, 1965). The dominance rank of male rhesus monkeys is determined largely by that of their mothers, while in some species the dominance hierarchy and aggressiveness are variable, depending on the degree of stress (such as crowding, or competition for food) in the social group (Kolata, 1976). If there were a high degree of consistency between different species, it might be reasonable to suppose that all primates, including the human species, were genetically programed for dominance relations, but, as Schaller said, "one of the most conspicuous behavioral differences between the apes is their group organization" (Schaller, 1965). There are such marked differences between nonhuman species that extrapolation from ape to human is quite unjustified. The chief proponents of this theory are members of a relatively new area of study, sociobiology. This important new field deals with the genetic basis of behavior. These few comments explain why this volume does not extensively discuss the social behavior of primates and other animals as a biologic basis for extrapolation to human behavior.

HUMAN STUDIES

The kinds of experiments described above cannot be done on human subjects. Nevertheless, genetic mutations and the side effects of medical treatments have provided us with natural experiments that to some degree parallel those done on experimental animals.

Feminization in Males

First, we should note that when human females have been exposed to excessive amounts of androgens before birth, their ovarian cyclicity at puberty has been normal (incidentally, this is also true of monkeys). Second, as regards behavior, some data are available on two genetic mutant types. We have already met the androgen insensitivity (*Tfm*) mutation. It results in failure to produce androgen receptors in chromosomally normal males. Male development is therefore arrested, the testes are undescended and external genitalia develop like those of females. The babies are usually brought up as girls, and their interests and games are typical of girls. On the basis of present-day evidence, it is impossible to decide whether their "feminine" characteristics (which include low scores in spatial skills tests) are due mainly to the hormone receptor abnormality or to the socialization they receive.

Masculinization in Females

The nearest approach to the opposite situation, masculinization of females, is caused by the mutation for adrenogenital syndrome (Chapters 2 and 4). This blocks the synthesis of cortisone and results in formation of excessive amounts of androgen. At birth this may cause a female baby to have an enlarged clitoris. There may be uncertainty as to her sex, and she may be brought up as a boy. Alternatively, if she is recognized early as female, it is possible to perform corrective surgery, if necessary, and also to give the compensatory hormone treatment needed for normal female development. A somewhat similar situation was created by the treatment of pregnant women with progestin to prevent miscarriage. This treatment is no longer practiced, but it, too, sometimes resulted in enlargement of the clitoris and created doubt as to a baby's sex. As the hormone treatment was only for a restricted period during pregnancy and did not continue after birth, hormone therapy during childhood was not necessary, even when corrective surgery was used. Money and Ehrhardt (1968) followed the development of 22 girls of one or the other of the prenatally androgenized groups, who received whatever corrective treatment was deemed necessary, and who were all raised as females. The authors found a strong tendency to "tomboy" behavior in this group.

Here also we must consider the possible effect of parents' fears and expectations. The clinical research of Money, Ehrhardt, and their colleagues has shown that gender identity (the way a person sees her/himself with regard to sex role) is not determined solely by sex chromosomes, the prenatal hormone pattern, postnatal hormone levels, or by any combination of these. It depends in large part on unambiguous identification of the child's sex and development of its gender role by the parents. If the parents have doubts about the sex of the child at birth, and these doubts are not completely resolved, they may transmit their uncertainty to the child. In the cases discussed, parents may have had doubts as to whether their children would behave according to accepted feminine patterns; or they may have made allowances for tomboyish behavior that they would have discouraged actively in a daughter whose sex had been unambiguous from the start.

Ehrhardt and Baker (1974) studied a second group of adrenogenital syndrome patients, in which the previous results were confirmed. The authors were reasonably sure that parental attitudes played no part in developing their daughters' tomboy behavior, but they admit that this possibility cannot be completely excluded. Another study, of prenatally treated individuals (Reinisch, 1977), is less open to these objections because none of the children were anatomically masculinized and the tests were double blind. A blind experiment is one in which the subject does not know how he or she is expected to perform. In a double-blind experiment, not only the subject, but the person carrying out the experiment (as opposed to the one who devised it) must be unable to tell how a particular subject is expected to perform. If the import of this is not immediately clear, the reader should refer to experiments on rats as well as on people (Rosenthal, 1963; Weisstein, 1976) in which the expectations of the person working with the subjects (whether rats or people) clearly influenced the results. Reinisch concluded that prenatal hormone exposure resulted in modification of personality.

Summary

While androgen level is known to affect development of the external genitalia in people, an effect on ovarian cyclicity has not been established. As far as human masculine and feminine behavior is concerned, some evidence suggests that it may be influenced by hormones. At the same time, there is clear evidence that behavior generally considered as masculine or feminine can result from socialization, quite independently of prenatal hormone levels (Money and Ehrhardt, 1972, p. 118). Even in the short term, behavior commonly resulting from changed hormone levels may be strikingly modified depending on the expectations aroused in the subject (Schachter and Singer, 1962). A fatalistic attitude about human behavior cannot be justified on the grounds that it is biologically determined, because it is unquestionably modifiable by environmental factors and learning.

CEREBRAL HEMISPHERES

Asymmetry of the Brain

The suggestive, but insufficient, evidence of hormonal effects on the human brain extends also to that part of the brain, the cerebrum, which is more highly developed than in any other animal. The evidence concerns cognitive skills in which the sexes tend to differ.

It is a well-known fact in academic circles that women score higher, on the average, in verbal or linguistic tests, while men tend to score higher in mathematics tests. The difference in the Scholastic Aptitude Tests taken in high school is about ten points in favor of women for the verbal and fifty points in favor of men for the mathematics test. It has also been established that the brain is not symmetric. The left and right cerebral hemispheres are different from one another, and although they may share functions, one side may perform a function better than the other. Thus, the verbal skills mentioned above are, to a large extent, mediated by the left side of the adult brain while the mathematical (and, as it turns out, spatial) skills are mediated by the right side. Further, experiments indicate that there are sex differences in the development of this asymmetry of the cerebral hemispheres.

Contralateral Relationships

Visual, tactile, and auditory stimuli are conveyed to the cerebral hemispheres mainly by nerves that cross over in the brain and end on the opposite side of the brain. In vision, most, but not all, of the nerve fibers cross over in the optic chiasm, and most of the stimuli received by the two eyes from the right side of the environment are recorded in the left hemisphere, and vice versa (Figure 10.1). This is called a contralateral relationship.

A difference between the perception of stimuli that can be easily verbalized and those that cannot has been documented. The easily verbalized (EV) stimuli include spoken and printed words, pictures of objects known to the observer, and shapes like cubes and spheres which, though identified either visually or by touch, can be accurately described in words. Stimuli that are difficult to verbalize (DV) are irregular three-dimensional shapes, abstract paintings, and random sound sequences or noises that the hearer cannot relate to everyday experience. Tests of right-handed adults show that the verbal type of stimulus is perceived better when presented to the right side (and therefore when transmitted to the left hemisphere), while the opposite is true of DV stimuli (Figure 10.2). With left-handed people, the situation is confused, because some of them do better when EV stimuli are presented to the left side (the opposite of right-handed people) while in some, there seems to be no advantage of one side

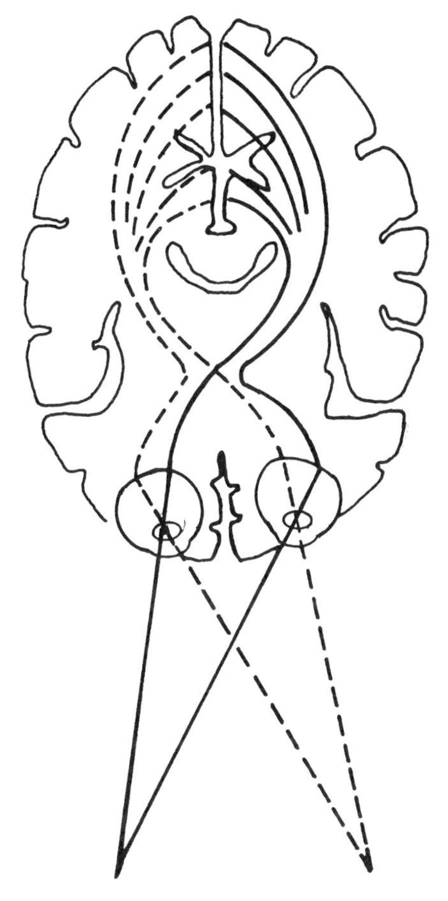

Figure 10.1. Section of the cerebral hemispheres showing the contralateral processing of visual stimuli. If an object lies to the left of the point on which the vision is fixed, stimuli are transmitted from it to the right hemisphere, and vice versa. (After Kimura, 1973)

over the other. For the right-handed, however, there is a clear functional asymmetry for the perception of both EV and DV stimuli (Kimura, 1973).

Sex Differences

Verbal Skills Some experiments that tested auditory reactions to EV stimuli in boys and girls suggested that girls develop the left hemisphere preference for verbal stimuli earlier than boys (Kimura, 1963). If true, this is consistent with the superior performance of women in the verbal Scholastic Aptitude Tests, and also with the fact that boys more often have reading difficulties than girls, and that these boys may fail to show a right ear, left hemisphere preference for verbal perception.

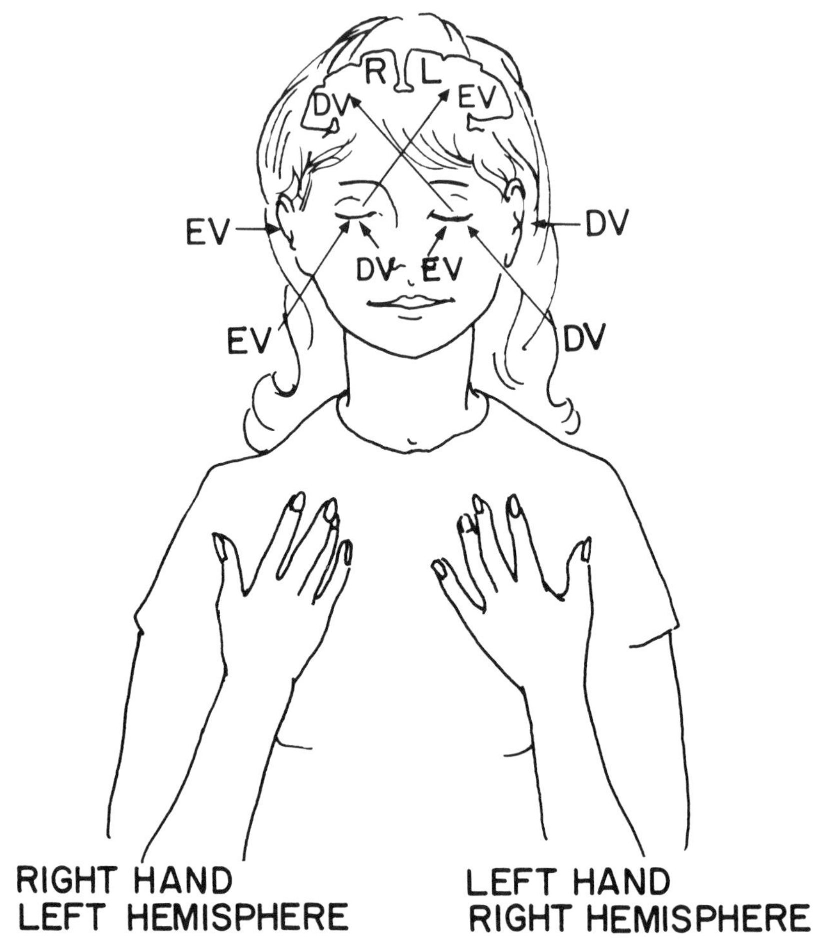

Figure 10.2. Diagram shows the preferred processing of easily verbalized (EV) and difficultly verbalized (DV) stimuli (visual, auditory, and tactile) in righthanded individuals.

Other points that may be relevant are: 1) that although male and female infants vocalize to the same extent, mothers seem to respond more to their girl babies, imitating their sounds (Goldberg and Lewis, 1969); and 2) that girls usually start to talk at an earlier age than boys (Darley and Winitz, 1961). Hutt (1972) raises the question whether the mother responds to a difference in the quality of her baby's sounds and whether this response may in turn lead to the earlier talking age of girls. The whole picture is complicated by the fact that girls develop faster than boys in many respects up to about age eleven, when boys catch up.

Nonverbal (Spatial) Skills Experiments have also been done to test sex differences in the asymmetry associated with nonverbal or spatial perception. Buffery (1970, 1971) and Witelson (1976) used different tests, in experiments that led to different results and contradictory conclusions. One wonders whether the different tests used, and what they test, are really equivalent. As to the interpretations, a somewhat similar conflict is noted by Star (1979).

Effects of Genes and Hormones The next question is the extent to which the differences described above can be ascribed to innate factors, on the one hand, or to training and socialization, on the other.

Bock and Kolakowski (1973) showed that the correlation between mother and son, and father and daughter, for sets of data on spatial ability tests, fitted those expected on the basis of a major sex-linked recessive gene for spatial skill, supplemented by autosomal modifiers.

Because of contradictions presented by test scores of 45,XO and androgen-insensitive individuals, Bock and Kolakowski postulated that in addition to the genetic effect, an androgen effect was necessary. Females with the karyotype 45,XO do not produce much androgen. This would explain why, though they resemble males in having only one X chromosome they do not show the male type of score distribution on tests. Androgen-insensitive (*Tfm*) individuals lack receptors that enable testosterone to stimulate its target organs. This explains why scores of these genetic males resemble those of females rather than those of normal males. Unfortunately, these authors dealt mainly with *Tfm* individuals raised as females, and were able to find only three *Tfm* individuals who were raised as males, for whom they found no consistent picture. It would be important to have data on such cases, because otherwise we cannot rule out the possibility that the difference between normal males and *Tfm* individuals reared as females (as well as 45,XO females) is due to socialization.

Brain Structure The left and right hemispheres do not differ in function only. There are also differences in structure. For instance, it is claimed that the part of the left temporal lobe known to be associated with speech is larger than the corresponding part of the right lobe (Geschwind, 1965).

Some differences in the brain structure of human males and females have also been noted (Levine, 1966). An example is an asymmetry of the vascular system in the brain which is found in female but not in male brains. A small sample of brains of 4-year-olds (five female and three male) gave indications of differences in asymmetry between the sexes. Even if these differences are general, however, they do not tell us that they existed initially, at birth or even earlier, before differential treatment of male and female infants becomes a factor to be reckoned with as a cause of differential development.

Brain Size At one time, much was made of the fact that the whole brain in women is, on the average, smaller than that of men. Using some shaky logic, it was inferred that women must be less intelligent than men. By extension of this reasoning, the most intelligent creatures would be those with the biggest brains, so that whales and elephants would be more than a match for people. It has subsequently been found that if the proportion of brain to body size is considered, the situation is reversed, and women average higher than men for this ratio. If various other ratios are considered, some favor women and some favor men. Needless to say, no conclusions about intelligence or cognitive skills can be drawn from these data (Shields, 1975).

Socialization There is abundant evidence of environmental influence on the human brain: all learned reactions depend on this. If girls and boys are treated differently from a very early age, this might be responsible for differences in brain stucture accompanied by differences in behavior and performance.

Hospital nurses treat newborn infants differently, according to sex, handling females more gently. Male babies are fussier and cry more, generally, so mothers pick them up more frequently and hold them longer than female babies during the first few months. Mothers vocalize more in response to their female babies' vocalization, although, as far as we know, there is no difference in the vocalization of boy and girl babies. Because everything we do, think, learn, or experience changes us at least a little, differences in treatment at this early age can start a trend. It may be that mothers feel more empathy for their daughters than for their sons. It would not be unreasonable to suppose that their response to the girls' vocalization stimulates the girl babies and gives them a head start over boys in developing communication skills. Likewise a case can be made, and has been made by Maccoby (1963), for believing that the difference in spatial skills may be due to socialization. Maccoby's studies contrast interestingly with those of Money and his associates. Maccoby compared girls who were good at mathematics and spatial tasks, but not outstanding in verbal tasks, with girls whose skills were verbal rather than spatial. She found differences in the way the two groups were treated by their mothers and she found that the first group tended to show traits conventionally considered "masculine," while the second group were more "feminine." Contrast this with the group described by Money and Ehrhardt (1972) showing the adrenogenital syndrome or progestin-induced modification of the external genitalia at birth. The members of this group were given corrective surgery and hormone treatment where necessary, and brought up as girls. Most of them showed "masculine or tomboyish characteristics," yet those with relatively high IQ scored higher on verbal than on nonverbal tests (Lewis, Money, and Epstein, 1968), with no evidence of the "masculine pattern of nonverbal superiority." Thus, masculine characteristics associated with prenatal hormonal influence were not accompanied by superiority in spatial skills, while similar characteristics associated with differences in upbringing were correlated with these skills.

In reconsidering the results from androgen-insensitive individuals, before this is assumed to be evidence of a hormonal effect (lacking in this case because of the lack of receptors), one should consider that from the very beginning the members of this group were brought up and socialized as girls.

A few studies of spatial skills have been made on people of other cultures. In one African culture, spatial skills were related to boyhood experiences (Nerlove, Monroe, and Monroe, 1971). Among Eskimos no sex difference was found in spatial ability (Berry, 1966; McArthur, 1967), and this has been attributed to greater similarity of men's and women's roles in Eskimo society.

SUMMARY

There is no question that there are structural as well as funtional asymmetries of the cerebral hemispheres. The relationship between the two is quite obscure. In our culture, the development and degree of functional asymmetry differs generally in females and males—women tend to be stronger on verbal tasks, usually mediated mainly by the left hemisphere, and men tend to be stronger on mathematical and spatial tasks in which the right hemisphere usually plays a larger part.

There are indications that the basis of this sex difference may be biologic and innate. First, there is some genetic evidence that rests on correlations between scores for spatial ability of mother and son, and father and daughter, indicating sex linkage. Spatial ability is, however, the only cognitive skill for which these correlations have been found: there are no comparable correlations for verbal ability. Then there is the fact that in other animals there are perinatal hormone influences that determine future behavior. It is possible that there are similar hormone effects, acting prenatally (as in monkeys) on human fetuses. On the other hand, there is evidence for a socializing effect on the kinds of human behavior that we have been discussing.

There is no reason why human sex differences of the kind considered in this chapter should be due exclusively to genetic determination or differences in hormone production, on the one hand, or to conditions of nurture, on the other. In fact, the biologic data discussed here and the studies of socialization indicate a complex interaction.

Much remains to be learned on sex differences in the brain. Without extended studies of brain structure in fetal stages and in the newborn, we cannot rule out an initial genetic difference between the sexes. We do not know whether different tests for nonverbal skills are really equivalent, nor which criteria of "masculinity" and "femininity" are likely to be critical for correlation with these skills.

Meanwhile, as long as such questions remain unresolved, the personal and sexual biases of experimenters provide in themselves an interesting field for study.

REFERENCES

Berry, J. W. 1966. Temne and Eskimo perceptual skills. Internat. J. Psychol. 1:207-229.

Bock, R. D., and D. Kolakowski. 1973. Further evidence of sex-linked major-gene influence on human spatial visualizing ability. Am. J. Hum. Genet. 25:1-14.

Buffery, A. W. H. 1970. Sex differences in the development of hand preference, cerebral dominance for speech and cognitive skill. Bull. Br. Psychol. Soc. 23:233.

Buffery, A. W. H. 1971. Sex differences in the development of hemispheric asymmetry of function in the human brain. Brain Res. 31:364-365.

Buffery, A. W. H., and J. A. Gray. 1972. Sex differences in the development of spatial and linguistic skills. *In* C. Ounsted and D. C. Taylor (eds.), Gender Differences: Their Ontogeny and Significance. Churchill Livingstone, Edinburgh, pp. 123-157.

Darley, F. L., and H. Winitz. 1961. Age of first word: Review of research. J. Speech Hearing Disorder 26:272-290.

Geschwind, N. 1978. Anatomical asymmetry as the basis for cerebral dominance. Fed. Proc. 37:2263-2266.

Goldberg, S., and M. Lewis. 1969. Play behavior in the year-old infant: Early sex differences. Child Dev. 40:21-31.

Goy, R. W. 1968. Organizing effects of androgen on the behavior of rhesus monkeys. *In* R. P. Michael (ed.), Endocrinology and Human Behavior, Oxford University Press, London, pp. 12-31.

Hutt, C. 1972. Neuroendocrinological, behavioral and intellectual aspects of sexual differentiation in human development. *In* C. Ounsted and D. C. Taylor (eds.), Gender Differences: Their Ontogeny and Significance. Churchill Livingstone, Edinburgh, pp. 73-121.

Kimura, D. 1963. Speech lateralization in young children as determined by an auditory test. J. Comp. Physiol. Psychol. 56:899-902.

Kimura, D. 1973. The asymmetry of the human brain. Sci. Am. 228(3):70-78.

Kolata, G. B. 1976. Primate behavior: Sex and the dominant male. Science 191: 55-56.

Levine, S. 1966. Sex differences in the brain. Sci. Am. 214(3):84-90.

Lewis, V. G., J. Money, and R. Epstein. 1968. Concordance of verbal and nonverbal ability in the adrenogenital syndrome. Johns Hopkins Med. J. 122: 192-195.

Maccoby, E. E. 1963. Women's intellect. *In* S. M. Farber and R. H. L. Wilson (eds.), The Potential of Women. McGraw Hill, New York, pp. 24-46.

Masica, D. N., J. Money, A. A. Ehrhardt, and V. G. Lewis. 1969. IQ, fetal sex hormones and cognitive patterns: Studies in the testicular feminizing syndrome of androgen insensitivity. Johns Hopkins Med. J. 124:34-43.

McArthur, R. 1967. Sex differences in field dependence for the Eskimo: Replication of Barry's findings. Internat. J. Psychol. 2:139-140.

Mead, M. 1935. Sex and Temperament in Three Primitive Societies. William Morrow, New York.
Money, J., and A. A. Ehrhardt. 1968. Prenatal hormone exposure: Possible effects on behavior in man. *In* R. P. Michael (ed.), Endocrinology and Human Behavior. Oxford University Press, London, pp. 32-48.
Nerlove, S. B., R. H. Monroe, and R. L. Monroe. 1971. Effects of environmental experience on spatial ability. A replication. J. Soc. Psychol. 84:3-10.
Phoenix, C. H. 1974. Prenatal testosterone in the nonhuman primate and its consequences for behavior. *In* R. C. Friedman, R. M. Richart, and R. L. Vande Wiele (eds.), Sex Differences in Behavior. John Wiley & Sons, New York, pp. 19-32.
Raisman, G., and P. M. Field. 1971. Sexual dimorphism in the preoptic area of the rat. Science 173:731-733.
Reinisch, J. M. 1977. Prenatal exposure of human foetuses to synthetic progestin and oestrogen affects personality. Nature 266:561-562.
Reynolds, V., and F. Reynolds. 1965. Chimpanzees of the Budongo Forest. *In* I. DeVore (ed.), Primate Behavior Holt, Rinehart & Winston, Inc., New York, pp. 368-424.
Rosenthal, R. 1963. On the social psychology of the psychological experiment: The experimenter's hypothesis as unintended determinant of experimental results. Am. Sci. 51:268-283.
Schachter, S., and J. E. Singer. 1962. Cognitive, social and physiological determinants of emotional state. Psychol. Rev. 69:379-399.
Schaller, G. B. 1965a. The behavior of the Mountain Gorilla. *In* I. DeVore (ed.), Primate Behavior. Holt, Rinehart & Winston, Inc., New York, pp. 324-367.
Schaller, G. B. 1965b. Behavioral comparisons of the apes. *In*: I. DeVore (ed.), Primate Behavior, pp. 474-481. Holt, Rinehart & Winston, Inc., New York.
Shields, S. A. 1975. Functionalism, Darwinism, and the psychology of women: A study in social myth. Am. Psychol. 30:739-754.
Star, S. L. 1979. Methods, limits and problems in research on consciousness. *In* R. Hubbard and M. Lowe (eds.), Genes and Gender, Vol. II. Gordian Press, Inc., New York, pp. 113-127.
Tiger, L. 1969. Men in Groups. Random House, New York.
Valenstein, E. S., W. Riss, and W. C. Young, 1955. Experimental and genetic factors in the organization of sexual behavior in male guinea pigs. J. Comp. Physiol. Psychol. 48:397-403.
Weisstein, N. 1976. Adventures of a women in science. Fed. Proc. 35:2226-2231.
Witelson, S. F. 1976. Sex and the single hemisphere: Specialization of the right hemisphere for spatial processing. Science 193:425-427.

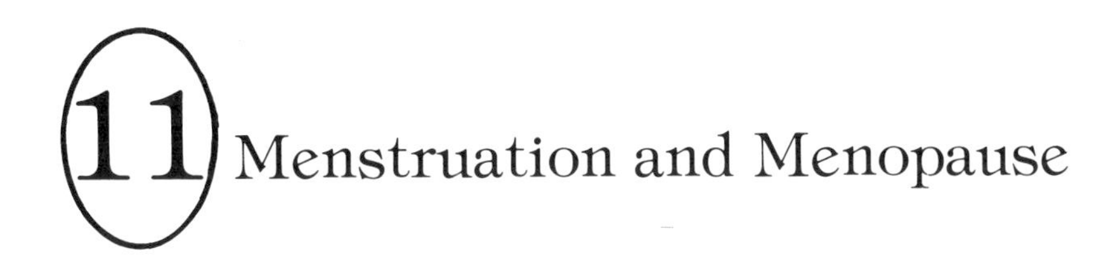

11 Menstruation and Menopause

DEFINITION AND DESCRIPTION

Menstrual bleeding is the appearance of unclotted blood in the vagina associated with hemorrhage in the inner wall of the uterus and its subsequent sloughing off. Menstruation, when well established, is cyclical, the average period being 28 days from the time of onset of bleeding to the onset of the following period. But the period is extremely variable in different women, and may be so in the same woman. The period may be most regular when women are between 25 and 35 years of age. Menstruation begins at puberty (about 12 to 13 years of age), and ends at menopause (about 50 years of age). The duration of bleeding is about 5 days, but ranges between 3 and 7 days. The amount of blood lost during menstruation is about 50 ml, but may vary between 10 ml and 200 ml. In most women, the amount of blood lost is not sufficient of itself to cause anemia.

The symptoms of the period range from barely noticeable to incapacitating. It is probable that many millions of woman-days are lost each year because of the malaise. Some women have no difficulties at all, some complain mildly of one or more of the following discomforts and pains: cramps, nausea, a dull aching in the abdomen, constipation, a feeling of heaviness, water retention, headache, backache, breast pain, irritability, tension, depression, and lethargy. The causes for these symptoms are extremely varied. For example, weight gain may involve water retention, which in turn may be related to salt retention (kidney function) and changes in some properties of the connective tissues of the body, all of which are possibly related in an obscure way to hormonal levels. Or some of the symptoms may be associated with conscious and/or unconscious conflicts that may originate in familial, social, or cultural problems. This is a field that urgently requires further analysis to reduce the widespread misery associated with menstruation in our society.

Menstrual pain is commonly associated with contractions of the smooth muscle of the uterus. Such smooth muscle activity is not consciously controlled, and lacks precise localization and awareness. A very encouraging advance in studies of menstrual pain is the recently developed knowledge bearing on the biochemical basis for the excessive pain. The biochemical basis appears to be certain substances called *prostaglandins,* specifically PGE and $PGF_{2\alpha}$ (Figure 11.1). These small molecules, thought to be produced locally, cause spastic contractions of smooth muscle. The administration of drugs that inhibit the synthesis of prostaglandins reduced or eliminated the menstrual symptoms. However, the effects of prostaglandins are so widespread that an extensive period of trial for the study of side effects and dosage would be prudent.

MENSTRUAL CYCLE

Menstruation signifies that many cyclic changes are taking place. These include cyclic hormonal changes in the anterior lobe of the hypophysis and the hypothalamus (discussed in Chapters 7 and 8) and in the target organs which they influence—ovaries, uterus, vagina, cervix, oviducts, and breasts.

Ovulation

The menstrual cycle may be thought of as a way of preparing the inner surface of the uterus for receiving the fertilized egg. Menstruation can also

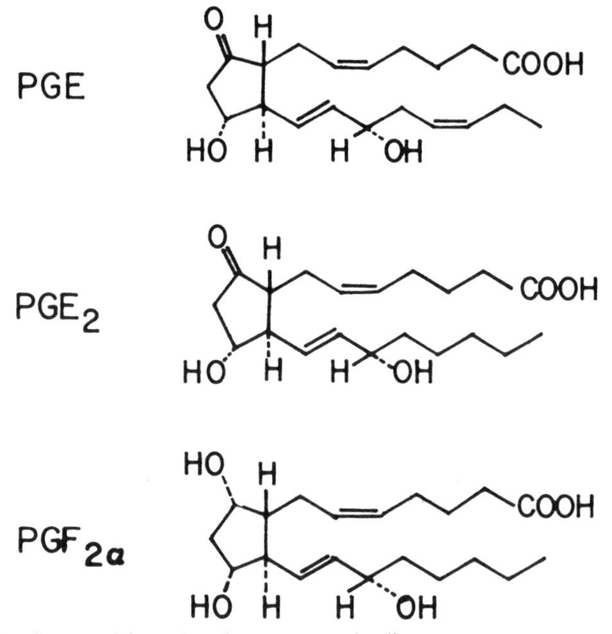

Figure 11.1. Structural formulae of some prostaglandins.

be regarded as revolving around the development of the egg and ovulation. From the few original gonadal cells in each embryo that wandered into the gonadal ridge from the embryonic yolk sac, arose millions of egg cells, which are gradually reduced in number to about 500,000 at puberty. Of this number, it is estimated that only about 500 mature and are ovulated in most women. Some stages in the development of the egg are shown in Figure 11.2. At first, the egg cell is surrounded by follicular cells. The latter become larger, multiply, and lay down the substance of an envelope around the egg cell called the *zona pellucida.* The follicular cells continue to multiply, and a cavity filled with follicular fluid appears. The follicular cells multiply even more, and the follicular cavity enlarges. Not just one, but a cluster of ten or more develop in each cycle. About this time, most of the cluster of follicles degenerate, leaving (usually) only one to reach maturity. The follicle bulges from the surface, and eventually ruptures, releasing the egg, which eventually reaches the uterus.

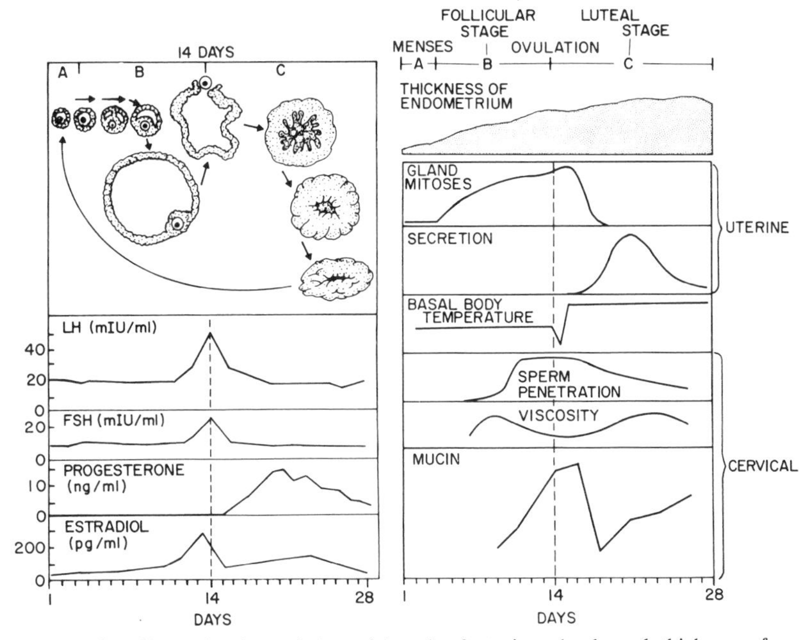

Figure 11.2. Chart showing relation of length of uterine glands and thickness of endometrium to the ovarian cycle and principal hormones involved (FSH and LH from the anterior lobe of the hypophysis and estradiol and progesterone secreted by the ovary). Also shown are changes in the mitotic rate of epithelial cells (labeled "gland mitoses") and secretory activity as estimated by the amount of secretion. Additionally shown here are changes in body temperature and in some properties of the cervical mucus, especially its viscosity, the amount of mucin, and its penetrability by sperm. (After Elias and Pauly, 1966; Bloom and Fawcett, 1975; Kletsky et al., 1975; Elstein, 1976; and Moghissi, 1973)

Luteinization

At the time of egg release, follicular cells that remain are transformed to luteal cells, and are joined by others from the internal layer surrounding the follicle, the theca. The corpus luteum is formed rapidly, and persists for some days. If the egg is not fertilized, the corpus luteum degenerates in 12 to 14 days, and becomes a *corpus albicans.* But if the egg is fertilized, the corpus luteum persists, increases markedly in volume, and is known as a *corpus luteum of pregnancy.* It is important to know that the follicular cells synthesize and secrete estrogenic steroids (and possibly some progesterone), and that the luteal cells secrete progesterone (and possibly some estrogen). These cyclical changes are illustrated in Figure 11.2 in the upper left diagram. In the lower left part of the figure are the blood curves for estradiol and progesterone. When the levels of both hormones have reached their peaks and begin to fall at the same time, menstruation takes place, or *bleeding from hormone withdrawal.* In Figure 11.2, A is the period of the menses, B is the follicular stage, and C is the luteal stage. Ovulation takes place at the end of the follicular stage or the beginning of the luteal phase. Development of these organs (follicle and corpus luteum) and secretion of the steroid hormones is triggered by the secretion of two hormones by the anterior lobe of the hypophysis, follicle-stimulating hormone (FSH) and luteinizing hormone (LH), and these anterior lobe hormones in their turn are triggered by another hormone, follicle-stimulating hormone/luteinizing hormone-releasing hormone (FSH/LH-RH) secreted by neuroendocrine cells of the hypothalamus.

Signs of Ovulation and the Need for Practical Tests

Apart from the external signs of menstrual cyclicity mentioned above, some few women have noted a sharp, sudden pain at about the time of ovulation. Others may have a slight vaginal bleeding for a day or two at the time of ovulation. But these signs of ovulation are not regular accompaniments of menstruation, nor are they clear enough to be used to identify the time of ovulation. There is no simple, certain method of ascertaining this time by tests that could be performed repeatedly at home. All the methods require laboratory controls, or are too variable. Vaginal smears, even if made practical, have no clear end-point. Some women have a slight rise in basal body temperature of less than 0.5°C (see Figure 11.2, middle right) or small differences in electrical conductivity of the vaginal fluid. Basal body temperature changes, which it may be practical to obtain in a home environment, are not sufficiently large or regular. The ascertaining of the viscosity of cervical mucus is not too practical, and might be of doubtful significance when contaminated with uterine and vaginal fluid, and the manipulations involved come close to being dangerously infectious. What is required is a simple, inexpensive, self-administered test that can be performed reliably and repeatedly by relatively untrained people.

Ovulation and Sexuality

Ovulation occurs at approximately 12 to 14 days preceding the onset of the next menstruation period (Figure 11.2). The preovulatory period varies more widely. The time of ovulation corresponds with the period of estrus in most other mammals, when they are also more receptive to copulation. In some women, heightened sexual feeling and desire occur at the mid-cycle, but in others this may occur later, during the premenstrual period.

Recent research on this subject has taken advantage of the development of extremely sensitive methods for the estimation of plasma steroid levels (testosterone, cortisol, progesterone and estradiol). It was found that there is no direct relationship between plasma estradiol levels and sexual arousal, intercourse frequency, and sexual gratification. There is however a relationship between sexual initiation, and sexual gratification of women and their average testosterone levels. In addition, intercourse frequency was related to the women's testosterone levels at their ovulatory peaks. It was suggested that the source of the testosterone is the adrenal cortex.

Endocrine Control of the Uterus and the Menstrual Cycle

The inner surface of the body of the uterus is rather flat, and consists of columnar cells that comprise the epithelial surface. Pits in the surface lead to the uterine glands, the lower parts of which are shown in Figure 11.3. These are supported by the connective tissue, which has many blood vessels. Together, the surface, the glands and the connective tissue with its

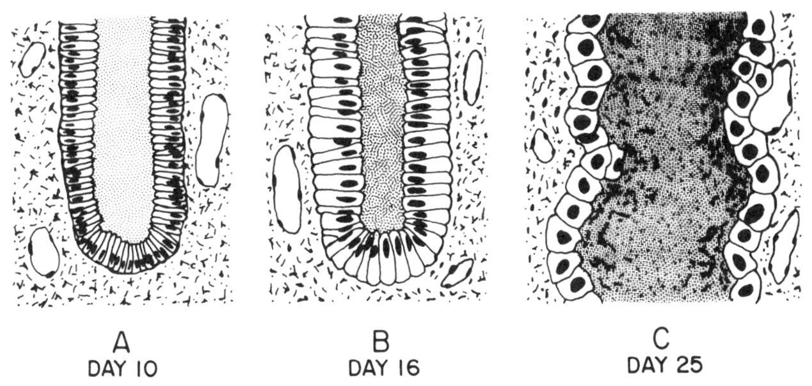

Figure 11.3. Drawings of parts of uterine glands showing some changes in the epithelium during the menstrual cycle. A) follicular stage (day 10) of the menstrual cycle; B) postovulation stage (day 16); C) late luteal stage (day 25). (After Hamilton, Boyd, and Mossman, 1972)

blood vessels constitute the *endometrium* (Figures 11.4 and 11.5). This is attached to a thick layer of smooth muscle called the *myometrium*. The endometrium undergoes cyclic changes during the menstrual period, while the myometrium shows little change except in contractility. In the endometrium are the terminal branches of the uterine artery, and these are key points in the cycle, whose external marking is bleeding. There are two kinds of blood vessels: 1) short straight arteries, which divide to form capillaries in the lamina basalis, and 2) long spiral arteries, which reach toward the surface epithelium where they break up into capillaries that exclusively supply the surface and the greater part of the glands with nutrients (Figures 11.5 and 11.6).

During the luteal phase, the endometrium is thickest. The glands are long, twisted, irregularly bulging, with enlarged cells and a dense secretion filling the distended opening (lumen) of the gland (Figure 11.3C). The spiral arteries are in their most elaborately coiled state, and terminate in a very large network of capillaries (Figures 11.5 and 11.6). The blood vessels are distended with blood and the connective tissue is edematous. In the last part of this phase (late luteal) some of these spiral arteries constrict and break, as do the capillaries nearby. The minute hemorrhages result in anoxia and death of the adjacent epithelium, and the endometrium locally affected sloughs off. The process takes place in small areas of the uterus throughout the period of bleeding, until the whole superficial part of the endometrium is shed.

During the follicular phase, the epithelium heals, and the surface of

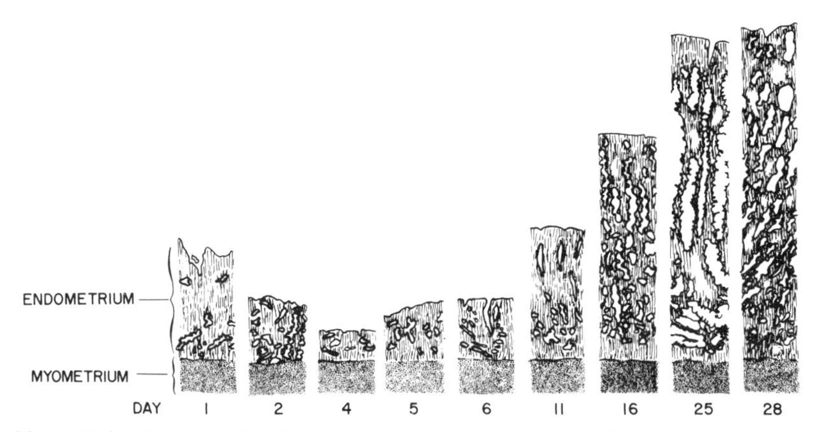

Figure 11.4. Drawings of sections of small sectors of the uterus to illustrate changes which take place in the endometrium at different times of the menstrual cycle. Menstruation, days 1-4; postmenstruation, day 5; follicular stages, days 6-11; postovulation, day 16; late luteal stages, days 25-28. The base of the endometrium is firmly and permanently attached to the myometrium at the base of each figure. Note changes in the length of the glands, their tortuosity, and the accumulation of secretion in the lumen of the gland as manifested by the diameter of the glands. (After Hamilton, Boyd, and Mossman, 1972; and Bloom and Fawcett, 1975)

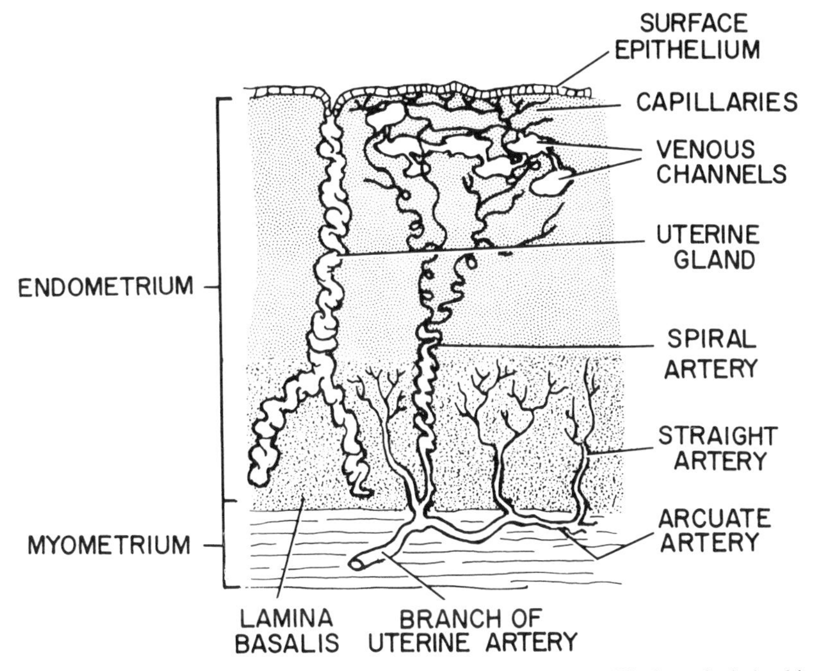

Figure 11.5. Components of endometrium in the late luteal phase. The branched gland is elongated, and its diameter is increased over that in earlier stages. (After Weiss and Greep, 1977)

the endometrium is completely covered with epithelium. At first the gland cells assume a less active appearance, and the secretion in the narrow lumen is not dense (Figure 11.3). The spiral arteries grow out and resupply the surface with capillaries. The endometrium continues to become thicker as the glands become larger and somewhat more elaborate, and the gland cells begin to show signs of increased activity (Figures 11.3C and 11.4). This is shown diagrammatically at the top of the right half of Figure 11.2, and below that in the curves where the gland mitoses and amount of secretion are plotted.

The processes of regeneration and reconstruction of the endometrium are related to the higher levels of plasma estrogens, which are the result of high levels of FSH secreted by the appropriate cells of the anterior lobe of the hypophysis. The growth process continues under the influence of progesterone to eventually reach the late luteal phase, thus completing the cycle.

Physiologic Changes During Menstruation

Vagina Changes also take place in the vagina during the menstrual cycle. In the vagina, the epithelium varies greatly in thickness, both as a

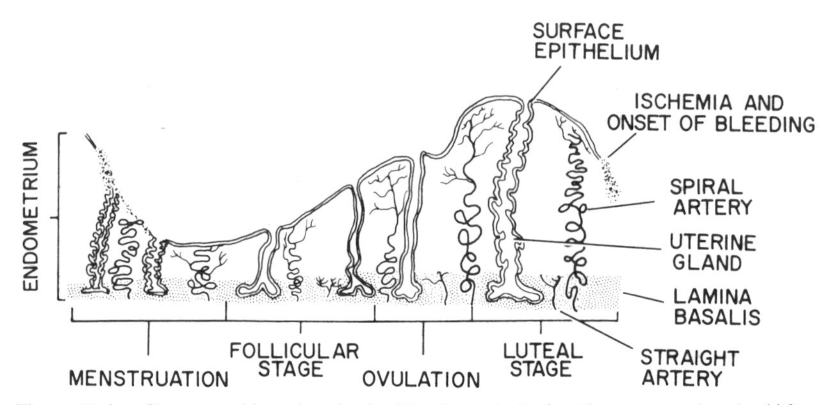

Figure 11.6. Changes taking place in the blood vessels during the menstrual cycle. (After Ramsey, 1977, and Barthelmez, 1957)

whole and in its several layers. The underlying connective tissue varies in consistency. The amount of vaginal fluid is always small, yet it varies in amount and in the kinds of cells that float in it. The external part of the cervix that dips into the vagina (ectocervix) is covered by the same kind of epithelial surface, and the epithelium shows the same kinds of changes as does the rest of the vaginal epithelium.

The wall of the vagina consists of several layers of cell types, all of which arise by cell division of the cells at or near the base of the epithelium. The process of cell replacement is the same as it is all over the skin, where the dividing cells move toward the surface and are shed. In the vagina, where the surface is moist, the cells which are shed float in the vaginal fluid. Beneath the epithelium is a very thin layer of elastic fibers, which rests on a dense layer of collagenous fibers. This layer binds, but loosely, the outer epithelium to the underlying layers of smooth muscle. Outside of this is a layer of dense connective tissue.

The epithelium is described as *stratified* (because of its many cell layers) and *squamous* (because of the flat cells on the surface). It consists of four layers of cells (Figure 11.7):

1. The deepest one is the basal layer, which is one cell thick; the cells here tend to divide rapidly and the progeny are pushed into the next layer.
2. The parabasal layer, which is two cells thick; these cells also divide rapidly and their progeny move into the next layer.
3. The intermediate layer has long, thin, wide cells and is about 20 cells thick. Their cytoplasm is filled with glycogen. All the cells of layers one to three are separated from each other by wide spaces, but remain joined to each other by numbers of intercellular bridges containing

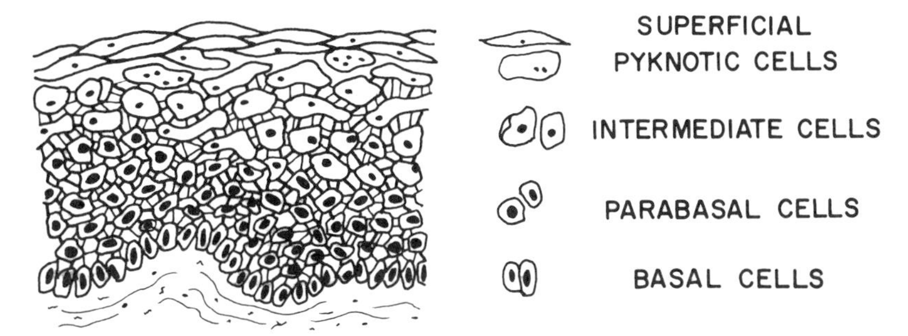

Figure 11.7. The structure of the stratified (many-layered), squamous (flat surface cells) epithelium of the vagina, and how it gives rise to the cells observed in vaginal smears. (After Steger and Hafez, 1978)

numerous protein filaments that run from cell to cell, and bind them together. It is probable that the lubricating fluid of the vagina originates as a kind of leakage of fluid from blood vessels in the connective tissue that percolates between the cells, and that this passage is facilitated by the wide intercellular spaces. The amount of fluid normally formed in this way is very small—about 2 g per day.

4. The cells of the superficial layer are even thinner than those in the third layer, and the layer may be ten cells thick. They contain little or no glycogen, but are rich in keratin, like the skin elsewhere in the body. Their nuclei are extremely dense and small, and frequently fragmented, and it is highly probable that their DNA is not active. These cells contain few or no intercellular bridges, and are readily shed from the surface of the epithelium.

The vaginal fluid, though small in amount, has a certain bacteriostatic action because of its high acidity. This high acidity (pH 3-4) is achieved by the enzymatic formation of lactic and acetic acids. The source of the acids is glycogen, which comes from the cells shed from the intermediate layer of the epithelium of the vagina. The enzymes causing the breakdown of glycogen come from bacteria that normally inhabit the vaginal fluid.

The height of the epithelium, as well as the thickness of each of the cell layers, varies with the time in the menstrual cycle. It is, for example, about two times thicker at the time of ovulation (end of the follicular stage) than just before the onset of menstruation (end of the luteal stage). The total thickness at the end of the follicular stage is about 0.25 mm.

Vaginal Smears These are used to diagnose certain infections, to aid in the efficiency of hormone treatment, to ascertain the time of ovulation, and to detect cancer at an early stage in parts of the genital tract. When cells appear in the smear with very basophilic cytoplasm and very large nucleoli, they are signs of rapid cell divisions such as are typical of cancer cells. This use of the smear was developed by Papanicolau, and the smear is commonly called the Pap smear. It is made by swabbing the vaginal fluid and rubbing cells from the wall of the vagina and/or the ectocervix, and transferring them to a glass slide on which they are stained by a specific procedure. Normally, in the follicular stage, the vaginal cells that are shed are chiefly from the superficial layer (flat, keratin-rich, glycogen-poor, and with shrunken or dark nuclei). In the luteal stage, the cells shed are chiefly cells of the intermediate layer—flat and glycogen-rich, with nuclei that are dark, but not as shrunken as those of the superficial layer (Figure 11.7, right).

Vaginal Infections One of the many adaptations of the genital tract of women is the relative infrequency of infections. In part, this is achieved because of the bacteriostatic acidity of the vaginal fluid; in part by the presence of immunoglobulins in the vaginal fluid; and in part, because of the presence in the vaginal wall of many plasma cells, which could assume the function of secretion of antibodies if challenged.

Nevertheless, the genital tract is vulnerable to many infections caused by viruses (*Herpes genitalis*), bacteria (*Hemophilus vaginalis* and other nonspecific bacteria), *Neisseria gonorrhoea* (the specific bacterium of gonorrhea), spirochaetes (*Trichomonas vaginalis*), *Trepanema pallidum* (the infectious agent of syphilis), and yeast or fungus (*Candida albicans* or *Monilia albicans*). In fact, with the indiscriminate use of antibiotics and the appearance of resistant strains, as well as carelessness and ignorance of available precautions, there are fears in medical circles of great epidemics of gonococcal infections, syphilis, and other venereal diseases.

Cervix Secretion by the cervix undergoes changes during the menstrual cycle though no such changes are identifiable in the epithelial cells or connective tissue. The cervix is the lower part of the uterus, and measures about 2.5 mm (1 inch) in length and thickness. At its widest, the diameter of the canal is about 6 mm. The relative size compared with the uterus throughout the life cycle is shown in Figure 11.8. The canal is filled with mucus secreted by the surface epithelium lining the canal. There are about 100 oblique folds or clefts of the surface epithelium, with many fine branches (Figure 11.9). The cervix is tough, being made up mostly of dense connective tissue, except in the last month of pregnancy when it becomes soft, distensible, and saturated with fluid, only to revert rapidly to its tough and dense consistency after parturition. The epithelium varies in the major parts of the cervix. In the ectocervix (Figure 11.8), the epithelial surface is like that of the vagina, and undergoes the same changes. In the canal, the epithelium consists of a single layer of tall columnar cells spread out as a sheet that covers the surface and the clefts (Figure 11.10). There are two kinds of columnar cells: cells with cilia and cells that synthesize mucigen and secrete it into the canal as mucus. Most of the cells

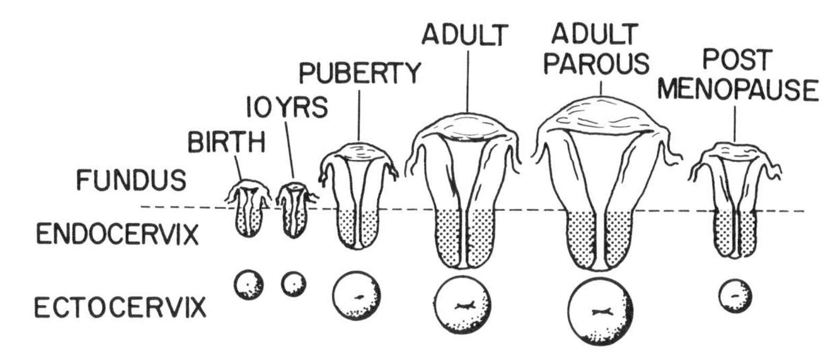

Figure 11.8. The fundus or body of the uterus and of the cervix show changes that take place during the lifetime. Above is a series of longitudinal views through the middle of the uterus, to show the position of the internal opening of the endocervix (at the level of the dotted line). Below this row is another of the ectocervix, as viewed from the vagina. (After Hafez, 1973)

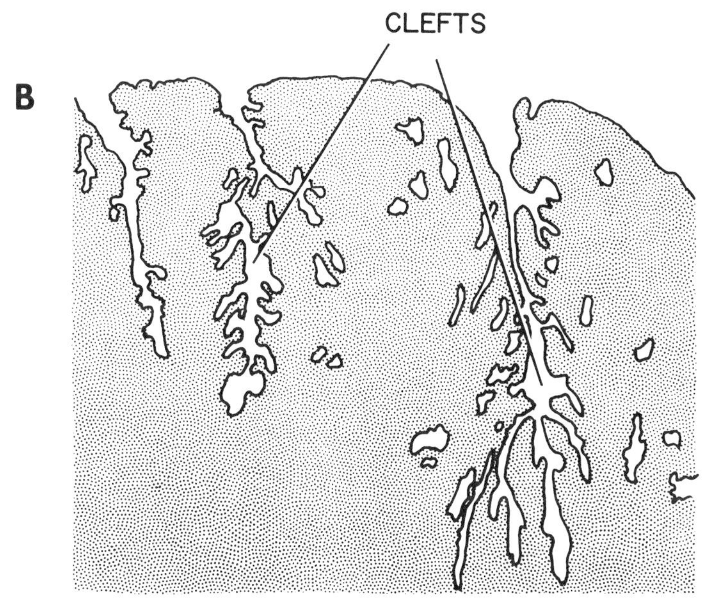

Figure 11.9. Reconstruction of the surface epithelium of the endocervix (black lines) and the clefts with their numerous minor folds. (After Fluhmann, 1961)

secrete mucus (Figure 11.10B). The total amount of mucus secreted is 20–60 mg/day. During the midcycle, nearly 1 g/day may be secreted. The mucus in the canal effectively acts as a plug preventing bacterial penetration to the body of the uterus and beyond. At the same time, it has certain viscosity properties that can be attributed to the mucin molecules themselves and the molecular aggregates they form. The net result is that they

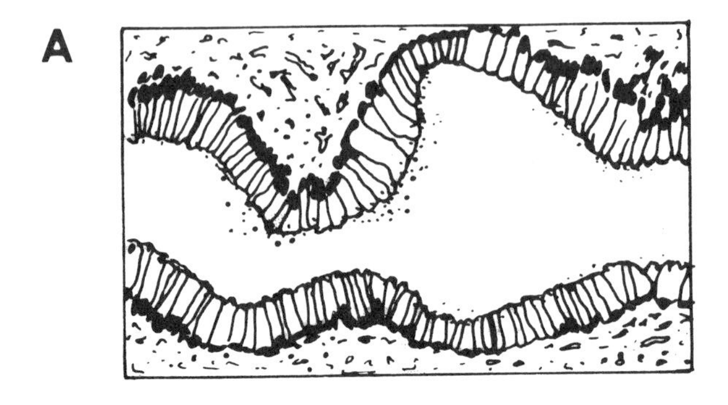

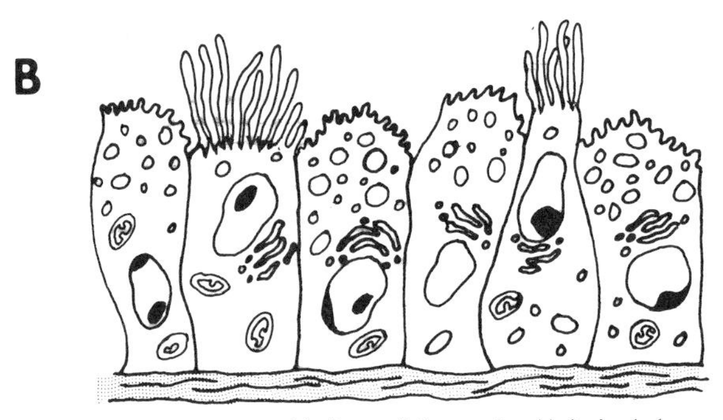

Figure 11.10. Some features of the epithelium of the cervix. A) A single layer of tall columnar cells. B) The epithelium consists of two cell types, ciliated and mucous. (After Weiss and Greep, 1977; and Elstein, 1976)

favor sperm penetration at about the time of ovulation, and prevent their passage at other times. This is shown diagrammatically in Figure 11.11. Sperm penetrability of cervical fluid can be ascertained in the laboratory and its variation is shown in Figure 11.2, in relation to the mucus content, the quantity produced, and the viscosity.

Sperm and the Cervix The head of the sperm is small (4.5 μm long × 2.5 μm wide × 1.5 μm thick), the chief component being the nucleus with all of its genetic information. The remainder of the sperm (about 45 μm) is the tail. Only at the attachment is an intermediate piece stuffed with mitochondria, which generate the energy required for sperm motility. The remainder of the tail is the motile portion—it oscillates at a definite frequency and whips the sperm forward about 2-3 mm per min. This rate of advance is not sufficient to account for their rapid appearance in the uterine tubes, and it is not clear how sperm reach their target so rapidly.

With a normal ejaculation, some 200-500 million sperm are deposited in the inner part of the vagina and on the cervix. The first portion of the

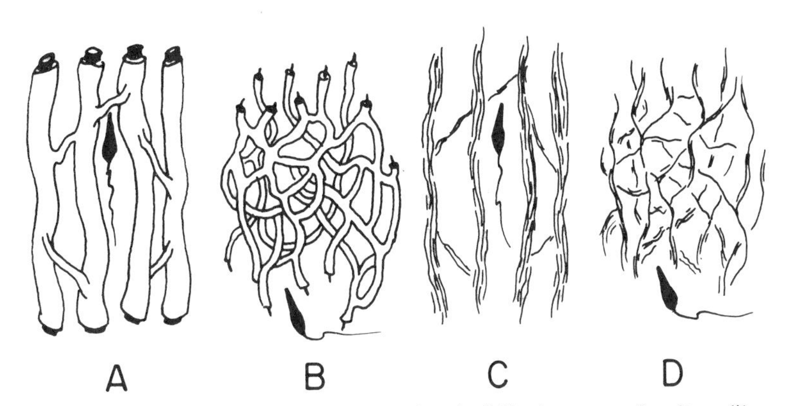

Figure 11.11. Structure of cervical mucus during the follicular stage, when it readily permits sperm passage (A), and during the luteal phase when it markedly reduces sperm passage (B). In both A and B, bound water of the mucin molecules is included in the structure. In C and D, only the mucin molecules are represented. (After Odeblad, 1973)

ejaculate contains approximately three-fourths of the sperm. The semen coagulates immediately and traps most of the sperm until seminal enzymes liquify the clot a few minutes later. Some sperm penetrate the cervix immediately in favorable conditions. The exact times when the sperm reach various parts of the genital tract are not known.

The chances that the ovum would be fertilized are greatly increased by the storage of sperm in the cervical crypts, by the favorable environment of the cervical mucus, by the orientation of sperm motility, and by the hormonally controlled cyclical variation in amount and consistency of the cervical mucus.

Oviduct The ovum spends seven days in its passage from the fimbriated end of the oviduct to the uterus, and is nourished during this time by the fluid in the lumen. At the same time, sperm passing in the opposite direction are matured by some component of the fluid and are supplied with an energy source in the oviductal fluid. The oviduct itself is made up of two chief layers: the surface epithelial layer and the underlying connective tissue and smooth muscle. The epithelium is columnar, and consists of two cell types—ciliated and mucous, both of which resemble similar cell types elsewhere in the body. The only cyclic change noted is in the height of the cells, which in the ovulatory phase is about twice that in the luteal phase (Figure 11.12). Correlated with this is an increase in the protein content of the oviductal fluid.

The oviduct performs functions in relation to the transport of sperm, ova, and the developing fertilized egg. Sperm are transported at least to the ampulla, but may pass into the peritoneal cavity. It is believed that in this passage, sperm may be capacitated or matured. Capacitation is fa-

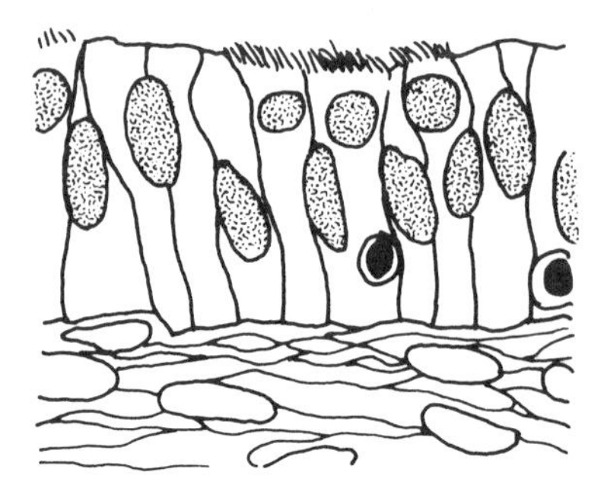

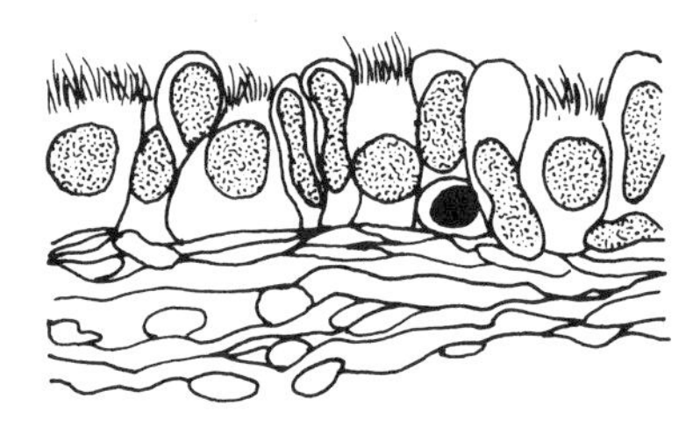

Figure 11.12. The epithelium of the oviduct consists of two cell types, ciliated and mucous. Both cell types increase in height during the follicular phase, and decrease in the luteal phase. (After Bloom and Fawcett, 1975)

vored by the bicarbonate of the oviductal fluid. The glucose that it also contains may supplement the energy sources of the sperm. The ova are transported through the fimbriated end to the ampulla where they may be fertilized. Though the movement of the ova may be facilitated in some way by ciliary movement, and by a more gross aspiration or suction action throughout the whole genital tract, more emphasis is placed now on the mechanical (peristaltic) movements along the course of the oviduct. The sweeping, rhythmic contractions of the fimbria, and the contractions of the smooth muscle in the mesosalpinx and ovarian ligaments, as well as the ciliary action of the fimbria on the cumulus oophorus may all be involved in recovery of the egg after ovulation and its guidance into the lumen of the oviduct. During its passage to the ampulla, the corona radiata cells of the cumulus and the zona pellucida are removed, possibly through the action of enzymes, some of which may arise from the acrosome in the head of the sperm. The denuded egg may be fertilized in the ampulla where it may remain for several hours. Then it passes into the isthmus where it is detained for one or two days before passing into the uterus. Here again the sugar derivatives (lactate and pyruvate) present in the oviductal fluid may supply the energy needs of the early, free-floating embryo.

Smooth muscle is oriented as an inner circular layer and an outer longitudinal layer. The oviduct seems to be actively mobile much of the time, with periodic bursts of activity of small or large amplitude. The greatest outbursts occur at the time of menstruation and during the early follicular stage.

Mechanisms of Control

The most obvious features of the mechanisms that control menstruation are the positive and negative feedback mechanisms involving the neuroendocrine portion of the hypothalamus, the anterior lobe of the hypophysis and the ovarian follicle and corpus luteum of the ovary, discussed in Chapter 7.

There is in addition a certain input into the hypothalamus from other parts of the central nervous system. While significant, this neural input is not essential for menstruation to take place. For example, in experimental animals complete severance of the genital tract from the parasympathetic and sympathetic nervous system has only a temporary effect on menstruation. After a brief cessation, cycles are resumed. But this does not exclude the possibility that the central nervous system through the hypothalamus might influence the severity or timing of the menstrual cycle in women. There are several examples of a tendency for women living or working together to synchronize their menstrual periods (see McClintock, 1971).

With so many variables involved, as to concentration, change in rates of hormone synthesis, and timing, it is not surprising that the menstrual cycle varies over a wide range, not only in the same woman but also as between different women. It is also not surprising that no method of testing can predict precisely when menstruation will set in, or how long or severe will be the bleeding, or indeed when ovulation will take place.

The cyclic hormonal variations appear perhaps more impressive when considered more broadly from the point of view of their relations to the life cycle of women from childhood through the five stages of sexual maturation, and the menopause and postmenopause. This is shown in Figure 11.13 for FSH, LH, estradiol, and progesterone.

Hormones and Target Cells Some hormones, like those considered in this chapter, have target cells. This concept requires that certain specific cells have a high sensitivity to certain specific hormone(s) and that all other cells be relatively insensitive to them. This property of target cells can be attributed to the specificity of receptor molecules on the surface of the target cells, which enables them to selectively bind with great avidity the specific hormones in the tissue fluid adjacent to the cell surface. The

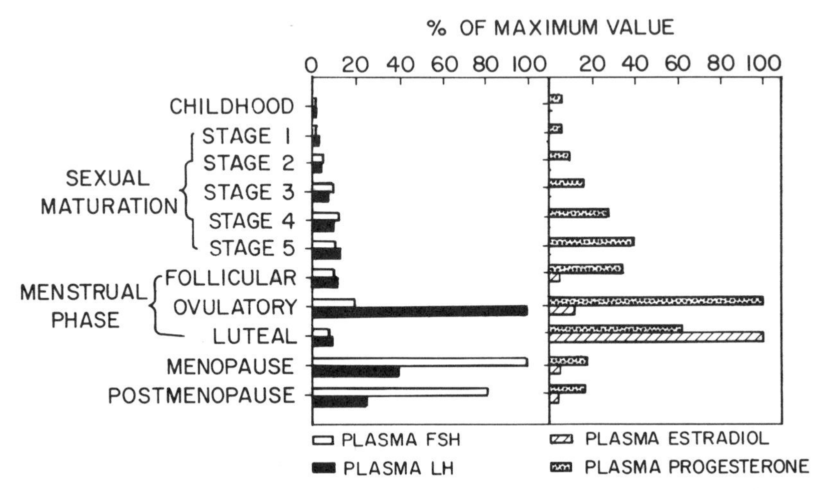

Figure 11.13. Relative plasma levels of FSH, LH, estradiol, and progesterone during the lifetime of women. Levels prior to and during puberty and during the menopause and postmenopause are discussed in appropriate sections of this chapter. Sexual maturation stages are discussed in Chapters 18 and 19. (After Collins and Newton, 1974)

number of hormone molecules in the tissue fluid is in equilibrium with the concentration of free hormone molecules in the plasma of the adjacent capillaries. This combined receptor and hormone molecule activates an enzyme in the cell membrane (adenyl cyclase) that releases cyclic adenyl monophosphate (cAMP) or cyclic guanyl monophosphate (cGMP). This activates a protein kinase specific for the target cell—which is to say that it catalyzes or accelerates protein synthesis. It may also accelerate ribosomal activity in the cytoplasm, again influencing protein synthesis. The cyclic monophosphate is then regarded as a second or intracellular messenger, if the hormone be considered the first.

Another method of action of another kind of hormone on the target cell is more direct. The hormone is bound to the cell surface through combining with a receptor. The bound hormone is transported to the nuclear membrane, and there combines with another protein, releasing the protein binder of the cytoplasmic membrane. In some way that is not understood, the hormone or a complex of the hormone and some other nuclear membrane component causes the DNA of specific genes to make messenger RNAs, which are transferred to the cytoplasm where in association with ribosomes and other nucleotides and other appropriate enzymes, specific proteins are synthesized. In addition, the genes controlling the synthesis of ribosomal RNA generate RNA molecules which are incorporated into ribosomes. These also pass into the cytoplasm and take part in protein synthesis.

Induced Variations in Menstruation

Women athletes in training are said to suffer relatively little from dysmenorrhea (difficult and painful menstruation). Moderate exercise has no effect on the menstrual cycle. Extreme exercise associated with rigorous athletic training eventually causes secondary amenorrhea (suppression of menses), but normal, regular menstrual cycles are resumed shortly after relaxing from the strict requirements of training. Menstruation is only a bar to sexual intercourse when it is brought irrelevantly into the situation through internalization of societal mores.

It seems "natural" to us that women should menstruate for up to 15% of their fertile years. Yet observation of other cultures shows that this assumption is a product of our cultural bias. In some cultures where women breastfeed their infants for prolonged periods, ovulation (and menstruation) may be inhibited over a stretch of 4 to 5 years. Short (1978) suggests that long periods of amenorrhea are just as natural as short periods (due to pregnancy) or none at all, and that treatment modifying hormone levels might be used for women's convenience to induce amenorrhea. Recent developments in the understanding of the action of prostaglandins mentioned above, encourage the hope that further research will provide safe ways to cope, not only with dysmenorrhea, but also with the symptoms of premenstrual tension and even, as Short suggests, the inconvenience of menstruation itself.

The fluctuations in hormone levels during the menstrual cycle are simulated by the use of contraceptive pills that commonly contain a mixture of estrogen and some substance or substances similar to progesterone. The dosages employed are such as to override the normal cyclical variations. They turn off the FSH/LH-RH of the hypothalamus, suppressing the production and release into the blood of FSH and LH. This leads to a cessation of follicular growth and maturation in the ovary, resulting in a cessation of ovulation. When the pill is not taken, the hormone levels fall, and the same changes take place in the uterine wall which lead to menstrual bleeding. Resumption of pill-taking raises blood hormone levels and the next cycle is initiated. Some women profit from this enforced regularity of hormone self-administration in that their bleeding periods become more or less predictable. In some women, the length of the bleeding time or the amount of blood lost is reduced; the flow may even stop altogether.

MENARCHE

The menarche, or the first menstruation, usually occurs at 12–13 years of age, but the age may range from 10–18 years. The age at menarche is a kind of measure of the rate of growth. It is very closely related to skeletal maturation. Menstrual cycles are usually irregular for 1–2 years, and in most of these early cycles, ovulation does not occur. At this time, the secondary sex characters are developed.

The amount of body fat might play an essential role in menarche, as well as in certain menstrual irregularities. It has been suggested that the fat content must exceed a certain level for menarche to occur. The argument is that such a condition would have increased survival value for the species. Pregnancy and survival of the infant would depend, among other things, on the energy supply available to the infant, and some calculations have shown that the amount of fat added during puberty might be sufficient for this purpose.

Data on twins indicate that there is a strong genetic component affecting the onset of menarche. The difference in age at menarche between pairs of identical twins is significantly lower than that between nonidentical twins or between sisters of different age. These two latter groups differ about equally, but less than the differences between pairs of unrelated girls of the same socioeconomic class. A common speculation is that the hypothalamic neuroendocrine center(s) of the male fetus are influenced by androgen some time before birth and become thereafter noncyclic. By contrast, in the female fetus (lacking such high levels of androgen), the hypothalamic center(s) is directed to assume a cyclic pattern that does not become manifest until puberty when the negative feedback systems become functional and initiate menstruation and the menstrual cycle. But exactly how the feedback systems are turned on is not known. It has been suggested that nutrition and body fat are important, but no single factor or sequence of changes can account for the onset of menarche.

Another hypothesis that has gained much support recently is that the threshold of sensitivity of the receptors in the hypothalamus (or of the nerve cells impinging on the neuroendocrine cells) is lowered, thereby starting off a genetically predetermined pattern that starts with the synthesis and secretion of follicle-stimulating releasing hormone, and ends with cyclical interrelated hormonal activities. It has been suggested that the change may follow a change in the number of receptors. Neither the onset and sequence of development of each trait, nor the relative rates of development of the several characters involved are explained by any known specific endocrinologic changes.

These data should be considered in conjunction with certain findings on synchrony of menstrual cycles. Studies showed a tendency for women living together in a dormitory to synchronize their menstrual periods (McClintock, 1971). It is not known whether the stimulus for this synchronization is visual, behavioral, or olfactory. It could be that the same environmental factors, whatever they are, that bring about this synchronization, could also be in effect with regard to initiation of menarche, at least in some girls.

Another set of environmental factors centers around the socioeconomic status of the girls. Girls of higher socioeconomic origins pass through menarche at an earlier age and mature earlier. The age of menarche was found to be lower in girls with parents in white-collar occupations than those with parents in blue-collar occupations. It seems also that menarche occurs earlier in girls of smaller families. And finally there is the trend, noted earlier, for menarche to take place at earlier ages in girls during the last 100 years in Western Europe and the USA, with the age reducing by about four months every ten years. Whether this accelerated time of onset of menarche represents a playing out of a genetic pattern developed in the course of evolution of the species, or reflects some of the environmental factors mentioned above, is not known.

The relative thickness of the vagina before and after menarche is shown in Figure 11.15. Many physical changes take place in the soma (muscle, bones, skin, etc.)—such as maturation of the breast—and in many physiologic functions during puberty and adolescence; these are treated at length in Chapters 17 and 18.

Very occasionally, female newborn babies menstruate. The uterine scene was prepared by high blood levels of maternal hormones that passed through the placenta into the fetal circulation. The sudden separation at birth from the maternal supply results in a rapid fall of hormone levels in the fetus, and this results in bleeding as does hormone withdrawal in women of menstruating age. Changes specific to the newborn are shown graphically in Figure 11.8 (ectocervix, endocervix and body of the uterus) and Figure 11.15. The prenatal growth of the uterus is shown in Figure 11.14.

MENOPAUSE

Menstruation continues in its irregularly cyclic way (except of course during pregnancy) until the climacteric, when cyclic functioning is reduced. During this period, bleeding may be profuse or painful, or short and irregular. The mammary glands, the uterus, and vagina begin to atrophy. This period of gradual reduction of ovarian function may continue for a short or long time, without any necessary reduction in sexual activity; eventually menstrual bleeding ceases entirely. After about a year, the woman has passed through the period of menopause. Thereafter, the woman is in the postmenopause. The average menopausal age is 50 years; but menopause may take place at any time between ages 35 and 60. The estrogen level in the blood, though markedly reduced and noncyclic because of the cessation of ovarian function is never zero; the hormone may originate from

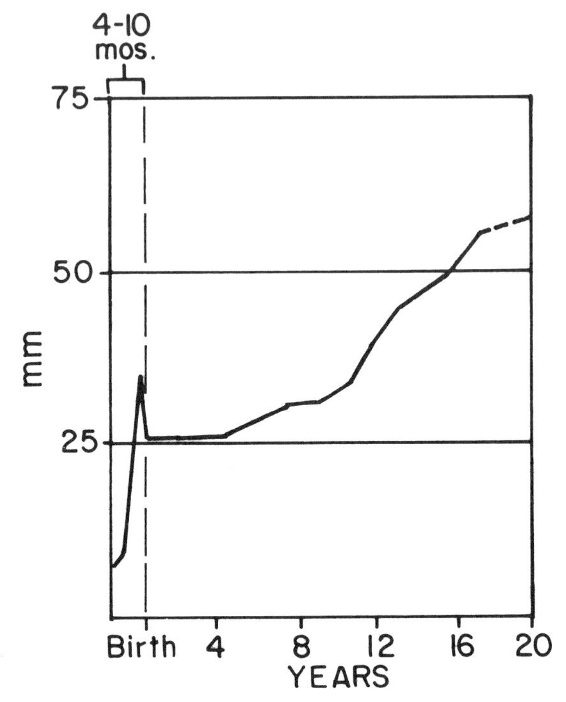

Figure 11.14. Prenatal and postnatal growth in length of the uterus. This shows the postnatal sharp regression and the slow growth of the uterus during childhood, with a gradual growth during adolescence until the adult size is reached at 20 years of age. (After Scammon, 1930)

some other source such as the adrenal cortex. The progesterone level is inappreciable. The marked reduction or absence of these hormones, and the consequent failure of the negative feedback action on the appropriate cells of the hypothalamus and anterior lobe of the hypophysis, commonly results in a high blood level of FSH and very high LH levels, both of which continue through the postmenopause (Figure 11.13).

The symptoms during the climacteric, in addition to heightened menstrual irregularities, may include one or more of the following: hot flushes, sweating, cardiac palpitations, osteoporosis, excitability, irritability, insomnia, and headache.

Early in the menopause some of these symptoms may be aggravated, such as the first three mentioned above. In later years, some of them disappear, while others become progressively more marked. The vagina begins to atrophy, becoming thinner, narrower, shorter, and less elastic (Figure 11.15). It also tends to be drier, which probably reflects a reduction in its vascular supply with coincidental reduced content of water and other components of the connective tissue underlying the epithelium. The

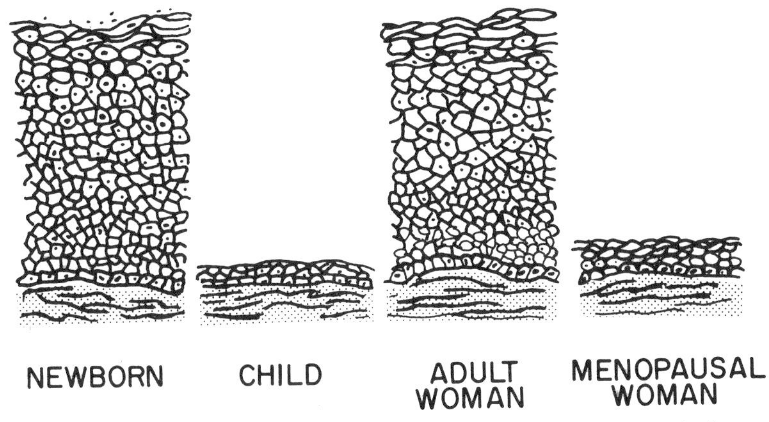

Figure 11.15. Relative heights of the vaginal epithelium in the newborn, the immature child, the adult (menstruating) woman, and the postmenopausal woman. (After Hafez and Evans, 1978)

reduced lubrication may cause pain during coitus unless some exogenous lubricant is applied. The vaginal wall may become ulcerated, and, in conjunction with decreased acidity of the surface film of water, more susceptible to infection. This is related to the fact that glycogen in the intermediate cells is reduced or absent. And, of course, the number of cells in mitosis in the basal and parabasal cells is reduced. The external genitalia shrink through loss of fat, vaginitis and itching may intervene, the vaginal wall may become sufficiently weaker structurally as to result in prolapse of the uterus. And, of course, all layers of the uterus atrophy. The cervical epithelium is somewhat atrophic and the cervical fluid is scanty and viscous. The epithelium of the oviduct is also atrophic, and the secretion of oviductal fluid stops. Late in the postmenopause, the ciliated cells of the oviduct lose their cilia. The skin becomes thinner, and loses some elasticity and turgor, through loss of water and other components of the subcutaneous connective tissue. The mucous membranes of the nasal and paranasal sinuses also atrophy, resulting frequently in rhinitis. The breasts shrink and droop as the fat and glandular tissue as well as tissue turgor decline (see Figure 13.3). The nipples become smaller and are less erectile. The bones may become structurally weaker and more subject to breakage, owing to the loss of calcium phosphate from the bones and a consequent increase in these inorganic constituents occurs in the plasma. This may also be related to the increased plasma level of alkaline phosphatase, an enzyme that may cause the removal of phosphate and calcium from bone. There may be an increase in the blood cholesterol level and a deposition of lipids in the wall of blood vessels leading to atheromatous plaques. This increase is less marked than in men of similar age.

During the menopause and postmenopause periods, changes take place in the relative dimensions and appearance of certain organs. These are indicated in Figure 11.9 (ectocervix, endocervix, and the body of the uterus); in Figure 11.15 (vagina) and the breast (Figure 13.2). Changes in plasma hormonal levels are illustrated in Figure 11.13.

Finally, the postmenopausal woman may become clinically depressed. The extent to which cultural attitudes enter this symptom is unknown but probably very important.

Very many of these symptoms are direct consequences of reduced levels (withdrawal) of estrogen, and can be ameliorated by estrogen replacement therapy. Administered cyclically, there may be a return of periodic bleeding. There is a controversy on the timing and duration of treatment with estrogen, as well as the kind of estrogenic compound given, especially in view of the possibility that prolonged estrogen treatment may cause cancer of the genital tract and undesirable side-effects.

This litany of ailments suffered by postmenopausal women is not very different from that suffered by men of comparable age. There is the similar atrophy of muscles and bones, and the similar trauma of joints. There are also very much the same motor, sensory, and psychologic defects of elderly people of both sexes. Most of the ailments may be attributed to the aging process, which is probably similar in both sexes. It is hoped that geriatric research will help to alleviate some of these all too common ailments. Despite these difficulties, sexual activity may continue throughout the postmenopausal period with as much (or more) enjoyment as in premenopausal times.

The postmenopausal period is a relatively new phenomenon in Western societies, for in earlier times few women survived beyond their childbearing years. That is to say, the very occurrence of the postmenopausal period has been made possible by better economic conditions, including better housing, nutrition, and public hygiene.

While Western societies have improved the bases of these biologic aspects of aging, they have not done so well in other social or cultural aspects. Only a small percentage of aged people can exist nowadays at a moderate level of comfort; most live a meager life from which many of the societal advances of the last two or three centuries have been withdrawn or are seriously curtailed. In many societies in the past, and in some which exist at present, postmenopausal women were and are honored for their judgment, their concern for the young, and their meaningful and responsible functions. The attainment of postmenopausal status in Western societies could carry with it certain freedoms and possibilities to develop economically, culturally and socially for the enrichment of both the individual and society, which are stalled by the niggardly treatment afforded most older people, especially women.

REFERENCES

Barthelmez, G. W. 1957. The form and the functions of the uterine blood vessels in the rhesus monkey. Contributions to Embryology, Carnegie Institution of Washington, Vol. 36, pp. 153–182.

Bloom, W., and D. W. Fawcett. 1975. A Textbook of Histology. 10th Ed. W. B. Saunders, Co., Philadelphia.

Bosch, J. J. van der Werff ten. 1966. Variations in activity of some endocrine organs. *In* J. J. van der Werff ten Bosch and A. Haak (eds.), Somatic Growth of the Child. H. E. Stenfert Kroese, N. V. Leiden, pp. 157–164.

Boston Women's Health Book Collective. 1973. Our Bodies, Our Selves. Simon & Schuster, New York.

Burgos, M. H., and C. E. Roig de Vargas-Lenares. 1978. Ultrastructure of the vaginal mucosa. *In* E. S. E. Hafez and T. N. Evans (eds.), The Human Vagina. Elsevier-North Holland Publishing Co., New York, pp. 63–93.

Cavalli-Sforza, L. L., and W. F. Bodmer. 1971. The Genetics of Human Populations. W. H. Freeman & Company, San Francisco.

Chantler, E., and E. Debruyne. 1977. Factors regulating the changes in cervical mucus in different hormonal stages. *In* M. Elstein and D. V. Parke (eds.), Mucus in Health and Disease, Advances in Biology and Medicine, Vol. 89. Plenum Press, New York, pp. 131–141.

Collins, W. P., and J. R. Newton. 1974. The ovarian cycle. *In* Biochemistry of Women: Clinical Concepts. A. S. Curry and J. V. Hewitt (eds.), CRC Press, Cleveland, pp. 1–22.

Cooke, C. W., and S. Dworkin. 1979. The Ms. Guide to a Woman's Health. Anchor Press, Garden City, N.Y.

Elias, H., and J. E. Pauly. 1966. Human Microanatomy. 3rd Ed. F. A. Davis Co., Philadelphia.

Elstein, N. 1976. Non-immunological factors in infertility. *In* The Cervix. J. A. Jordan and A. Singer (eds.), W. B. Saunders Co., Philadelphia, pp. 175–184.

Elstein, N. 1978. Functions and physical properties of mucus in the female genital tract. Br. Med. Bull. 34:83–88.

Fluhmann, C. F. 1961. The Cervix Uteri and Its Diseases. W. B. Saunders Co., Philadelphia.

Hafez, E. S. E. 1973. Histology and microstructure of the cervical epithelial secretory system. *In* M. Elstein, K. S. Moghissi, and R. Borth (eds.), Cervical Mucous in Human Reproduction. Scriptor, Copenhagen, pp. 23–32.

Hamilton, W. J., J. D. Boyd, and H. W. Mossman. 1972. Human Embryology. Williams & Wilkins, Co., Baltimore.

Johnson, J. D., and C. W. Foley. 1974. The Oviduct and Its Functions. Academic Press, New York.

Keebler, C. M., and J. W. Reagan. 1975. A Manual of Cytotechnology. 4th Ed. American Society of Clinical Pathologists, Chicago.

Kletzky, O. A., R. M. Nakamura, I. H. Thorneycroft, and D. R. Mischell, Jr. 1975. Log normal distribution of gonadotrophins and ovarian steroid values in the normal menstrual cycle. Am. J. Obstet. Gynecol. 121:688–694.

Lein, N. 1979. The Cycling Female: Her Menstrual Rhythm. W. H. Freeman & Co., San Francisco.

Lennane, K. J., and J. R. Lennane 1973. Alleged psychogenic disorders in women—A possible manifestation of sexual prejudice. N. Eng. J. Med. 288:228–292.

Marx, J. 1979. Dysmenorrhea: Basic research leads to a rational therapy. Science 205:175-176.

Maximov, A. A., and W. Bloom. 1957. A Textbook of Histology. 7th Ed. W. B. Saunders Co., Philadelphia.

McClintock, M. 1971. Menstrual synchrony and suppression. Nature 229:244-245.

Moghissi, K. S. 1973. Sperm migration through the human cervix. *In* R. J. Blandau and K. Moghissi (eds.), The Biology of the Cervix. University of Chicago Press, Chicago, pp. 306-327.

Nilson, O., and S. Reinius. 1969. Light and electron microscopic structure of the oviduct. *In* E. S. E. Hafez and R. J. Blandau (eds.), The Mammalian Oviduct. University of Chicago Press, Chicago, pp. 57-83.

Odeblad, E. 1973. Biophysical techniques of assessing cervical mucus and microstructure of cervical epithelium. *In* M. Elstein, K. S. Moghissi, and R. Borth (eds.), Cervical Mucous in Human Reproduction. Scriptor, Copenhagen, pp. 58-74.

Odeblad, E. 1977. Physical properties of cervical mucus. *In* M. Elstein and D. V. Parke (eds.), Mucus in Health and Disease, Advances in Biology and Medicine, Vol. 89. Plenum Press, New York, pp. 217-225.

Pauerstein, C. J. 1975. Seminar on tubal physiology and biochemistry. Gynecol. Invest. 6:100-264.

Persky, H., N. Charney, H. I. Lief, C. P. O'Brien, W. R. Miller, and D. Strauss. 1978. The relationship of plasma estradiol level to sexual behavior in young women. Psychosomat. Med. 40:523-535.

Persky, H., H. I. Lief, D. Strauss, W. R. Miller, and C. P. O'Brien. 1978. Plasma testosterone levels and sexual behavior of couples. Arch. Sexual Behav. 7: 157-173.

Platzer, W., S. Poisel, and E. S. E. Hafez. 1978. Functional anatomy of the human vagina. *In* E. S. E. Hafez and T. N. E. Evans (eds.), The Human Vagina. Elsevier-North Holland Publishing Co., New York, pp. 39-53.

Ramsey, E. M. 1977. Vascular anatomy. *In* R. M. Wynn, Biology of the Uterus. 2nd Ed. Plenum Press, New York, pp. 59-76.

Scammon, R. E. 1930. The measurement of the body in children. *In* J. A. Harris, C. M. Jackson, D. G. Paterson, and R. E. Scammon (eds.), The Measurement of Man, pp. 171-215. University of Minnesota Press, Minneapolis.

Schachter, H. 1977. Control of biochemical parameters in glycoprotein production. *In* M. Elstein and D. V. Parke (eds.), Mucus in Health and Disease, Advances in Biology and Medicine, Vol. 89. Plenum Press, New York, pp. 103-129.

Schaller, G. B. 1965. Behavior comparisons of the apes. *In* I. DeVore (ed.), Primate Behavior. Holt, Rinehart & Winston, Inc., New York, pp. 474-481.

Schmidt, E. H., and F. K. Beller. 1978. Biochemistry of the vagina. *In* E. S. E. Hafez and T. N. Evans (eds.), The Human Vagina. Elsevier-North Holland Publishing Co., New York, pp. 139-149.

Singer, A., and J. A. Jordan. 1976. The anatomy of the cervix. *In* J. A. Jordan and A. Singer (eds.), The Cervix. W. B. Saunders Co., Philadelphia, pp. 13-36.

Steger, R. W., and E. S. E. Hafez. 1978. Age-associated changes in the vagina. *In* E. S. E. Hafez and T. N. Evans (eds.), The Human Vagina. Elsevier-North Holland Publishing Co., New York, pp. 95-106.

Tanner, J. M. 1960. Genetics of human growth. *In* J. M. Tanner (ed.), Symposia of the Society for the Study of Human Biology, Vol. III. Pergamon Press, New York, pp. 43-58.

Weideger, P. 1976. Menstruation and Menopause. A. A. Knopf, New York.

Weiss, L., and R. O. Greep. 1977. Histology. 4th Ed. Harper & Row, New York.

12 Orgasm

It is difficult to define the word "orgasm" objectively. The rhythmic body movements of one or both members may culminate in seizure-like thrusting movements, which are accompanied by altered systemic changes in respiration, cardiac action and blood pressure, and sweating. This leaves out of consideration the intensity of the drive toward completion of the act of coitus, the frequently pleasurable act of sharing, in fact or in fantasy, the buildup and release of tension, or the ecstatic sense of achievement. Many of these factors cannot be described adequately or analyzed objectively. Some of them have been studied in great detail, and are presented in this chapter, which is based almost entirely on the pioneering books by Masters and Johnson.

EROGENOUS ZONES

These are the parts of the body which, when stimulated by touch or pressure, cause and intensify sexual arousal or activity. These are primarily the external genitalia and their surroundings, the breasts, especially the nipples, the mouth, the thighs and buttocks, the hands and feet, the eyes, ears, and nose, the anus, and in fact almost all parts of the body, depending on the situation, and the previous experiences, the culture, and the variations in sensitivity in different parts of the body in different individuals.

ORGASM IN WOMEN

The orgasm, as observed clinically, is the same following coitus or masturbation. Many of the features of the orgasm are common to women and men, though the timing and intensity may vary. The observations are considered for convenience as occurring in four stages: 1) the excitement phase, 2) the plateau phase, 3) the orgasmic phase, and 4) the resolution phase. These phases are continuous in time, and their intensity and extent may vary in different individuals, depending in part on prior events. They include systemic, or extragenital, events and genital events.

Excitement Phase

This may last for minutes to hours. The nipples become erect, the breast begins to enlarge, and the areola swells (Figure 12.1). A little later, the skin over the anterior part of the ventral abdominal wall (the epigastrium) may become flushed, and the area affected may spread over the breasts. The heart rate increases, as does the blood pressure. Muscle tension increases in many of the voluntary muscles of the body. The clitoris swells, especially the glans, increasing in length and diameter. At the same time, the wall of the vagina becomes moistened (and lubricated) by fluid that originates from the distended blood vessels of the wall of the vagina and passes through the connective tissue and epithelium of the vaginal wall (Figure 12.2). The vascular engorgement is accompanied by expansion of the inner part of the vagina, both in length and in diameter. Perhaps as a consequence of these changes, the uterus tips to a more elevated position. The labia majora swell in women who have already had children (multipara) and flatten in those who have not had children (nullipara), vasocongestion occurring in both, with slight separation accompanying their elevation. The labia minora are also engorged and slightly swollen (Figure 12.3).

Plateau Phase

The areola and the breast continue to swell, and the erect nipples also enlarge (Figure 12.1). The cutaneous flush is well developed, and may spread to other parts of the body. Muscle tone increases further, and

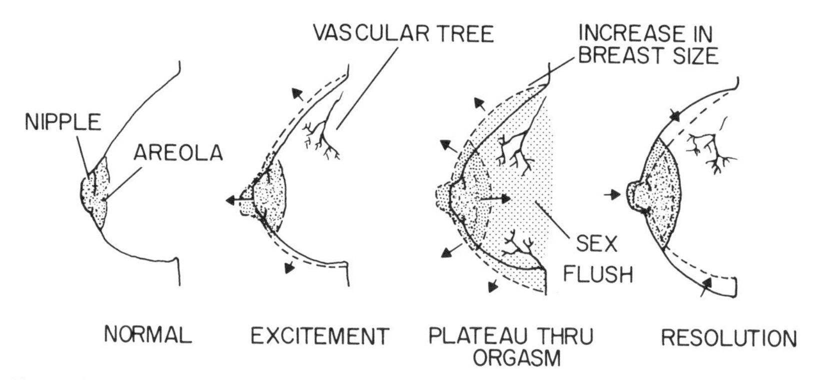

Figure 12.1. Changes in the female breast that take place during the sexual response cycle. (After Masters and Johnson, 1966)

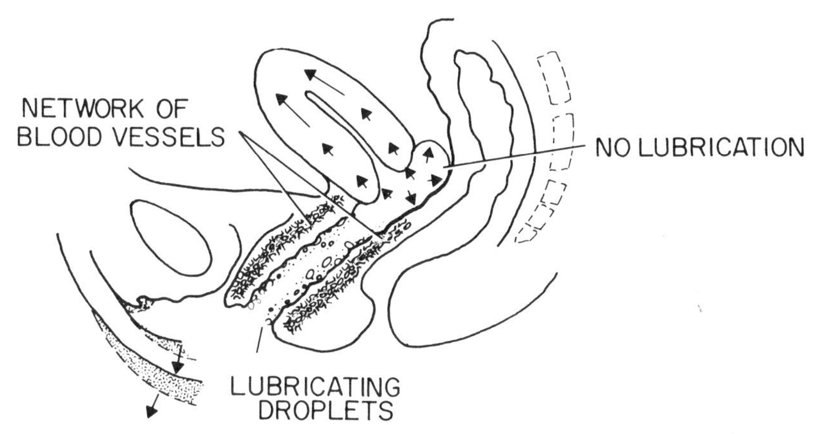

Figure 12.2. Midsagittal view of the female pelvis during the excitement phase illustrates the finding that the major source of vaginal lubricant is from leakage from the rich network of blood vessels surrounding it. (After Masters and Johnson, 1966)

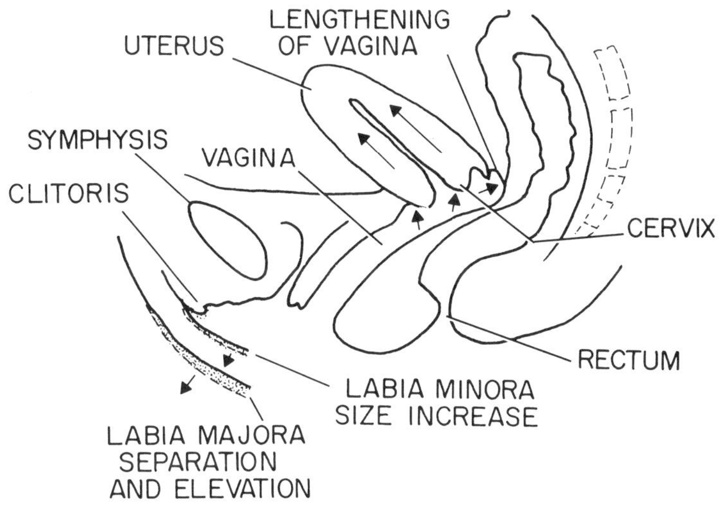

Figure 12.3. Midsagittal view of the female pelvis showing changes that take place during the excitement phase. In this figure, as in most of those which follow, the vagina is not really encompassing an inner space. During most of the time its walls are collapsed; it is only during coitus when the erect penis is inserted that its walls are separated. (After Masters and Johnson, 1966)

there may be involuntary and semispastic contractions of facial, abdominal, and intercostal muscles. The cardiac rate increases, as well as the blood pressure. The respiratory rate may increase markedly late in the plateau phase.

The clitoris continues to be engorged and swollen, and as a consequence, is retracted against the anterior wall of the symphysis pubis. The outer third of the vagina is markedly swollen and is known as the orgasmic platform. The swelling is accompanied by a reduction in the diameter of the vagina. The rest of the vagina continues to elongate and increase in diameter (Figures 12.4 and 12.5B).

There is secretion of a drop or two from the bulbourethral glands, but lubrication is not aided much by this. Perhaps associated with the increased diameter and length of the vagina is the continued anterior tipping and elevation of the uterus. The labia majora swell markedly. The labia minora also swell markedly, due to venous engorgement and accordingly change in color from a bright red to a deep wine color. Because of this swelling which extends to the coverings of the glans, and despite the swelling and erection of the glans and shaft of the clitoris, the glans becomes completely hidden from view (Figure 12.5A).

Orgasmic Phase

When the cutaneous flush occurs at all, it reaches its maximum extent during this phase. The breast changes continue (Figure 12.1). Involuntary (spastic) contractions of muscle groups occur, resulting among other things in the thrusting of the pelvic area irregularly and spasmodically. The rectal sphincter contracts spasmodically simultaneously with contractions of the outer part of the vagina (orgasmic platform). Involuntary contractions of the perineal and lower abdominal areas occur.

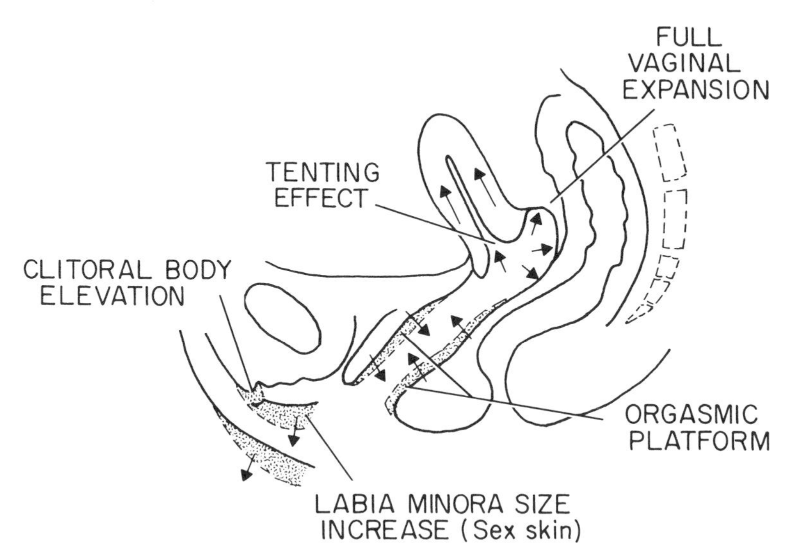

Figure 12.4. The origin of the orgasmic platform and the enlarged labia and swollen clitoris in the plateau phase. (After Masters and Johnson, 1966)

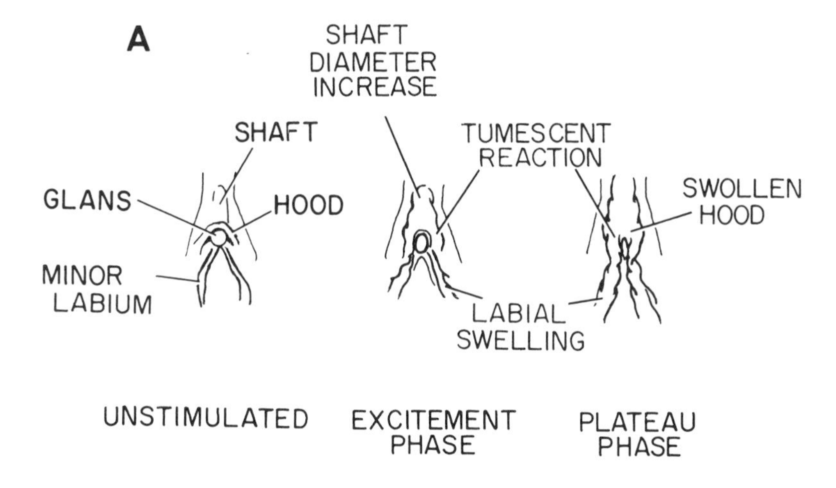

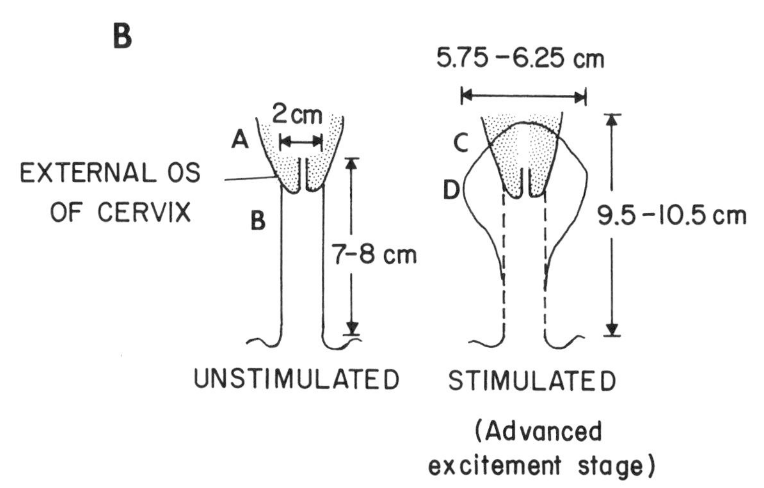

Figure 12.5. A) Changes in the size and position of the clitoris during the excitement and plateau phases in the female sexual response cycle. B) Changes in the size and shape of the vagina during the height of the excitement stage. (After Masters and Johnson, 1966)

Cardiac rate, blood pressure, and respiratory rate reach their peaks. Observation of the clitoris is not feasible because it is covered by swollen membranes filled with blood. The orgasmic platform contracts five to twelve times at 0.8-sec intervals. The uterus also contracts rhythmically, the contractions beginning at the fundus and progressing toward the vagina. The labia majora and minora remain swollen (Figure 12.6).

Resolution Phase

The areolae and nipples revert to their pre-excitement phase, and the breasts return to their normal size and shape (Figure 12.1). The cutaneous flush disappears rapidly, lingering longest in the areas of first appearance. Voluntary muscle contractions subside, but more slowly than the vascular changes. Heart rate, blood pressure, and respiratory rate return to normal levels. Some women show a film of perspiration over the chest, back and thighs, and a more obvious layer of sweat on the forehead, upper lip and armpits. The amount of perspiration, when apparent, is unrelated to the amount of muscular effort.

The clitoris rapidly returns to its normal position, and slowly becomes less swollen. The vagina also returns to normal, but more slowly. The first reaction is a rapid return to normal of the outer third of the vagina, the orgasmic platform, a relaxation of the remainder of the vaginal walls with a shortening of the vagina and decrease in its diameter, until the normal collapsed state is attained. The external cervical

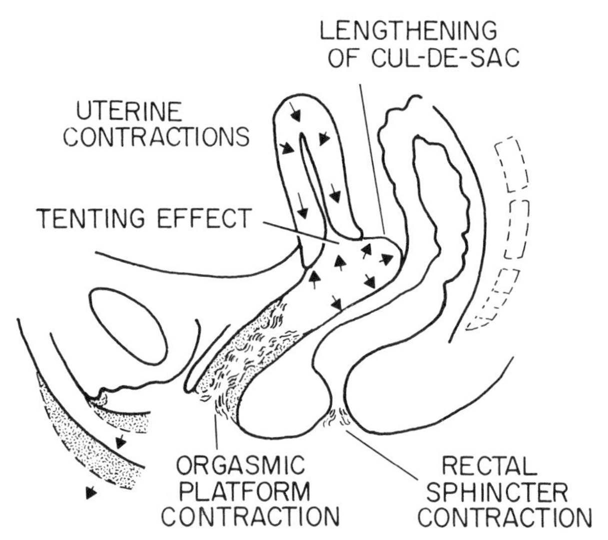

Figure 12.6. Midsagittal view of the female pelvis shows changes which take place during the orgasmic phase. (After Masters and Johnson, 1966)

opening remains gaping for up to 30 min (*tenting*). The uterus also returns to its normal descended position, during which time the cervix dips into the portion of the vagina containing the most of the seminal fluid (Figure 12.7). Both labia return to the preexcitement size, position, and color (Figure 12.8).

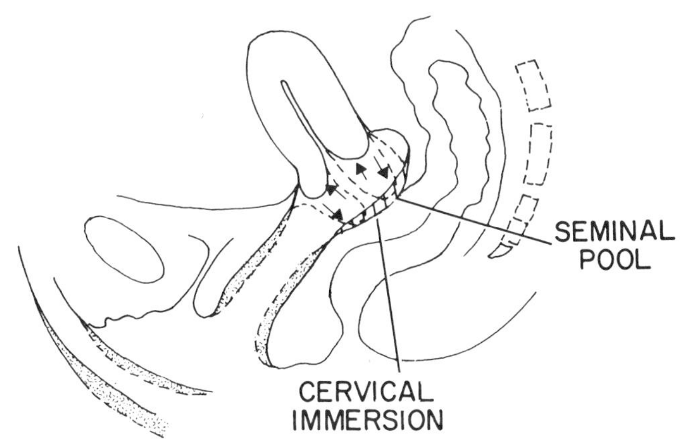

Figure 12.7. Midsagittal view of the female pelvis shows how the orgasmic platform aids in confining the seminal pool into which the cervix dips during the orgasmic phase. (After Masters and Johnson, 1966)

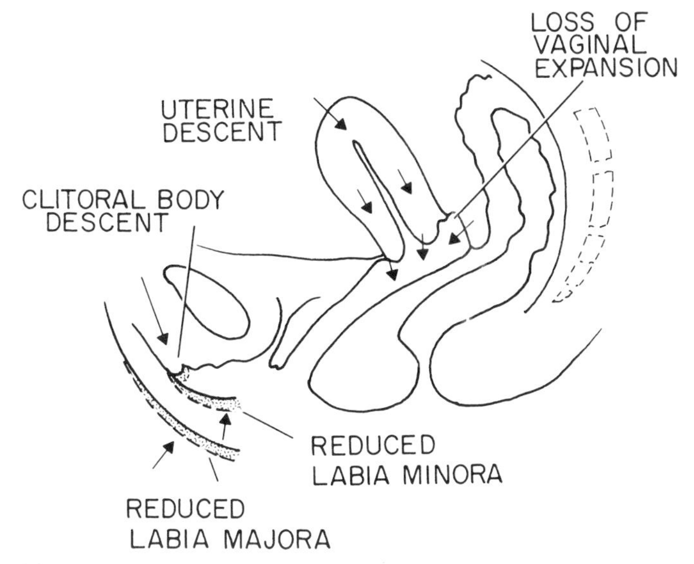

Figure 12.8. Midsagittal view of the female pelvis shows the return to the unstimulated state during the resolution phase. (After Masters and Johnson, 1966)

FEMALE ERECTILE TISSUE

Clitoris

The radiating center of the orgasm in women is the clitoris. This is the chief erectile organ of females and corresponds to the penis of the male. The other major erectile organs of females are the bilateral corpora cavernosa or the cavernous bodies or bulbs, which lie in the walls of the labia minora.

The clitoris comprises three parts: The glans, the body, and the crura. The first two parts are about 2–2.5 cm long, while the crus is about 5 cm on each side of the pubis; but these dimensions vary enormously from person to person. Each crus is enclosed by a very tough layer of dense connective tissue, the tunica albuginea. This envelops the crus on each side and is attached to the pubic bone. The tunica extends around the body of the clitoris, and is partially fused with the membrane of the opposite side. The erectile tissues enclosed by the tunica albuginea are fully fused in the glans. The glans and body of the clitoris are shielded to a large extent by a fold of the labia minora. This fold is essentially loose connective tissue and is continuous with the skin of the labia minora; it is called the prepuce.

Labia Minora

The labia minora consist of loose connective tissue, and contain a great interlacing network of large, probably venous blood vessels. These swell as they become filled with blood, so that the labia behave like erectile tissue.

When not stimulated, the glans and body of the clitoris are directed posteriorly and somewhat dorsally, mostly covered by the prepuce, only about one-tenth of the tissue being visible externally. During erection, none of the clitoris is visible externally, since the clitoris hugs the pubic bone and is covered entirely by the swollen prepuce (Figure 12.5A).

Each crus is partly enclosed by a thin muscle, the *ischiocavernosus*; each cavernous bulb is also enclosed by a thin layer of muscle, the *bulbocavernosus*. These muscles are probably important during erection, since by contracting around the emergent veins, they constrict the veins and promote engorgement and swelling of the erectile tissue. During the orgasm, their spastic contractions and relaxation are expressions of the orgasm, and their subsequent relaxation helps the blood vessels to empty, with the subsidence of the erection.

Vasculature

Microscopically, the most prominent feature of the clitoris is the presence of numerous large blood vessel spaces, the blood sinuses, which are closely enmeshed and interconnected. Their thin walls, called *septae*, are lined by endothelial cells, and these are backed by connective tissue. In the unstimulated state, the blood sinuses are mostly collapsed, and contain little blood. During erection, the sinuses are expanded with blood delivered from the helicine arteries, which are derived from branches of the internal pudendal artery and the internal iliac arteries. The septal walls are nourished by arterial blood from another set of very fine vessels. Blood from the sinuses empties into a plexus of veins which, when joined to veins from a second plexus of the glans, form the deep dorsal vein. This passes under the pubic ligament (which stretches between the pubic bones near the symphysis pubis) and joins the venous plexus of the pelvis. Both venous plexuses are important in erection, during which they are engorged with blood. Characteristic of the main arteries entering the clitoris and of the main veins are slight thickenings of their inner layers, which are also important for erection. When these pads are thickened in the vein, they obstruct venous flow and promote venous congestion and favor erection. When these pads are thickened in the artery, they reduce blood flow; conversely, when the arterial pads become thinner, they result in increased blood flow and favor erection.

Innervation

The innervation of the clitoris is very extensive. The sympathetic component arises from the first and second lumbar segments, and includes sensory and motor fibers. The fibers synapse with cells in the sympathetic ganglia and join the hypogastric plexus in the pelvis. The parasympathetic component arises from the second to the fourth lumbar segment as the nervi erigentes. The cells of origin lie in the dorsal root ganglia (sensory) and the ventral gray matter (motor). The nerve fibers enter into the hypogastric plexus also and there synapse with cells in the dispersed ganglia of the plexus. By this point mostly postganglionic fibers are distributed to the pelvic structures, the perineal region, and more particularly the clitoris, the cavernous bulbs and surrounding region. In the erectile tissues they accompany the deep dorsal vein of the clitoris and are distributed peripherally as these vessels branch. Although this seems to be the general pattern of nerve distribution, the exact nerve cells of origin or their connections and the exact course of the fibers involved in the action of any specific part of the erectile tissues are not known.

As expected, afferent and efferent nerve endings are prominent. All types of nerve endings present in nonsexual skin are present in the skin of the clitoris and neighboring erogenous zones, since the same sensory modalities are common to both regions. These are: touch, pain, cold, warmth, and physical displacement. There do not seem to be any sensory bodies specialized as "genital" bodies. Deep sensory (afferent) endings are found mainly as muscle endings in muscles (in the bulbocavernosus and ischiocavernosus muscles) and as Pacinian corpuscles or bodies. These are also to be found in other sites and are commonly assumed to monitor blood flow through the region. They are sensitive to physical displacements or distortions as small as 1μ. They are very common in the nerve plexuses that surround the blood vessels of the erectile tissue, especially at the points where the blood vessels branch. The efferent endings are associated with smooth muscle of the arteries and veins of erectile tissues and with the end plates of the muscles named above. The clitoris, more often than not, is richer in nerves, both afferent and efferent than other parts of the external genitalia. By contrast, there are very few endings in the vagina of most women. This is in accord with the greater importance of the clitoris in sexual arousal of most women. In one exceptional woman of a series studied for this purpose, however, there were few afferent endings in the clitoris, but many in the labia minora. In the sheep and ox more nerve fibers were counted at the base of the clitoris than were counted at the base of the penis. It is possible that the same is also true for women and men.

The Clitoris and the Orgasm

The basic physiologic factor in erection is the engorgement with venous blood of the erectile tissues (clitoris and cavernous bulb) and veins of neighboring parts of the external genitalia as well as of pelvic tissues. This results in the leakage of plasma components into the surrounding connective tissues and the consequent swelling associated with edema. Rapidly after orgasm, or more slowly in the absence of orgasm, the veins and erectile tissues empty their excess blood, the edemal swelling subsides as the excess water of the connective tissues reenters the circulating blood, and the genital and pelvic tissues return to their unstimulated condition.

Clitoral erection may be induced through fantasy, by breast stimulation, rectal stimulation, clitoral and mons veneris stimulation and by vaginal coitus, alone or in combination. The clitoris is hardly ever stimulated directly by the penis during vaginal coitus. The clitoris is stimulated mainly indirectly through intermittent displacement of the labia minora, which surround the vaginal opening almost completely. As the labia are attached loosely to the prepuce, there is rhythmic tugging on the prepuce which slides over the clitoris and stimulates it rhythmically and gently. The mechanism of clitoral erection and of bulbar enlargement has not been worked out in detail.

ORGASM IN MEN

Excitement Phase

The nipples may be erect. Voluntary muscles are tensed, especially those of the abdomen and thorax. The testicles are partially elevated by shortening of the spermatic cord, and the scrotal skin may show a slight thickening and wrinkling, as the scrotal sac is raised by contraction of the cremasteric muscles. The penis is elongated and engorged with blood to an increased diameter (Figure 12.9).

Plateau Phase

Nipples, if erect, may swell. Some men develop a cutaneous flush, beginning in the anterior part of the ventral abdominal wall (the epigastrium), and the area involved may spread anteriorly to include the head and neck as well as the extremities. Voluntary muscle tension is increased, and there may be semispastic contraction of facial, abdominal, and thoracic wall musculature. The heart rate, blood pressure and (late in the phase) respiratory rate all increase.

The coronal ridge of the glans is more prominent in the erect penis. The testes are also congested and elevated. Two or three drops of mucoid fluid originating from the bulbourethral glands may be released from the urethral orifice—sometimes they contain motile sperm (Figure 12.10).

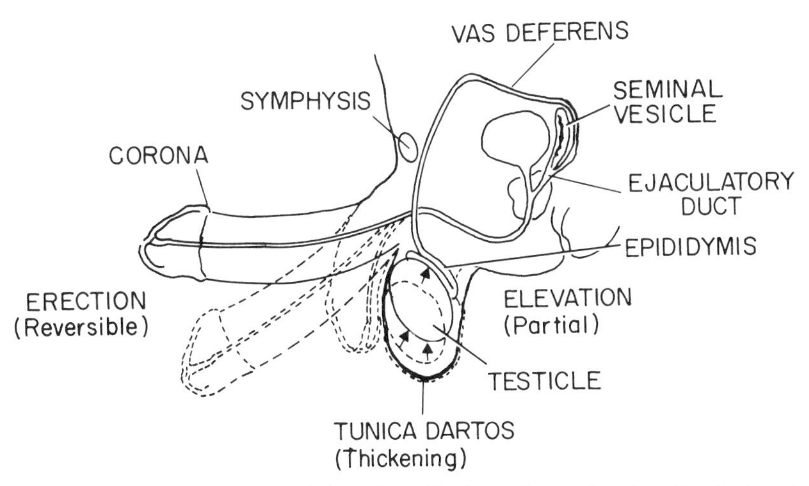

Figure 12.9. Near sagittal view of the male pelvis shows changes in external and internal relations during excitement phase. (After Masters and Johnson, 1966)

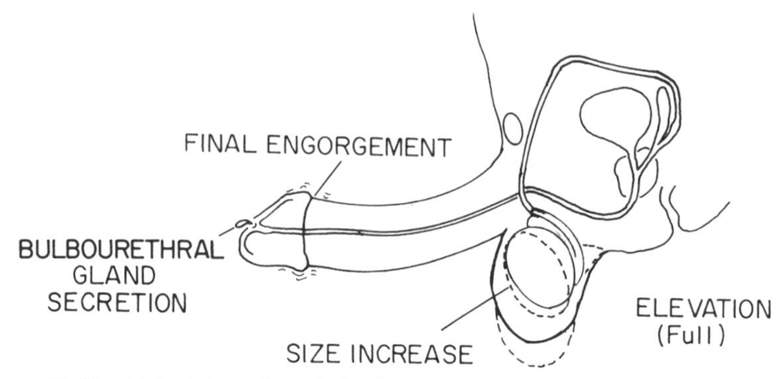

Figure 12.10. Male plateau phase. (After Masters and Johnson, 1966)

Orgasmic Phase

The cutaneous flush, when it occurs, is fully developed. Involuntary contractions and spasms of muscle groups occur, accompanied by involuntary contractions of the anal sphincter at more or less regular intervals of 0.8 sec. Cardiac rate, blood pressure, and respiratory rate reach their peaks.

The smooth muscle of the secondary sex organs (seminal vesicle, seminiferous duct, ejaculatory duct and prostate gland) contract, expelling their contributions to the semen. The contractions of these muscles develop the sensation of impending ejaculation which initiates the process. The sphincters of the urinary bladder are contracted, thus preventing the backward flow of semen into the urinary bladder.

The penile urethra contracts rhythmically throughout its length, at about 0.8-sec intervals, and expels the semen. The first three or four contractions are accompanied by more forceful thrusts, which thereafter become more irregular and weaker, as the urethral contractions become weaker and less frequent (Figure 12.11).

Resolution Phase

Slow or rapid return of nipples to their preexcitation state accompanies the rapid disappearance of the cutaneous flush, if these changes took place in earlier stages. Muscle relaxation takes place, as the cardiac rate, blood pressure, and respiratory rate return to preexcitation levels.

The penis becomes less rigid and less swollen, and returns to its unexcited state slowly. The vascular swelling of the testicles and of the scrotal sac, as well as of the organs in the floor of the pelvis, regresses and the normal relations are reestablished, ending with the full descent of the testes in their scrotal sac (Figure 12.12).

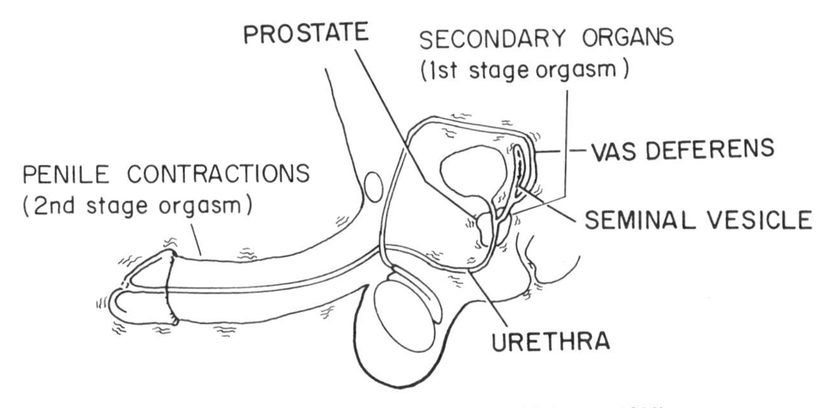

Figure 12.11. Male orgasmic phase. (After Masters and Johnson, 1966)

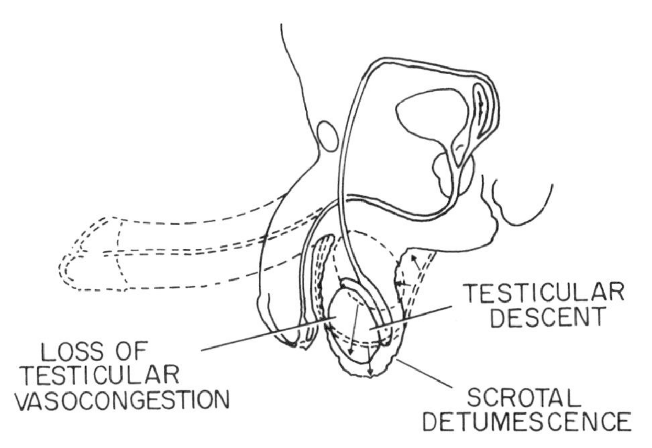

Figure 12.12. Male resolution phase. (After Masters and Johnson, 1966)

COMPARISON OF ORGASM IN WOMEN AND MEN

The orgasm in women is described in greater detail than in men. But enough has been given to make clear the similarities of at least the clinical observations of the orgasm in the two sexes. This is as true of the systemic or extragenital events as of the genital events. This is not really surprising in view of the similar embryologic origin of the various tissues and organs involved, and of their similar neurologic connections (sensory and motor, voluntary and autonomic) with the central nervous system.

Although the relaxation and release from vascular congestion take place similarly in women and men, there are subsequent differences. Women return to normal somewhat more slowly, and it may be for this reason that they may reenter the plateau phase at a stage somewhat advanced over that of the first encounter. By contrast, men enter a refractory period before they enter on a second encounter. Women can have many successive orgasms until they are physically exhausted, whereas the number of successive orgasms of men is markedly restricted.

The psychologic-emotional aspects of the orgasm presumably are more or less similar in the two sexes. The major differences that affect the pleasure level achieved by individuals of the two sexes are probably cultural and psychologic. In a rapidly changing world, these differences are to some extent being moderated to the extent where it is no longer "necessary" or desirable for the man to express aggression and dominance sexually, or for the woman to be passive during coitus. The elements of pain and fear of pregnancy need not intrude into what can be an unalloyed pleasure. Sharing of the pleasure and pride through verbal and nonverbal communication before, during, and after the whole coital act would then contribute to and intensify the pleasure.

NEED FOR FURTHER STUDIES

Women of the present day are the first generation in the history of womankind to be able to separate sexuality and sexual pleasure from the restraints of pregnancy and of the menstrual cycle. They can and some do engage in lovemaking from the time of puberty to the time of menopause and after. Understanding the orgasm can lead only to a greater enjoyment of sexuality. This is not to say that there is enough understanding of the orgasm. In fact, the clinical observations in women are far in advance of the understanding of the neurologic bases for the sensory and motor aspects involved, not to speak of the role of the central nervous system. Only by extension of a great amount of work on laboratory animals do some of these features become understandable. Even less clear are the roles played by the mind: the drives, the emotions, the dream world, and the fantasy world.

REFERENCES

Freud, S. 1905. Three essays on the theory of sexuality. II, Infantile Sexuality. *In* The Complete Psychological Works of Sigmund Freud, Vol. VII, pp. 173-206, 1953. Hogarth Press, London.

Lowry, T. P., and T. S. Lowry. 1976. The Clitoris. Warren H. Green, Inc., St. Louis.

Masters, W. H., and V. E. Johnson. 1966. Human Sexual Response. Little, Brown & Co., Boston.

Masters, W. H., and V. E. Johnson. 1970. Human Sexual Inadequacy. Little, Brown & Co., Boston.

13 Pregnancy and Contraception

To be a mother is, or can be, a matter of choice in most Western countries, since sex and reproduction are now more readily separable. Also a matter of choice is the limitation on the number of children a woman has, as well as their spacing, and even the sex of the prospective child. Choice has become possible in recent years with the introduction and spread of contraceptive techniques, which are discussed in general terms at the end of this chapter. Most women and men are not even aware of the possible motivations for having children, much less of their own reasons. Such awareness is the other side of the coin of choice. Motivations of this sort are summarized by Berelson (1974).

PREGNANCY

Pregnancy Tests

The first sign of possible pregnancy is the cessation of menstruation, brought about by persistence of the corpus luteum and its continuous secretion of progesterone and estrogen. A second sign is a positive test for urinary chorionic gonadotropin.

After conception, the corpus luteum does not degenerate, as occurs in the normal menstrual cycle, but persists and grows in diameter from about 1.5 cm to as large as 3 cm in the third month of pregnancy, after which it degenerates. It continues to secrete progesterone, which inhibits the development of new Graafian follicles and thus prevents further ovulations. The hormone also prevents menstruation, and promotes the continuance and growth of the endometrium. This hormone from the corpus luteum, as well as others from the anterior lobe of the hypophysis, is supplemented by progesterone and human chorionic gonadotropin (HCG) from the placenta, which is described in the next chapter. HCG is detectable in maternal blood as early as 8 to 10 days after fertilization. These placental hormones (and others to be discussed shortly) are synthesized and secreted by trophoblasts. Trophoblasts are extraembryonic cells that are embedded in the maternal endometrium and form part of the placenta. They are derived from the outer layer of the developing embryo in a very early (blastula) stage.

Human chorionic gonadotropin is excreted in the maternal urine, and its presence is detected by an immunologic method that is rapid, sensitive, and convenient, so that it can be performed in the doctor's office or at home. In this procedure, purified gonadotropic hormone is used to produce antibodies against it in a rabbit. These are concentrated, and a drop of a clear solution of the antibody is added to a drop of test urine. If the urine contains human chorionic gonadotropin (the antigen), it combines with the antibody to produce a cloudy precipitate, which constitutes a positive test. Or the antibody could be attached to red blood cells or to microscopic latex particles; in the presence of HCG (in the urine) the red blood cells or latex particles aggregate to form a precipitate.

Physical Changes During Pregnancy

Uterus The muscle layers of the uterus increase in bulk about twenty-five times during pregnancy, and the negligible internal capacity of the nonpregnant uterus increases grossly at term to accommodate the fetus, the fetal membranes, and the amniotic fluid. These increases are not simply an effect of stretching, as if it were a balloon, because the internal pressure remains virtually the same. The tremendous enlargement is achieved perhaps partly by an increase in number of the uterine muscle cells (perhaps due to increased progesterone in the maternal plasma), but mostly because of the increase in size of the uterine muscle cells due to higher levels of estrogen in the maternal circulation. This is accompanied by increased formation of connective tissue and vascularity. Growth of the uterine tissues ceases about ten days before term.

There is no externally visible or palpable indication of pregnancy until the sixth week, when a skillful obstetrician can palpate the enlarging uterus. While there is a considerable variation in different women, the first external sign is visible at about 4 months. At about 6 months the uterus reaches the navel, and at about 9 months the uterus reaches the xiphoid process which is attached to the lower portion of the sternum or breastbone. During the last month, as the head (or in rare instances the buttocks) of the fetus drops into the pelvis, the external form of the mother's abdominal wall is reshaped to accommodate the altered position of the fetus in the uterus (Figure 13.1). The increased size of the uterus compels some portions of the gastrointestinal tract to be pushed into unaccustomed positions; it also raises the diaphragm and thus affects the heart and respiratory movements. The added weight bearing down on the pelvis may reduce distention of the rectum and urinary bladder, and may

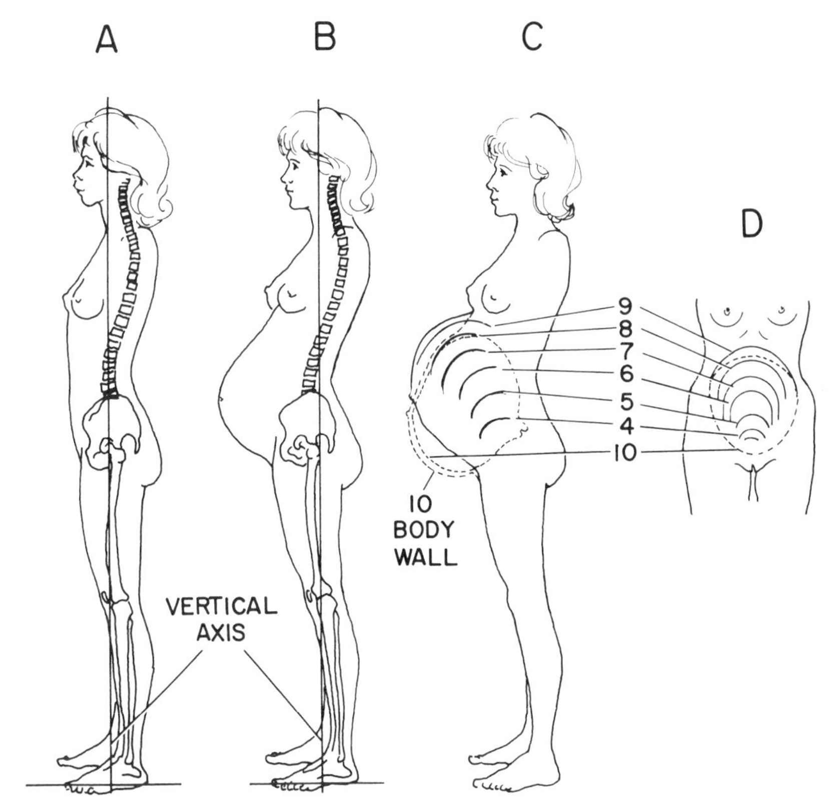

Figure 13.1. Outline diagrams of nonpregnant (A) and pregnant (B) woman show adjustments in posture associated with enlarged abdomen with changes occurring from the ankles up to the shoulders and neck. Without these adjustments, the woman would fall forward. In C are shown the extent of the enlargement of the abdomen from the third lunar month of pregnancy. During the last month, when the fetal head is engaged in the pelvis, there is a notable fall in the abdominal enlargement. (After Corliss, 1976, and Greenhill, 1966)

compress pelvic blood vessels, particularly veins. These physical changes contribute to constipation (associated with the retention of the feces and the longer time for water reabsorption to take place), frequent urination, and edema in the legs because of the venous compression in the pelvis.

Cervix and Vagina Changes in the cervix and vagina are not as drastic as those in the uterus. The cervix becomes softer and dilates, in part perhaps because of the action of relaxin secreted by the corpus luteum and the placenta. The lumen is occluded by a thick, viscous plug, which serves to protect against infection. The walls are more richly vascularized, with consequent edema and more collagen is deposited. The vagina is edematous, and the lumen is increasingly moist. The surface fluid contains flecks of a white, curdlike pasty material which consists of desquamated cells and bacteria. The epithelium is thickened, the surface becomes rougher, and the cells contain more glycogen. The muscle cells increase in size.

Breasts The breasts also undergo changes in their external form and internal structure during pregnancy. They are larger, and more full, with a more rounded upper surface (Fig. 13.2). The nipples and areolae are also larger. After lactation, the breast returns to a size larger than before pregnancy, while the lower surface is more pendant so that the nipple appears to be at a lower level. It and the areola are reduced and approach the prepregnant state. The internal structural changes during pregnancy involving the fat tissue, connective tissue, blood vessels, ducts, and secretory gland cells are described in Chapter 15.

Skin In the second half of pregnancy, stretch marks appear in about half the women, on the abdomen, breasts and buttocks. These are reddish streaks. After parturition, the reddish marks are replaced by silvery lines, which are permanent. The pigmentation of the skin of the breasts and genitalia is more marked during pregnancy; and irregular pigmented patches may appear on the forehead and cheeks, but these disappear after parturition. Growth of hair is also affected (see Chapter 17).

Physiologic and Biochemical Changes

Many physiologic and biochemical changes also take place in the pregnant mother. Most notable is the increase in weight. Nearly half of the weight increase is to be attributed to fetus, placenta, and amniotic fluid. The remainder is attributed to the uterus and the breasts of the mother, the increased blood volume and tissue fluid of the pregnant woman, and the storage chiefly of fat (Figure 13.3). The amount of fat

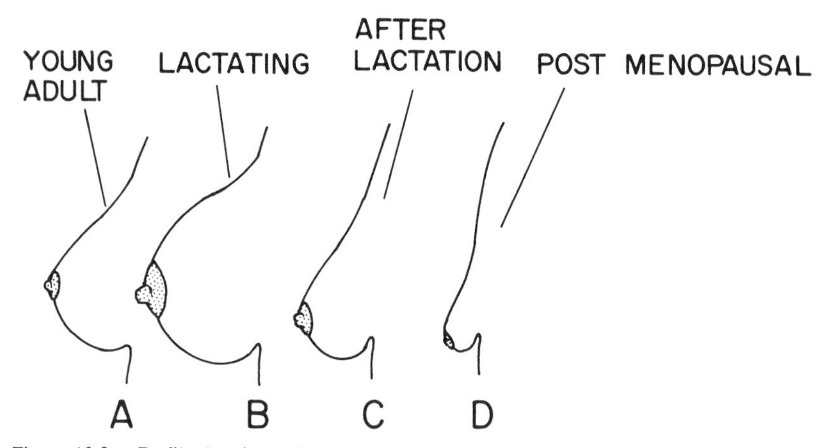

Figure 13.2. Profile drawings of the breast at different stages. (After Corliss, 1976)

added is sufficient to sustain fetal growth during the last trimester of pregnancy, and to subsidize lactation. Both factors are still important in marginal economies, such as those in India, West Africa, and Southeast Asia. For many Westerners, at the present time (though not for the vast numbers at or below the poverty level), they are obsolete relics of natural selection. The amount of water (and electrolytes) added is quite considerable—6.8 liters in women without edema, 7.2–9.8 liters in women with varying degrees of edema.

The plasma volume begins to rise at the end of 3 months, reaches a peak at about 34 weeks, and then falls slowly to term. The increase is considerable (630–1940 ml, depending on fetal size). There is also an increase in the red blood cell volume, which may reach 400 ml or about 18%. As the increase in the plasma volume is greater than the increase in red blood cells, the woman is said to have a *physiologic anemia of pregnancy.* The plasma albumin concentration falls even more than immunoglobulins in-

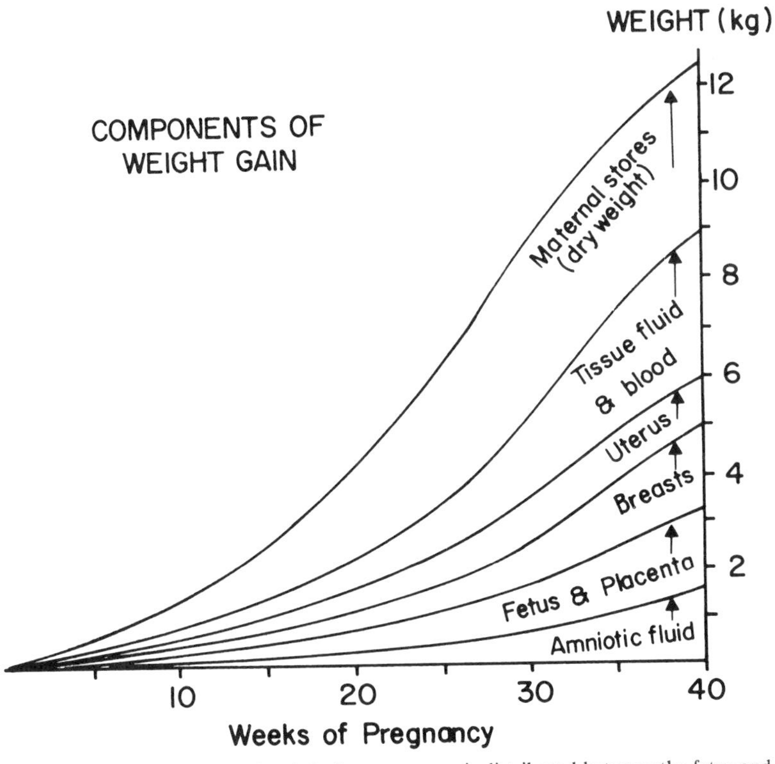

Figure 13.3. Weight accumulated during pregnancy is distributed between the fetus and its membranes and the mother. Most of the maternal stores are considered to be fat tissue. (After Hytten and Thomson, 1968)

crease, and this results in a reduction in the plasma osmotic pressure, which can contribute to edema of the legs. Cardiac output increases significantly during pregnancy, both because of the increased heart (pulse) rate and the larger stroke volume. The consequence is that more oxygen is returned to the heart with each cycle during pregnancy. This, along with an increase in respiration rate and alveolar ventilation in the lungs, supplements the effect of the increased number of red blood cells. In addition, the increased cardiac efficiency tends to compensate for the increased venous resistance to blood flow in the pelvic veins. There is also increased blood flow in various organs of the body—the uterus, from 200 ml/min at 28 weeks to about 500 ml/min at term; the kidney, from a nonpregnant level of 900 ml/min to 1200 ml/min at 10 weeks, with a slow return to nonpregnant levels during the last 10 weeks; the forearm, about 2.5 times the nonpregnant level at the end of pregnancy; also the blood flow in the breast and the gastrointestinal tract increases notably.

Renal function improves, and the digestive system is unusually efficient. Perhaps owing to the high progesterone level and its nonspecific effect in causing relaxation of the smooth muscle of the gastrointestinal tract, there may be periodic relaxation of the esophageal sphincter followed by heartburn. Perhaps also for the same reason, fecal material is not moved along as rapidly as usual, which is another cause of constipation.

Nutrition

Adequate nutrition during pregnancy is of great importance, because it is a major factor in determining the birth weight of the newborn. The survival rate of newborns with low birth weight is lower than that of babies of normal weight. Moreover, there is the possibility of irreversible fetal brain damage, such as has been described in other species when the mother was underfed. Adequate nutrition is especially important for adolescents who are pregnant, for their diet must be sufficient for their own continued growth, as well as the growth of the fetus and its membranes. The nutrients of greatest concern are protein and iron. It is not only a question of poverty, it is also a question of adequate and suitable prenatal care and education. Birth weight is not determined solely by nutritional factors—it is also determined by parental genotype, race, and age. Despite the number of variables concerned, most women deliver their baby at 266 days ($\pm$11 days) postfertilization, when the time of fertilization is presumed to be 2 weeks after the onset of the last menstrual period preceding pregnancy.

The changes in mood, interpersonal relationships, fears, and anticipations are to a large extent socially, economically, and culturally imposed. These aspects, which are of enormous interest, are not treated in this book but are analyzed in books on psychology, psychiatry, and social science.

Delivery

Parturition is the expulsion of the fetus and placenta. The factors that trigger the whole complicated process are not clear—it may be initiated by 1) the fetus, 2) a preset aging process of the placenta, or 3) a sharp change in the maternal hormone balance (including oxytocin) and some one or more of the prostaglandins. Both of the latter may cause uterine muscles to contract. The delivery of the baby is divided into three stages:

1. From the beginning of labor to the complete dilation of the cervix, when it is flush with the vagina. Labor begins when the persistent, rhythmic contractions of the uterine muscle exceed three per half hour, accompanied by progressive dilation of the cervix.
2. Complete cervical dilation and the appearance of the fetus.
3. Delivery of the baby, placenta, and fetal membranes.

There is a growing movement to engage both prospective parents in the delivery of the child, and several internationally practiced methods are available. As a part of this tendency, there is also a desire for birthing at home, or in a more informal setting if in a hospital. This also implicates the services of midwives, who were almost entirely replaced by doctors in Western societies until recently. The prospective parents learn enough about the process to make choices and cooperate intelligently with doctors or midwives during the delivery process. This includes, for the mother, dietary control, exercise, breathing exercises to prevent overventilation, and instruction on relaxation. The prospective parents also come to share in the tasks of caring for the baby, and anticipate the pleasure of closer bonding with the baby. Sometimes medical complications require the active intervention of an obstetrician, for instance, in the performance of a cesarian section, an operation in which the fetus is removed through an incision in the ventral abdominal wall and the uterus.

After the fetus is expelled, there is a short interval before uterine contractions are resumed. The placenta is then expelled with very little bleeding. The reason is that the very strong contraction of the uterine muscle closes the numerous broken arteries and veins, greatly reducing hemorrhage. There may be some oozing of blood for 2 to 4 weeks.

Puerperium

This shedding of the placenta closes the period of labor, and begins the puerperium—the healing and return to the nonpregnant state of the uterus. The uterus at term weighs about 1 kg. Within a week, it regresses to about 500 g; within two weeks, 350 g; within 8 weeks, 50–70 g, which is close to normal. The endometrium is regenerated from small islets of uterine glands which remain in the basal layer after expulsion of the placenta. The bits of tissue not expelled are removed by lysosomal activity released by leucocytes and phagocytes that have wandered in. Menstruation usually recurs after about 3 months, though nursing may prolong the period of amennorhea; but ovulation may take place much earlier and another pregnancy may occur.

The cervix regains its shape in 1 day, and the internal os (opening) is nearly closed in 2 weeks. The external os appears as a transverse slit in 4 weeks. At the end of 3 to 6 weeks, the muscle and connective tissues of the pelvic region regain their normal tone, and the surface circular folds of the vagina reappear. Glycogen disappears from the vaginal epithelium at 2 weeks postpartum, and reappears when ovulation is resumed. The epithelium of the oviduct shows minimal changes and remains in the luteal condition throughout pregnancy.

All the various physiologic adaptations to pregnancy discussed above are reduced to normal levels in the course of time. These include weight, stored fat, blood volume and cardiac output, red blood cell count, and water and electrolyte content (and, with them, edema).

LACTATION AND THE RESUMPTION OF OVULATION

When the placenta is shed, the hormones it produced are lost and are no longer synthesized. The basophil cells of the anterior lobe of the hypophysis, which secrete prolactin, are reduced in number. The cyclic behavior of the hypothalamus is resumed, and the ovarian and menstrual cycles are also resumed. The time when the normal cycles are resumed varies in different women, and according to whether lactation was engaged in or not, illustrated in Figure 13.4. Normal cycles are resumed even faster after abortion, because of the absence of lactation.

CONTRACEPTION

For a variety of reasons, many women seek some form of contraception in order to continue their sex life without the threat of pregnancy. The ideal contraceptive should be certain to prevent fertilization without damage to either partner or the subsequent offspring (planned or unplanned) at some later time, and should be convenient and reversible. No presently available method of contraception is entirely acceptable. In general terms, the aim of contraception is to prevent ovulation, fertilization and/or implantation of the developing embryo. While abortion is commonly used to prevent a childbirth, it is not included in the term *contraception*.

The effectiveness of the various methods of contraception listed below

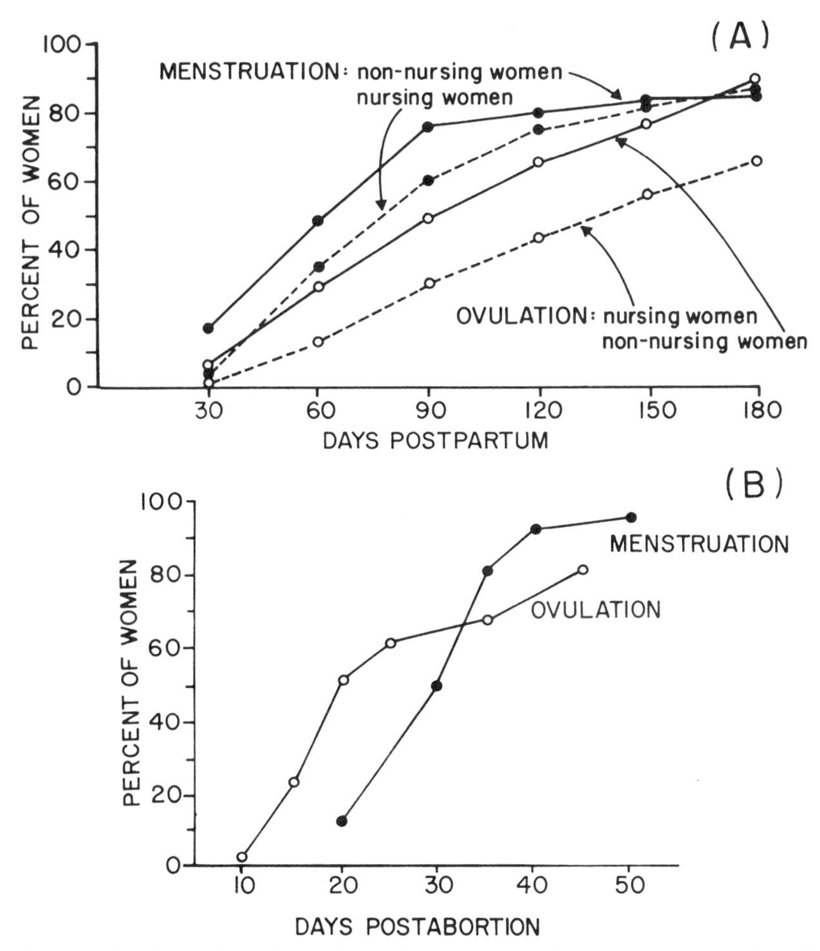

Figure 13.4. Resumption of ovulation and menstruation after pregnancy or abortion. A) breastfeeding versus nonbreastfeeding mothers. B) after abortion the hormonal changes associated with pregnancy are suddenly cut off. (After Vorherr, 1974)

is compared by means of two statistics. The theoretical effectiveness (TE) is indicated by the number of pregnancies during the first year of use per one hundred nonsterile women who have been screened to ascertain that the method is used correctly and consistently. The actual effectiveness (AE) is the average experience among women who are motivated but who may have lapses of some sort. The AE is indicated by the number of pregnancies during the first year of use per 100 nonsterile women (in this case unscreened) using the method. Because these statistics really measure the failure, rather than the success of the method, the index of actual effectiveness is almost always greater than the TE, and the greater the effectiveness, the lower is the index.

Coitus Interruptus (Withdrawal of the Penis Prior to Ejaculation)

Timing is of the essence, and it may not be sufficiently well controlled to prevent some semen from escaping before the penis is withdrawn, so an amount sufficient to result in pregnancy might be deposited in the woman's external organs. The psychologic effects of this method may be quite frustrating. TE—9; AE—20-25 (i.e., the method failed in up to 25% of women).

Postcoital Douche to Dilute and Kill Remaining Sperm

The rapidity with which sperm penetrate into the cervix and cervical clefts accounts for the poor record of this method. AE—40.

Prolonged Lactation

While menstruation is postponed for a variable time in different women, ovulation may return at any time and even precede the return of menstruation (see Figure 13.4). TE—25; AE—40.

Condom

Sperm are prevented from entering the vagina except when imperfect or friable condoms are used or sometimes when withdrawal is overly delayed. The aesthetics of the situation before and during coitus may also interfere with the full enjoyment of sexual intercourse. TE—3; AE—10. When a spermicidal agent is also used, the effectiveness index decreases to <1 and <5, respectively.

Vaginal Diaphragm

A diaphragm with a spermicidal jelly or cream for better sealing off the cervix from the vagina, and thus acting as an additional barrier to the passage of sperm into the cervix is a reliable method requiring some foresight and high motivation. Very rarely, poor fitting by a physician or health worker or incorrect insertion may cause a failure of this simple method. TE—3; AE—17.

Spermicides without Diaphragm

This method is less effective than the preceding one. Failures may be caused by an inadequate amount or quality of the agent—which can be contained in jellies, creams, foams, suppositories—or by shortening the prescribed time for the agent to be thoroughly distributed. TE—3; AE—22.

Rhythm Method (Calendar and Temperature)

The first is based on the individual woman's menstrual history over several months; the second is based on basal body temperature changes during menstruation (see Figure 11.2 and legend). Abstinence is prescribed for a few days before and after ovulation in the first method, and for the preovulatory period and some days after ovulation in the second. The well-known irregularity of menses and the difficulty of reading correctly the fractional temperature changes are responsible for many failures. Other difficulties involved are accuracy in interpreting the cyclic chart, and the psychologic factors of self-control and motivation. TE—13; AE—21 (calendar rhythm only).

Oral Contraceptives

In the dosages used, the artificial steroids suppress ovulation. There are two distinct regimens, combined and sequential. Under the combined regimen, 20 or 21 tablets containing a progestin and estrogen are taken from the fifth to the twenty-fourth or twenty-fifth day of the menstrual cycle. Under the sequential regimen, tablets containing estrogen only are taken for 15 or 16 days, after which tablets containing progestin and estrogen are taken for 5 days. Suspension of medication usually results in "menstrual" or withdrawal bleeding. On the fifth day of the new cycle, self-medication is resumed. This method is almost completely effective when the regimen is adhered to, but there are a great variety of side effects. These include disturbances in the function of the liver and thyroid gland. Early in their use, there may be nausea, vomiting, breast enlargement, headache, dizziness, weight gain, and breakthrough bleeding. A very serious complication, thromboembolic disease, is caused by blood clots that form in the veins and are carried to the heart, lungs, and brain with possibly very serious consequences. Jaundice, hypertension, obesity, depression, and lack of libido can also be caused in susceptible individuals. Research has not yet ascertained whether the pill causes cancer if continued for many years. One or more of the complaints in this formidable list of possible side effects may afflict any one woman, but, statistically considered, the number of women actually affected is small. TE—0.34; AE—4-10.

The fluctuations in hormone levels during the menstrual cycle are simulated by the use of contraceptive pills. The dosages employed are such as to override the normal cyclical variations. They are effective by turning off the FSH/LH-RH of the hypothalamus, thus suppressing the production and release into the blood of FSH and LH. This leads to a cessation of follicular growth and maturation in the ovary, resulting in a cessation of ovulation. When the pill is not taken, the hormone levels fall, and the same changes take place in the uterine wall, which leads to menstrual bleeding. Resumption of pill-taking raises blood hormone levels and initiates the next cycle. Some women profit from this enforced regularity of hormone self-administration in that their bleeding periods become more or less predictable. In some women, the length of the bleeding time or the amount of blood lost is reduced; the flow may even stop altogether.

Intrauterine Devices (IUDs)

Of various sizes, shapes, and materials, they are inserted into the uterus. Like oral contraceptives, IUDs are effective, safe, and acceptable for most couples; both methods also share the advantage that their use is dissociated from coitus. Although it is not certain how they work, they may act by causing the appearance of many leucocytes in the uterine fluid or perhaps by the release of prostaglandin. When these leucocytes degenerate, they release substances that may be toxic to sperm and/or preimplanted embryos. Prostaglandins in excess may also interfere with implantation. This method is about as effective as the use of the diaphragm, but less effective than the use of oral contraceptives. There are some side effects. These are mostly local but may be serious: bleeding or spotting, pain including cramps and backache, pelvic inflammatory disease involving infection, and very rarely, uterine perforation. TE—1-3; AE—5.

Surgical Intervention

This usually irreversible form of birth control involves cutting, ligation, and removal of a portion of the oviduct (tubal sterilization) or of the vas deferens (vasectomy). Some individuals are psychologically affected by the realization that the birth control achieved may be permanent or a perceived sense of lack of virility or femininity accompanying loss of fertility. Tubal ligation: TE—0.04; AE—0.04. Vasectomy: TE—0.15; AE—0.15+. Hysterectomy (the removal of the uterus, with the ovaries remaining intact): TE—0.0001; AE—0.0001.

Sterilization should not be entered into lightly, since it is considered to be irreversible and permanent (even though the operation is indeed reversible in some people). Accordingly, following numerous exposures of sterilizations of poor and minority women under unethical conditions, the laws have been tightened and attempts made to enforce them. Women are assured of informed choice and consent procedures. The partner need not consent. Women must be 21 or over and mentally competent, if the expenditure of federal funds is involved. Also, when federal funds are involved the government requires that there be counseling in the minority patient's language with a friend or relative present, followed by a waiting period of 1 month.

Use of sterilization has been escalating during recent years, especially among women, who outnumber men who have this operation performed. Undoubtedly this reflects in large part the fact that many parents with one or two children do not want any more.

The surgical procedures for men and women are simple, and can be performed under local anesthesia. In vasectomy and tubal ligation, the vas deferens or oviduct, respectively, is exposed and ligated or cauterized, thus blocking the passage of sperm and preventing fertilization. In the woman, menstruation and ovulation continue, but the egg cannot be fertilized. The minute figure for theoretical and actual use effectiveness represent the rare instances when the blockage was incomplete or undone following the recovery period.

These and other methods of contraception are being studied in the greatest detail in the laboratory and clinic in order to reduce the side effects of methods now in use and otherwise to improve their effectiveness, and to broaden the effects to include postcoital treatments. At present these consist of high doses of estrogens administered the morning after coitus or a very limited time after that.

Abortion

Abortion is used in some cultures as a birth control measure. In some Western cultures, abortion is considered a matter of choice by the woman pregnant. In this country abortions are legal and may not be prohibited by the state during the first third of pregnancy, and can be decided on by the woman involved and her doctor. From then up to 24 to 28 weeks of pregnancy, the states can establish regulations to make abortions safe in ways reasonably related to health, but cannot prohibit them. After this period, the fetus may be viable after separation from the mother, and the states can regulate abortions restrictively. This decision of the Supreme Court was handed down in 1973 and was followed by a dramatic fall in mortality following abortion, presumably because of improved surgical conditions. The benefits were to a great extent shortlived, since Congress through the Hyde Amendment cut off federal (Medicaid) funds for abortion at any time except when the life of the woman is endangered, or in cases of conception after rape or incest. But abortion during the first 6 months of pregnancy remains legal, if the patient can pay for it. The constitutionality of the Hyde Amendment has been challenged but has been upheld by the Supreme Court.

The medical procedure for abortion varies according to the time of conception. Up to 3 days, abortion may be ensured by taking a large dose of an estrogen (the "morning-after pill"). Up to 12 weeks, the vacuum suction method may be used. This involves dilation of the cervix, insertion of a sterile tube into the body of the uterus, and connection of the free end to a vacuum bottle. The procedure is rapid (5 to 7 min) and painless except for cramps. Another procedure is cervical dilation and curettage (D & C), in which the uterine wall is scraped with a metal loop at the end of a long, thin handle inserted through the dilated cervix, and fetal tissues are removed. This is regarded as major surgery, and the patient may be hospitalized for 1 or 2 days.

After 16 weeks of pregnancy, the saline method can be used. Some amniotic fluid is removed through the ventral abdominal wall, and is replaced by an equal volume of concentrated salt solution. Uterine contractions as strong as those of labor begin a few hours later, and after 8 to 15 hours of labor the fetus is expelled. Hypertonic urea may be used instead of saline. Prostaglandins may also be used to accelerate the process, though not without risk of side effects.

After 20 weeks of pregnancy, the fetus is removed through a small abdominal incision usually below the pubic hair line. This is considered major surgery. The patient is given local anesthesia for suction abortions, sleeping pills postoperatively for saline abortions, and general anesthesia for D & C and abdominal incisions.

REFERENCES

Assali, N. S. (ed.). 1968. Biology of Gestation, Vol. III. The Fetus and the Neonate. Academic Press, New York.

Assali, N. S., P. V. Dilts, Jr., A. A. Pleutl, T. H. Kirschbaum, and S. J. Gross. 1968. Physiology of the placenta. *In* Biology of Gestation, N. S. Assali (ed.), Vol. I. Academic Press, New York, pp. 185-289.

Basel, J. A. 1977. Factors that affect nutritional requirements in adolescents. *In* Current Concepts in Nutrition, Vol. 5, Nutritional Disorders of American Women. John Wiley & Sons, New York, pp. 53-65.

Berelson, D. 1974. The value of children: A taxonomical essay. *In* N. B. Talbot (ed.), Raising Children in Modern America. Little, Brown & Co., Boston, pp. 11-24.

Boyd, J. D., and W. J. Hamilton. 1970. The Human Placenta. Macmillan Press Ltd., London.

Briggs, M. H., and M. Briggs. 1976. Biochemical Contraception. Academic Press, New York.

Corliss, C. E. 1976. Patten's Human Embryology. McGraw Hill Book Co., New York.

Coursin, B. D. 1974. Overview of the problem. *In* M. Winick (ed.), Nutrition and Fetal Development. John Wiley & Sons, New York, pp. 1-25.

García, C-R., and D. L. Rosenfeld. 1977. Human Fertility: The Regulation of Reproduction. F. A. Davis Co., Philadelphia.

Greenhill, J. P. 1966. Obstetrics. 13th Ed. W. B. Saunders Co., Philadelphia.

Hatcher, R. A., G. K. Stewart, F. Stewart, F. Guest, P. Stratton, and A. H. Wright. 1979. Contraceptive Technology 1978-1979. 9th Ed. John Wiley & Sons, New York.

Hytten, F. E., and A. M. Thomson. 1968. Maternal physiological adjustments. *In*

Biology of Gestation, N. S. Assali (ed.), Vol. 1. Academic Press, New York, pp. 449-479.

James, V. H. T., and C. M. Andre. 1974. Androgen metabolism in the human female. *In* A. S. Curry and J. V. Hewitt (eds.), Biochemistry of Women: Clinical Concepts. CRC Press, Cleveland, Ohio, pp. 23-39.

Parfitt, R. R. 1980. The Birth Primer. Signet, New York.

Phillips, M. G. 1976. Food for the Teenager during Pregnancy. U.S. Department of Health, Education and Welfare, Publication No. (HSA) 76 5611. Superintendent of Documents. U.S. Government Printing Office, Washington, D.C.

Rhodes, P. 1971. Birth Control. Oxford University Press, New York.

Rosso, P. 1977. Maternal nutrition, nutrient exchange, and fetal growth. *In* Current Concepts in Nutrition, Vol. V. Nutritional Disorders of American Women. John Wiley & Sons, New York, pp. 3-25.

Sloane, E. 1980. Biology of Women. John Wiley & Sons, New York.

Swanson, H. D. 1974. Human Reproduction. Biology and Social Change. Oxford University Press, New York.

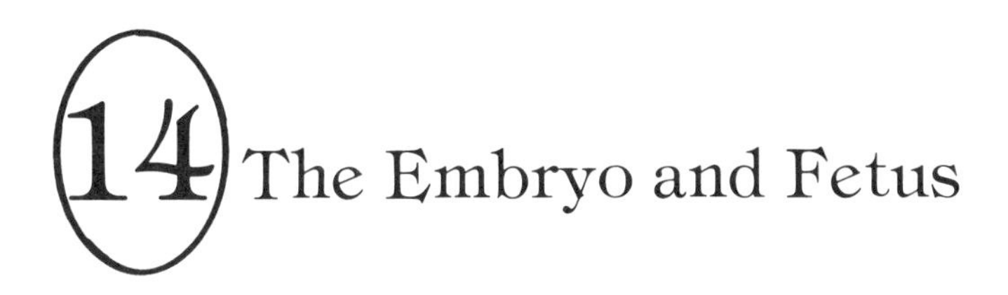

14 The Embryo and Fetus

This chapter discusses the course of development of the human embryo and fetus from fertilization of the egg to the birth of the baby, and the development and fate of the embryonic and fetal membranes.

PRENATAL DEVELOPMENT

The Egg

The ovaries arise early in embryonic life, and are populated by oogonia or germ cells that arise from the yolk sac (see Chapter 2). These multiply so rapidly that there may be nearly 7,000,000 in the 5-month fetus. The number declines to about 2,000,000 at birth. The number continues to decline until at puberty only about 500,000 remain. During this period, some oogonia go through the initial stages of the first meiotic division (see Chapter 4). After puberty, the first meiotic division is completed with the formation of the secondary oocyte and the first polar body.

During the reproductive period, some ten to twenty follicles begin to develop periodically under the cyclic influence of estrogen. But usually only one continues to develop to maturity to a ripe follicle and release its egg. The egg passes the fimbria of the Fallopian tube, and reaches the ampulla of the oviduct where it may be fertilized. Altogether, about 500 egg cells reach this level of development during the reproductive period of a woman. The rest degenerate and their remains are removed from the ovary; but a few persist even through the menopause.

The egg and adherent cells (corona) enter the Fallopian tube, and move down the oviduct during the next 4 days. It is thought to be fertilizable by sperm during the first 2 days after release from the ovary. After 4 days, it passes into the uterus where it dies if it is not fertilized, and passes out of the body with the menstrual fluid.

The Sperm

The sperm-bearing seminal fluid or ejaculate is around 3-5 ml, and contains some 200-500 million sperm. They had matured and become potentially motile in the vas deferens. The seminal fluid accumulates in a kind of vaginal pool into which dips the external os of the cervix. The sperm pass rapidly (in minutes) up the cervical canal and the lumen of the uterus to the ampulla of the oviduct where they meet the egg. The exact mechanism of sperm transport is not known, but it is thought that contraction of smooth muscle of the uterus and oviduct play an important role. The sperm penetrate the corona, and separate and disperse the coronal cells by releasing enzymes which digest the sticky cement, and penetrate the denuded zona pellucida. Timing is essential in this complicated process, for the egg and sperm are effectively viable for only a limited time, although other factors are also important, as for example, the competence of egg and sperm, the number of sperm reaching the oviduct, and the adequacy of the enzymes released by the sperm. Just before the union of the pronuclei takes place, the egg nucleus rapidly goes through its second meiotic division (see Chapter 4 and Figure 14.1).

Fertilization

The nucleus of the egg rapidly goes through its second meiotic division, the second polar body is separated (Figure 14.1). The male and female pronuclei approach each other. The DNA of each nucleus is replicated: chromosomes are formed and are aligned on the spindle, where they split longitudinally during metaphase. Then, as in mitosis, each set moves toward a pole and is enclosed in a nuclear membrane. When the cytoplasm divides, each cell contains a complete set of paternal and maternal chromosomes (the diploid number), and the genetic sex of the embryo is fixed.

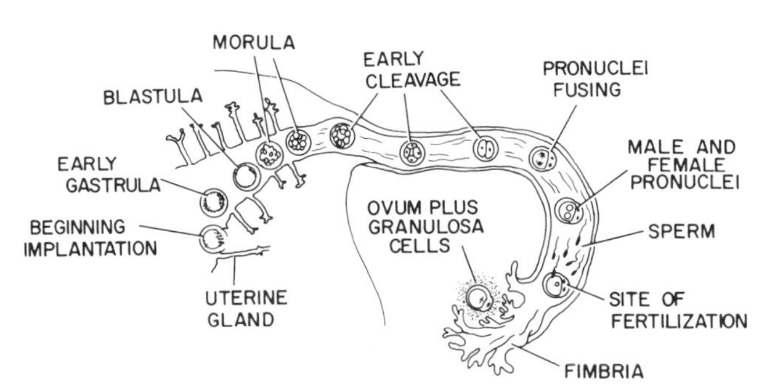

Figure 14.1. Course of development of the human egg as it passes from the ruptured ovarian follicle through the oviduct to implant in the endometrium of the uterus. (After Arey, 1954)

The Embryo

Ten hours later, the embryo consists of four cells. At 3 days, the embryo consists of about 36 cells, which are packed closely as a "solid" ball of cells called the *morula.* At 4 days, the free-floating embryo reaches the uterus. At 6 days, the embryo consists of about 150 cells and is called the *blastula.* At this stage fluid accumulates between the outer cells, which become flattened, and the inner cell mass (Figures 14.1, 14.2A, and 14.2B). The inner cell mass develops into the *embryo.* The flattened outer cells are now called the *trophoblast.* This forms the *chorion,* which develops into the *placenta.* At about this time the zona pellucida becomes thinner until it is completely digested away.

Implantation

Between 7 and 8 days, the trophoblast impinges on the surface of the epithelium of the uterus and remains attached long enough for the cells of

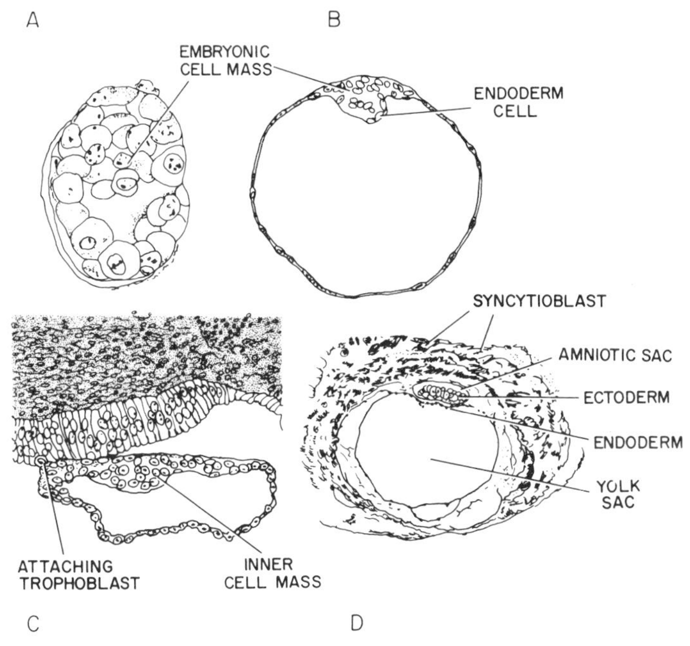

Figure 14.2. Sections of some early stages of known age in the development of the human and macaque embryo. A) Human 58-cell blastula (blastocyst) recovered from the uterine cavity. A small part of the zona pellucida remains at the bottom and left, a small polar body on the upper right. B) A 9-day macaque blastocyst floating in the uterine cavity, nearly ready for implantation. C) Another 9-day macaque blastocyst just after implantation. D) A human embryo with an estimated age of 12 days, nearly embedded in the uterine wall. (After Hamilton, Boyd, and Mossman, 1972)

the outer layer to grow between the epithelial cells and into the underlying connective tissue (Figure 14.2C). Usually this connection is made on the upper, dorsal part of the uterus, near the midsagittal plane, but on rare occasions contact is made on other sites, many of which are so dangerous to the fetus and/or mother as to require special attention by the obstetrician. Some of the sites of ectopic pregnancy are shown in Figure 14.3. The cells of the inner cell mass are rearranged to form two flat contiguous layers—one of which forms the ectoderm and the other the yolk sac (Figures 14.2D and 14.4). The yolk sac of human embryos never contains yolk, but develops a very fine network of capillaries containing primitive blood cells from which all the cells of the blood arise later. The yolk sac is

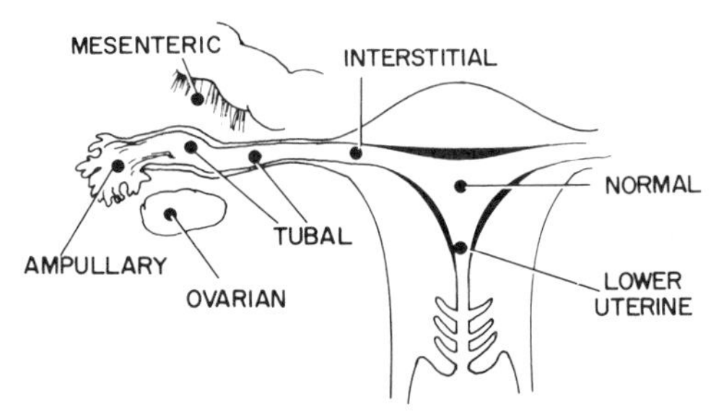

Figure 14.3. Normal and some abnormal sites of implantation. Other sites include the peritoneal wall of the abdominal cavity, including the mesentery. (After Hamilton, Boyd, and Mossman, 1972)

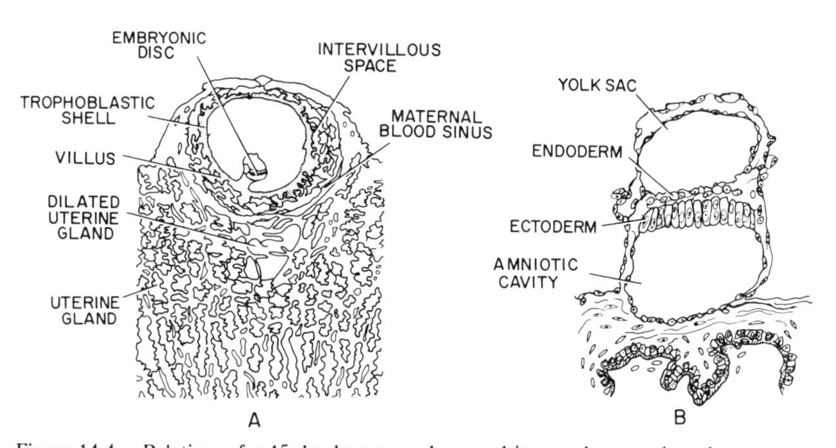

Figure 14.4. Relations of a 15-day human embryo and its membranes when they are completely enclosed in the hypertrophied endometrium. In A), the trophoblastic shell is penetrated by numerous intervillous spaces, which are also invaded by chorionic villi. The embryonic disk is attached to the chorionic vesicle and its components by the connecting stalk. In B), a human embryo of about the same stage is shown. (After Hamilton, Boyd, and Mossman, 1972)

also the site of origin of the sex cells which migrate by their own ameboid activity to the genital ridge where they proliferate. The yolk sac develops in later stages into the greater part of the gastrointestinal tract, as well as a vestigial organ, the allantois, whose remnants are shed with the placenta. At about 2 weeks, the embryo appears as a flat disk and is about 1.5 mm long. Cell columns extend from the proliferated trophoblastic cell mass into the loose connective tissue of the maternal endometrium (Figure 14.4).

Embryonic Development

Week 3. At 3 weeks, the embryonic heart consists of a microscopic tube bent on itself. The nervous system consists of differentiating cells beginning to fold over into what later becomes a spinal cord. The first condensations of embryonic mesenchyme appear adjacent to the neural groove described above. These are called *somites* and from them originate the segmental vertebrae, ribs, and intercostal muscles. The body stalk is now clearly established.

Week 4. At 4 weeks, the crown-rump (CR) length of the embryo is 5-6 mm. (See Figure 14.5 for methods of measuring human embryos.) The heart rudiments are very prominent, and the circulation of blood in the embryo begins. All forty somites are clearly marked out and arm buds are present. The head is large, and contains the earliest beginnings of eyes, ears, mouth and a nervous system. The neural tube has closed over. The trachea and lung buds are visible. The rudiments of many parts of the gastrointestinal tract appear (common bile duct, gall bladder, pancreas,

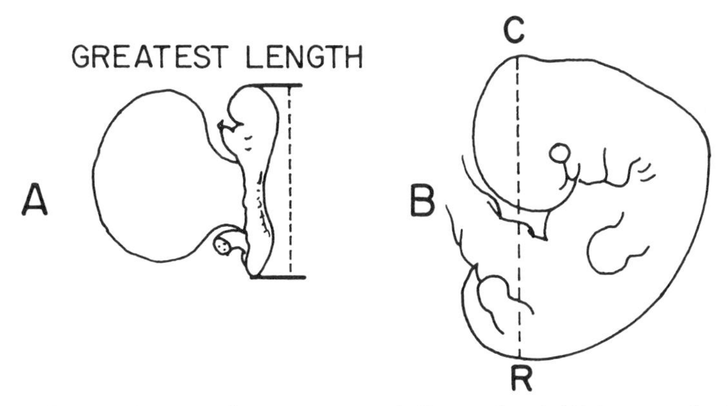

Figure 14.5. How human embryos are measured. Greatest length (A) is measured only in very early embryos, before the head and rump curl under. Crown-rump length (B) is measured in all other embryos and fetuses, and is the distance between the farthest points on the head and rump when the embryo or fetus is in its natural curved position. (After Corliss, 1976)

and the main parts of the intestine (fore-, mid-, and hind-gut)). The leg buds are present. The thyroid gland precursor and the precursor of the future anterior lobe of the hypophysis make their first appearance. The first embryonic kidney (pronephros) has already formed and degenerated, and the second embryonic kidney (mesonephros) is partly formed, and ready to be transformed later into parts of the genital tract. (See Figures 2.3, 2.6, and 2.7).

Week 6. At 6 weeks, the embryo is about 12 mm long (CR). The thin-walled cerebral hemispheres are enlarged as is also the midbrain. The large eyes are widely separated and nerve fibers are beginning to enter the primitive optic nerve. The ear has developed further and external auditory structures are visible. Olfactory organs are beginning to approach their definitive site. The adrenal cortex is being organized, and the posterior lobe of the hypophysis appears. The arms are divided into three segments (arm, forearm, and hand rudiments) and the indications of fingers are visible. The legs lag, though the three major segments are now visible. The somites have fused, and the muscle mass is spreading ventrally. The milk streak is present (see Chapter 15). The liver is engaged in making blood cells. The muscles of the arms and body can contract. Figure 14.6 shows these early stages.

Fetal Development

Week 8. At 8 weeks, the embryo is about 25 mm long (CR) and attains the status of *fetus* (Figure 14.7). The separation of these two stages of development is entirely arbitrary. It is intended to designate the time when the beginnings of all the major organs of the body are present. It should be emphasized that this is far from the actual realization both structurally and functionally of these potential structures—the realization requiring further growth and maturation. In fact, the embryo or fetus retains its dependency status until about 6 or 7 months *in utero,* after which, it may lead a life independent of the mother, though heavily dependent on machinery. To refer to the embryo or fetus at any time *before birth* as a "baby" confuses the developmental stages. To refer to any embryonic structure in adult terms such as the "heart" or the "cerebral cortex," without qualification confuses the functional capabilities of different stages. For example, the heart would be incapable of supporting the embryo or fetus, even though the heart beats. The same is true of all other organ systems. They are only *potentially competent* organs for the time being, not actually or functionally competent, as they are in the newborn child.

At this transitional time, the large head is the most prominent feature. Well-developed nerve cells are visible in the cerebral cortex, and the olfactory lobes are visible. Taste buds begin to appear, and eyes are

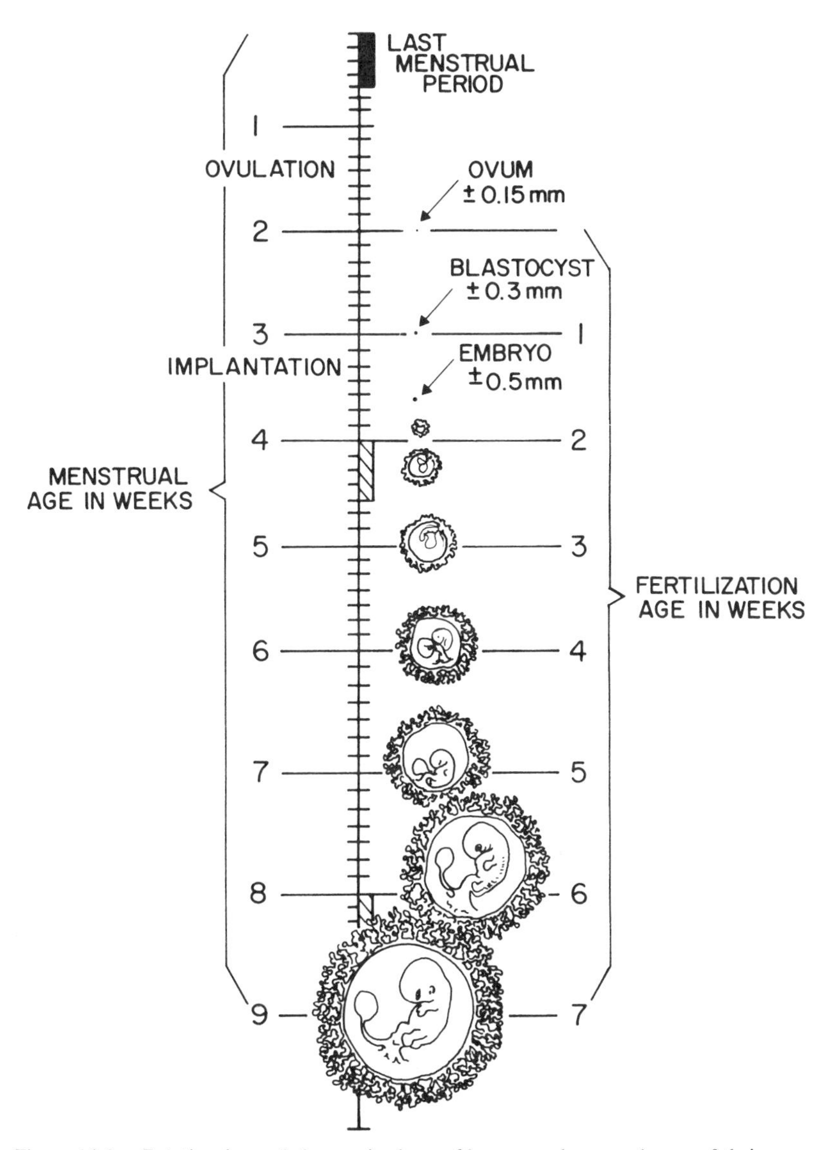

Figure 14.6. Relative size and changes in shape of human embryos and some of their membranes in relation to the menstrual age, which is two weeks earlier than the presumed actual age of fertilization. (After Corliss, 1976)

beginning to approach the definitive position in the adult. The adrenal medulla is now enclosed by the adrenal cortex. The first center of bone formation forms in the cartilage model of the upper arm and with it the bone marrow. Muscles have developed throughout the body, and the first

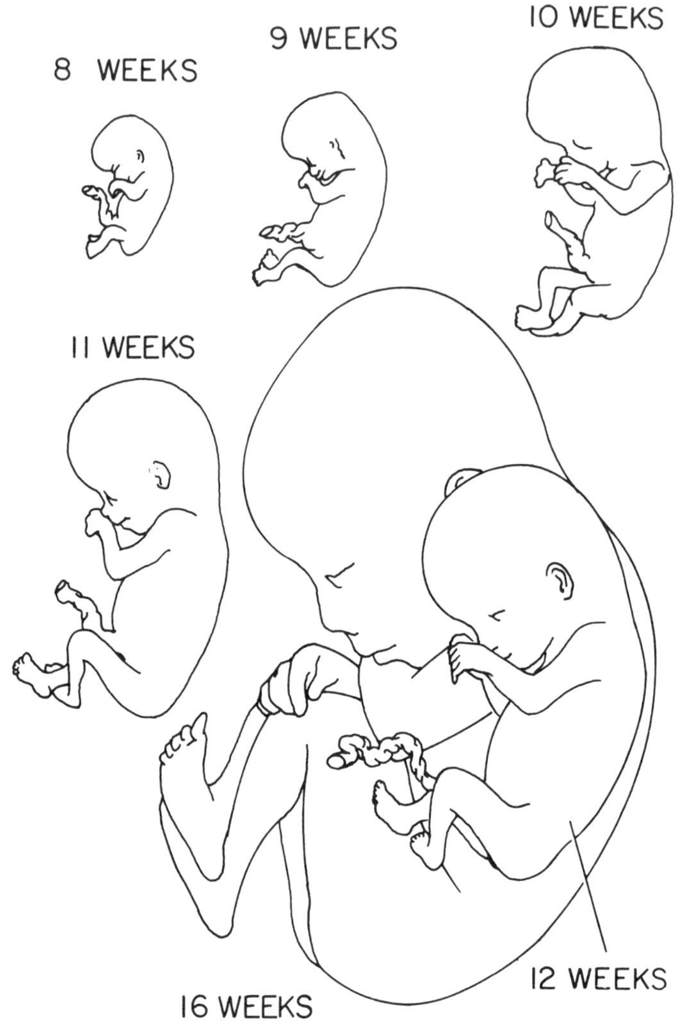

Figure 14.7. Outline drawings of human embryos and fetuses of known age, showing the progressive growth in size and change in shape. (After Corliss, 1976)

reflex movement can be elicited by stroking the upper lip with a very fine hair. Two weeks later some simple reflexes can be detected. The genital organs have developed as male or female.

Week 12. At 3 months, the fetus is about 56 mm long (CR) (Figure 14.7). The major subdivisions of the brain are now clear, and the spinal cord attains its definitive structure (white and gray matter, dorsal and ventral horns). The fetus is capable of many movements (kicking, turning feet, curling and fanning toes, making fists, bending wrists, moving thumb in opposition to fingers, opening mouth, swallowing). Bone formation is more extensive and blood cell formation takes place in bone marrow. The external sex organs are readily distinguishable.

Week 16. At 4 months, the fetus is about 112 mm long (CR) (Figure 14.7). The eyes, ears, and nose approximate their relative positions in the adult. The various parts of the brain are much more developed, especially the cerebrum and cerebellum. The general sense organs are developing. Muscles are more highly developed, and spontaneous muscular movements are strong enough to be felt by the mother. The surface epithelial layers of the skin are thicker, as is the underlying connective tissue. Hair appears on the head. The definitive kidney is structurally and functionally developed.

Week 20. During the fifth and sixth month of pregnancy, the fetus grows and matures both in external form and internal structure and function (Figure 14.8). Most of the skeletal parts are calcified. Nails appear on the nail beds of the fingers. The beating of the heart may be heard with a stethoscope placed on the mother's abdomen. The fetus alternates between quiet and active periods and may hiccup. The nerve fibers of the spinal cord begin to be myelinated. The surface layer of the skin becomes cornified, and is covered with a whitish, waxy paste. Very fine hairs appear over the body, which are mostly shed before birth. At this time the fetus appears lean and wrinkled.

Week 28. From the seventh month to birth, the cerebral cortex is further enlarged, cerebral fissures and convolutions appear, and myelinization of the nerve fibers begins. Fat is deposited in the subcutaneous connective tissue, and the body appears rounded. Hair on the head may grow longish, and the fetus may begin to suck on a thumb.

The changes in size of the developing embryo and fetus are presented graphically in Figures 14.9 and 14.10. Also, Figures 14.7 and 14.8 show progressive changes in size and shape of developing embryos and fetuses.

Review

With birth, the infant has achieved independence from the mother, in the sense that the baby can be nurtured and protected by any other motivated adult, male or female, or group of adults. It is worthwhile to review the steps by which this relatively independent status has been achieved. At first, oxygen and nutrients reach the free-floating egg, and metabolic products and carbon dioxide leave the egg, by diffusion from the fluid in the lumen of the oviduct and uterus. This continues, aided by the development of villi on the surface of the chorion, throughout the morula and blastula stages (Figures 14.1, 14.6, and 14.11). Shortly after implantation, the luminal fluid as a medium of exchange is replaced by the mater-

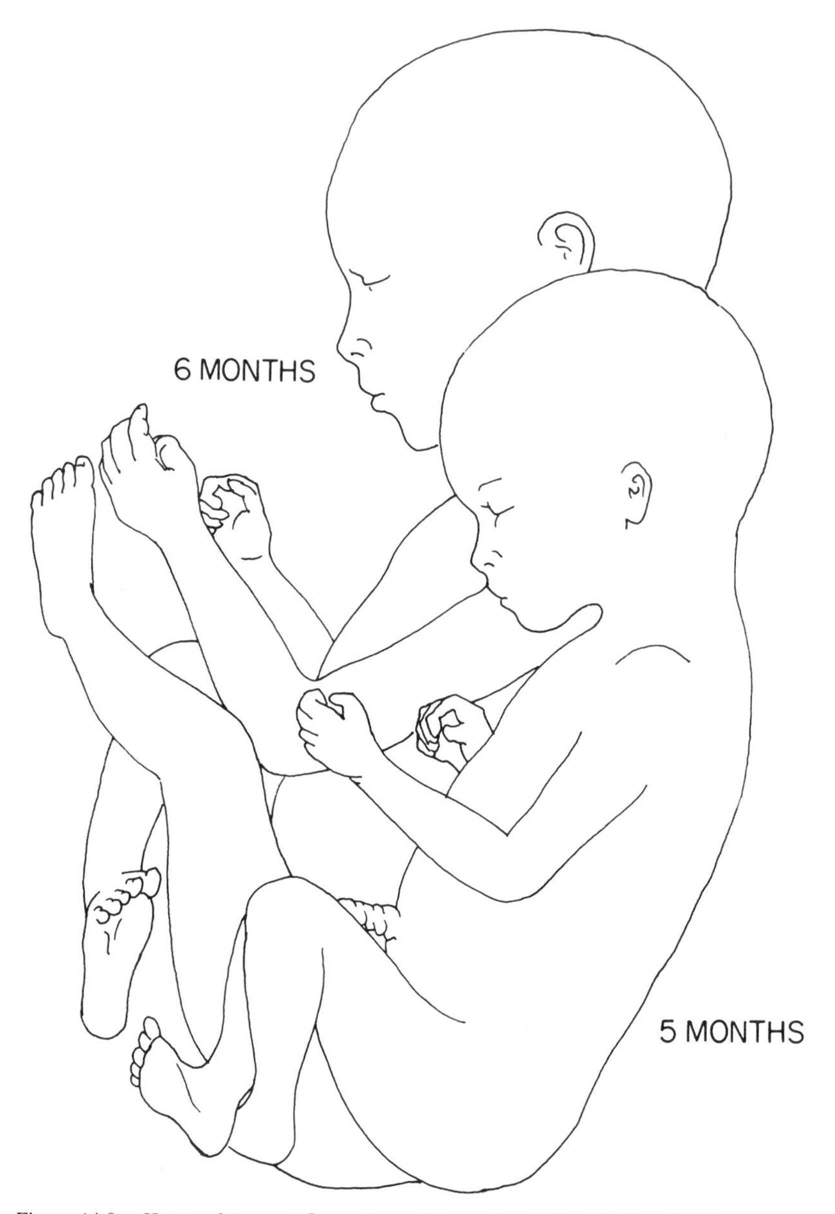

Figure 14.8. Human fetuses at 5 months and 6 months. (After Corliss, 1976)

nal tissue fluid that surrounds the rapidly growing chorionic villi and by gland secretions. The villi increase markedly the surface area through which exchanges take place. As growth continues and the separate fetal and maternal blood circulation is established in the chorion and early pla-

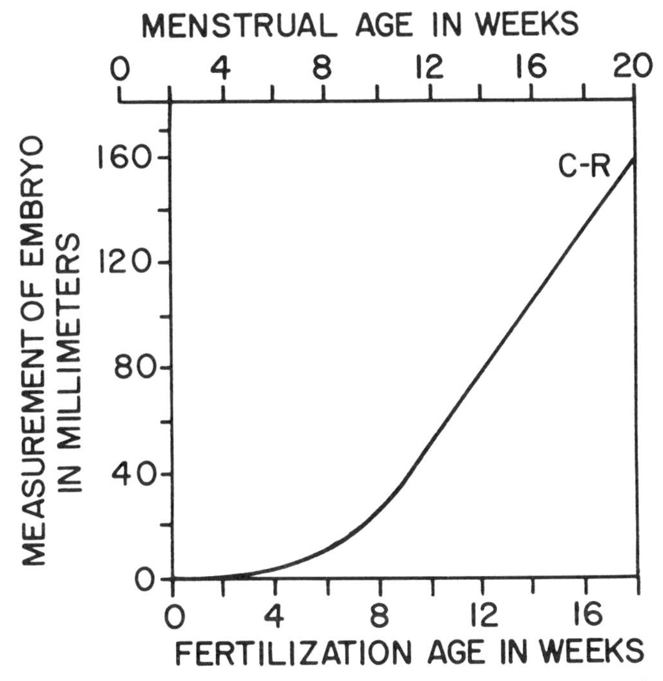

Figure 14.9. Increase in crown-rump length of human embryos and fetuses during the first 18 weeks of life. (After Corliss, 1976)

centa, the transfer of substances is accelerated. Later, the distance between the circulating fetal and maternal blood streams becomes extremely attenuated, further facilitating the transfer in either direction of the mostly small molecules from the one to the other (see later, Figure 14.14.)

In the embryo, the rudiments or buds arise on a definite, preset schedule, and each one further develops its own, predetermined pattern of organization. And eventually, the organs are fully prepared to function, without actually having achieved this condition. For example, the fetus near term is able to perform simple actions, the most essential ones being to feel, see, swallow, and suckle; as a consequence, the newborn is able to partake of nourishment. The fetus is able to make respiratory movements in the uterus; as a consequence, the newborn is able to breath air, though not without difficulty because of the fluid in the terminal bronchi and alveoli and the mucus in the trachea and bronchi. In addition, the pulmonary alveoli are not all developed and expanded because of the immaturity of the alveolar surface. The fetus is able to make and secrete digestive enzymes (though not stimulated to do so until after birth), because the protein-synthesizing systems of each cell type have differentiated and developed. The same applies to the presence of certain hor-

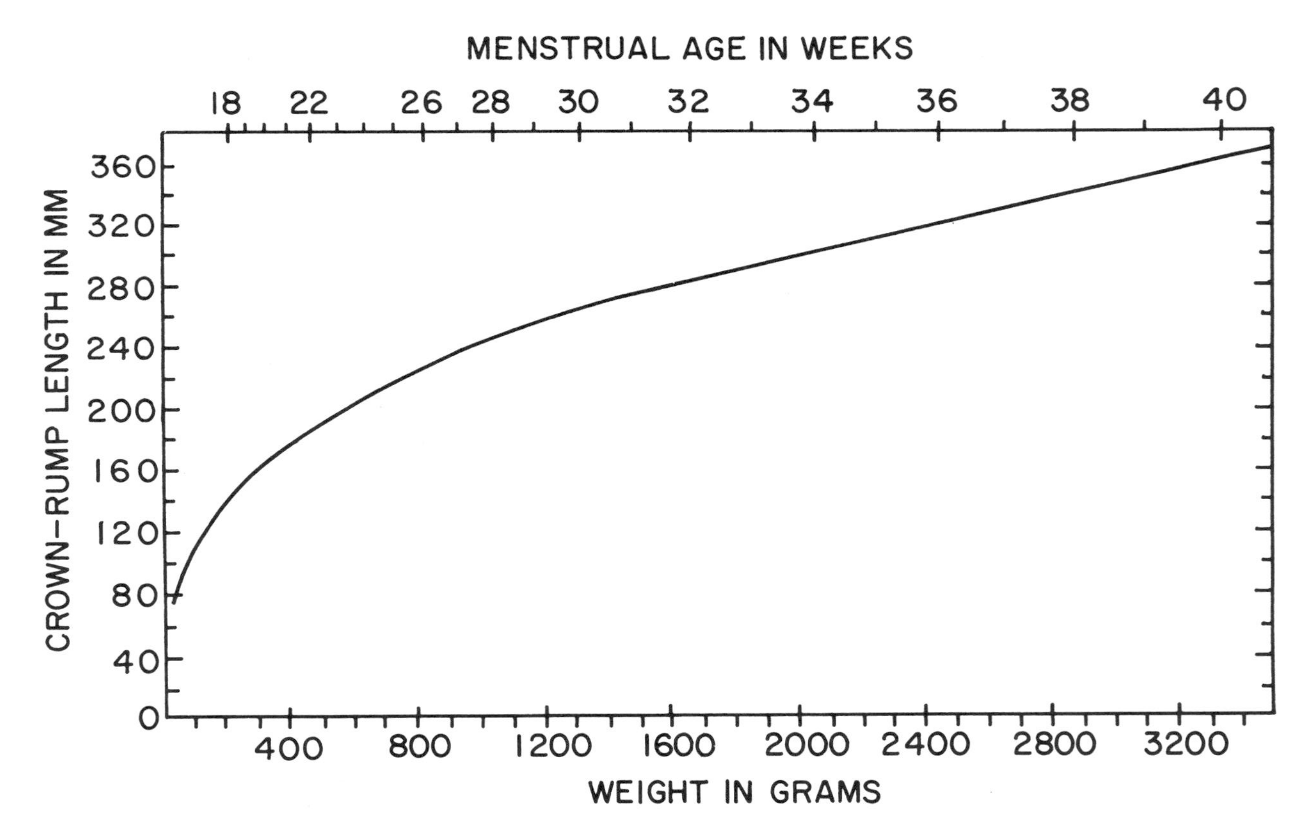

Figure 14.10. Relation of weight of human embryos and fetuses in relation to their height (length) and menstrual age. (After Streeter, 1921)

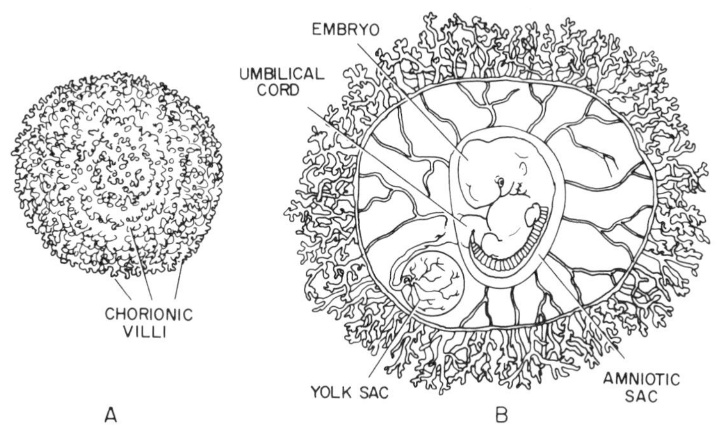

Figure 14.11. The rich field of chorionic villi as viewed from the outside (A) of a human embryo about 8 weeks old and with part removed (B) to show relations of the chorion and its villi to the embryo in the amniotic sac and the yolk sac at an earlier stage, about 5 weeks after fertilization. (After Hamilton, Boyd, and Mossman, 1972)

mones in the newborn, as well as the development of others at puberty. The kidneys excrete into the amniotic sac enormous volumes of "urine," which is reabsorbed into the fetal circulation and exchanged in the placenta whence it enters the maternal circulation from which it is eliminated as maternal urine. Though functionally useless in the fetus, the end result is that the kidney of the newborn is ready to function as an excretory organ of an independent organism. The central nervous system and the sympathetic and parasympathetic nervous system are rather incompletely developed at birth. And at the last moments, when the newborn baby takes it first breaths of air, the circulatory pattern is readjusted to initiate the functionally mature circulatory system through the lungs. These instances of development are cited to call attention to the remarkable adaptive quality of integrative growth of embryo and fetus. This kind of growth is not without its dangers: at certain early stages, when fetuses are particularly sensitive, they may be affected by some noxious agent, and the basic plan may be altered, resulting in functional disturbances or structural malformations.

Birth Defects

Considered statistically, about 7% of all live births have structural or functional defects recognizable by early childhood. Less than half of these are detectable at birth.

Pregnant women risk exposure to noxious agents that may cause selective growth retardation, functional defects, malformations, or death of the embryo or fetus. Such agents are called *teratogens.* These substances, their mechanism of action and the abnormality are the subject matter of teratology. The susceptibility of the fetus to teratogens is not a simple matter, but depends on the interaction of the genetic makeup and some one or several environmental factors.

There are few if any examples of proven sensitive periods in humans, because the exposures of pregnant women are usually chronic (rather than acute or single), the women may not be aware of the precise time or dose, and the timing of the genetic program that acts throughout development is poorly known. In general, it is acceptable to consider the human fetus as usually resistant during early stages of development (approximately the first 17 days after fertilization), as very susceptible during early periods of organogenesis (18-30 days after fertilization), and as becoming increasingly resistant as the fetal period progresses (35 or 60 days after fertilization to birth). The fetal variability and conditions of exposure to a suspected teratogenic agent in humans are so uncertain as to justify replacing the term *sensitive period* with the term *termination period*, which is less precise but more correct. The use of the expression *termination period* reflects the view that one can ascertain from the facts of development the *latest* possible time when a factor could cause malformations, because the main structural features of the tissues and organs have already been laid down and are insensitive to the damaging agent.

In fact, the causes of most developmental defects in the human fetus are unknown, as shown in Table 14.1. Among the drugs and chemicals known to have caused fetal malformations are androgenic hormones, folic acid antagonists (such as aminopterins), diethylstilbestrol, corticosteroids, thalidomides, certain organic mercurials, and certain pesticides (such as parathion). Diethylstilbestrol (DES) has some long delayed effects on the offspring of women treated with this drug, including a slightly increased incidence of cervical cancer. Smoking of tobacco has not been shown to be teratogenic, but it does have the effect of promoting the incidence of small babies. Maternal ingestion of alcohol or caffeine may also affect the health of the baby.

Table 14.1 Causes of developmental defects

Known genetic causes	20%
Chromosomal abnormalities	3-5%
Virus infections and syphilis	2-3%
Radiation	1%
Maternal imbalances or metabolic errors	1-2%
Drugs and chemicals	2-3%
Unknown	65-70%

EMBRYONIC AND FETAL MEMBRANES

In general, embryonic and fetal membranes are tissues and structures developed from the fertilized egg that do not form part of the embryo proper. They include the amniotic membrane (which encloses the amniotic cavity), the chorionic membrane (which encloses the chorionic space) and its derivatives, which comprise the fetal portion of the placenta, the yolk sac and allantoic duct and the umbilical cord.

When the blastula is attached to the body of the uterus at about 7½ days, it consists of an inner cell mass separated by fluid from an outer cell mass (Figure 14.2). The latter consists of flattened trophoblast cells that react immediately on coming in contact with the surface cells of the uterus. The trophoblasts penetrate between the epithelial cells, and penetrate the underlying connective tissue where they multiply and spread rapidly. The trophoblasts ingest cells of the maternal tissues that have degenerated and it is presumed that some of the digestion products pass out of the cells and are utilized as nutrients by the growing embryo. These substances may contribute to the very local edema, some of whose components, derived from maternal intracellular fluid and plasma, also may become nutrients for the embryo (Figure 14.5).

When the embryo is about 11–12 days old, the trophoblasts that have grown out as a more or less solid sheet of cells are now arranged as interconnected cords between which are spaces called *lacunae* (Figure 14.5A). The adjacent maternal capillaries become very wide, and are penetrated by the trophoblasts, and their blood escapes. The blood then enters the lacunae and presumably is part of the nutrient supply of the growing embryo. The more complete view of nutrition of early embryos includes the highly coiled spiral arteries that branch into many capillaries, some of which ooze their contents into the lacunar spaces. Other parts of the lacunar system become continuous with neighboring veins, and then a very slow circulation of maternal blood flows through the region in the lacunar spaces, thus replenishing and contributing to the nutrient supply of the embryo, which is now about 8 mm long (CR) and about 5 weeks old.

The Placenta

At term, the placenta measures 15-20 cm in diameter and is about 3 cm thick. Its volume is about 500 ml, of which the greater part consists of fetal villi (Figures 14.12 and 14.13). It is estimated that about 500 ml of maternal blood passes through the intervillous spaces every minute, and that about 400 ml circulates through the capillaries of the fetal villi every minute, with normally no gross mixture of maternal and fetal blood cells. Between the two circulating blood streams is the placental barrier. This comprises the outer epithelial covering of the villi, the connective tissue of

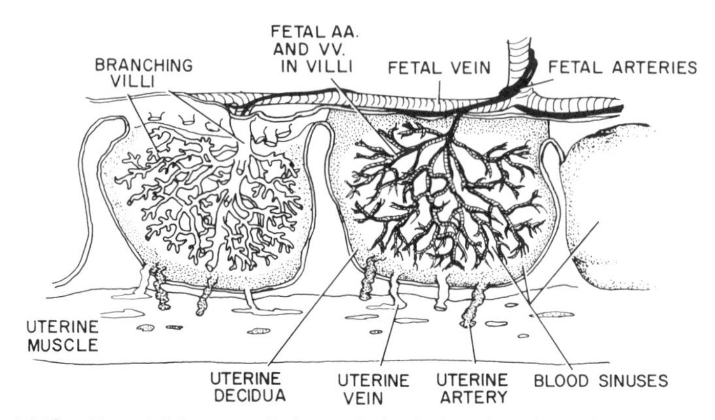

Figure 14.12. Essential features of placental circulation after about the eleventh week of gestation. AA, arteries; VV, veins. (After Ramsey, 1973; and Hamilton, Boyd, and Mossman, 1972)

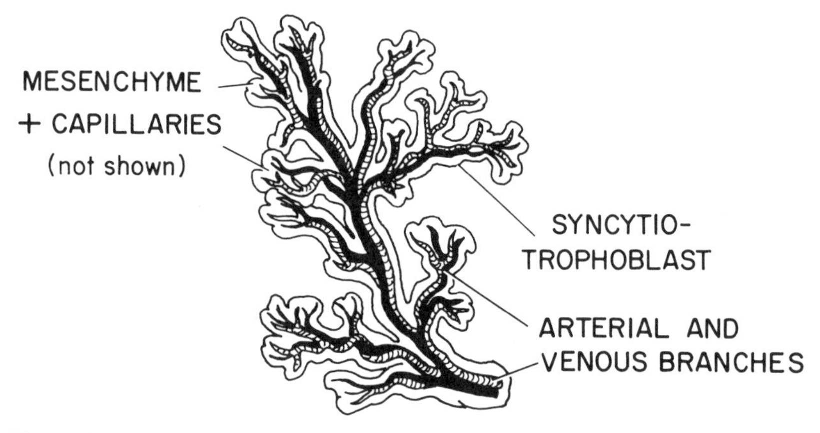

Figure 14.13. Placental villus which consists of connective tissue richly supplied with small and very fine capillaries. The whole is completely enclosed in a syncytiotrophoblast, which separates the inner contents of the villus from the maternal blood in the intervillous space. The contents of the villus are shown at higher magnification in Figure 14.14. (After Corliss, 1976)

the villi, and the endothelium of the capillaries in the villi, which together vary from 25 μm in early stages to 2 μm in very late stages (Figure 14.14). The surface of the villi exposed to the maternal blood is further increased by the numerous submicroscopic cell processes that extend out from the cell surface. It has been estimated that the surface area of the villi is as large as 4–14 m^2. This improves the efficiency of the transfer of substances

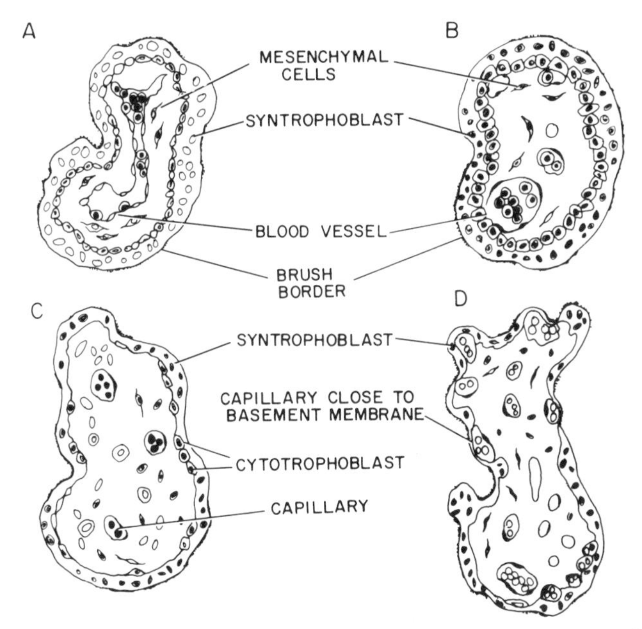

Figure 14.14. Sections of placental villi, in order of gradually increasing age. (After Corliss, 1976)

between maternal and fetal circulatory systems, especially in conjunction with the high rate of circulation of blood in the fetal and maternal portions of the placenta.

The placenta has many functions. The preeminent function of the placenta is the transfer of essential nutrients and other substances such as oxygen from the maternal to the fetal circulation. These include small molecules such as glucose, amino acids, lipids, vitamins, electrolytes, and water. An equally important function is the passage of fetal metabolic products and perhaps other substances to the maternal circulation. Also transferred from mother to fetus are many macromolecules that are important for the protection of the infant against bacteria. These include antibodies, agglutinins, precipitins and bacteriolysins. The transfer of antibodies is not an unmixed blessing. It can be responsible for serious damage to the fetus if it is Rh positive and the mother is Rh negative. Also an essential function is the synthesis and secretion by trophoblasts of several hormones: 1) human chorionic gonadotropin (HCG); 2) human chorionic somatotropin (HCS); 3) human placental lactogen (HPL); and 4) human chorionic thyrotropin (HCT), as well as several steroids, chiefly 17-β-estradiol and 17-OH-progesterone.

The Umbilical Cord

The umbilical cord arises early in embryonic life as the connecting body stalk, and is well formed at 12 days (before somites have appeared). In these early stages, the umbilical cord is very short, but at 43 mm length (CR) (about 10 weeks), the fetus can float freely in the amniotic cavity. During the interval, the connective tissue cells secrete a material rich in hyaluronic acid that gives the umbilical cord its special appearance and consistency. The umbilical cord carries the blood vessels that pass between the fetus and the placenta. It also bears portions of the yolk sac and duct and the vitelline duct, which are shed when the umbilical cord falls off about a week after it is tied. Circulation through the umbilical cord stops after the birth of the baby with the collapse of the umbilical vein. One day after the cord is cut, it is shrunken, but soft. During the next 2 days, it becomes brittle and dry. Two days later, it appears dark brown and semitranslucent. On or about postnatal day 7, it is shrivelled and separates.

The Amniotic Sac

The amniotic sac is present at the time of implantation, about 7½ days, when the embryo consists of two sheets of cells—one columnar and the other cuboidal (Figures 14.2D and 14.5B). The columnar cell layer continues as a thin, cellular membrane around the embryo and encloses the amniotic sac. At first, the amnion is attached to the chorion by some very delicate mesenchyme, but soon the two membranes are separated except at their attachment at the connecting stalk. This gives rise to the umbilical cord. At first the amniotic fluid is formed by the transfer of fluid from the chorionic sac. At about 4 ½ months, the amniotic and chorionic membranes fuse, and now nutrients and waste products are exchanged between the blood vessels of the maternal part of the placenta and the lumina of the uterine glands, the trophoblasts, mesenchyme, amniotic membrane and the amniotic fluid. Later, as the embryo grows, the main source of amniotic fluid is the fetal kidney, supplemented by fluid from the lungs. When the fetus begins to swallow amniotic fluid, the constituents pass from the fetal intestine into the fetal circulation and then to the placental blood vessels from where they pass into the maternal placental sinuses and the maternal circulation. At 6 months, there may be a liter of amniotic fluid. This volume is reduced to about 500 ml at 7 to 9 months, and remains at that level until parturition, when the amniotic sac ruptures and releases the fluid. Though the volume of the amniotic fluid during the last months is relatively constant, the constituents are continually changing. It has been estimated that every hour about one-third of the amniotic fluid is removed and replaced by an equivalent volume of fluid originating in the fetus. In the course of a day, this is equivalent to the movement of nearly 20 liters of fluid, a veritably continuous rinsing flow of fluid.

The amniotic fluid has another function essential for the survival and development of the embryo. Floating in the fluid, almost weightless, the musculosupportive tisues of the embryo, including the skeleton, develop to the point where they are functionally supportive of all the cells of the body, so that they retain their relations in the new gravitationally strong environment of the newborn. In earlier stages, when the embryo is more than 90 percent water and is very soft, it would be deformed if implanted on any surface without the cushioning effect of the amniotic fluid.

The Yolk Sac

The yolk sac arises early in the human embryo (about 7½ days) from the layer of embryonic cuboidal cells, which are destined to form much of the intestine. As the head and tail parts of the embryo form, the sac is restricted to the middle part of the embryo. The connective tissue of the connecting stalk contains within it part of the yolk sac that does not exceed 5 mm in diameter. The embryonic part becomes narrower, and is known as the yolk stalk, part of which is also included in the connecting stalk (Figures 14.6 and 14.11). The primitive sex cells arise in the walls of the yolk sac and migrate to the gonadal ridges of the embryo. Slightly later, blood islands appear in the walls of the yolk sac, and interconnect to form a capillary network that is connected to the blood vessels of the embryo proper. The blood islands contain the primitive blood cells that later populate the liver and still later the bone marrow. These primitive blood cells thus are the origin of the cells of the circulating blood of the fetus and adult, including the red blood cells and the granular and nongranular white blood cells, and most of the cells of the red bone marrow, lymph nodes and lymphoid tissue, tonsils, and spleen. Thus, though the yolk sac in human embryos is small and does not contribute to the nutrition of the embryo and fetus, it performs functions essential for survival. Long after these functions are over, the yolk stalk and sac are disposed of when the umbilical cord atrophies about a week after the birth of the infant.

Allantoic Duct

This vestigial organ arises from the hindgut and is incorporated in the umbilical cord. It degenerates at about 5 months, and the vestigial remnants are shed at birth.

REFERENCES

Arey, L. B. 1954. Developmental Anatomy. 6th Ed. W. B. Saunders Co., Philadelphia.

Bodmer, W. F. and L. L. Cavalli-Sforza. 1976. Genetics, Evolution, and Man. W. H. Freeman, San Francisco.

Brinster, R. L . 1973. Nutrition and metabolism of the ovum, zygote, and blastocyst. *In* R. O. Greep (ed.), Handbook of Physiology. Section VII. Endocrinology. Vol. II. Female Reproductive System. American Physiological Society, Washington, D.C., pp. 165-185.

Carsten, M. E. 1968. Regulation of myometrial composition, growth and activity. *In* N. S. Assali (ed.), Biology of Gestation, Vol. I. Academic Press, New York, pp. 355-425.

Flanagan, G. L. 1962. The First Nine Months of Life. Simon & Schuster, New York.

Miller-Catchpole, R., E. A. McGrew, and H. R. Catchpole. 1979. The appearance of tumor cells in the milk of C_3H mice in successive pregnancies. Anat. Rec.193.625.

Ramsey, E. M. 1973. Placental vasculature and circulation. *In* R. O. Greep (ed.), Handbook of Physiology. Section VII. Endocrinology. Vol. II. Female Reproductive System. American Physiological Society, Washington, D.C., pp. 323-337.

Schardein, J. L. 1976. Drugs as Teratogens. CRC Press, Inc., Cleveland.

Steinbeck, H., and F. Neuman. 1972. Aspects of steroidal influence on fetal development. *In* Drugs and Fetal Development, Advances in Experimental Medicine and Biology, Vol. 28. M. A. Klingberg, A. Abramovici, and J. Chemke (eds.), pp. 227-242.

Streeter, G. L. 1921. Weight, sitting height, head size, foot length, and menstrual age of the human embryo. *In*, Contributions to Embryology. Carnegie Institution of Washington (Washington, D.C.), Vol. 11, No. 55, pp. 142-170.

Vorherr, H. 1968. The pregnant uterus: Process of labor, puerperium, and lactation. *In* N. S. Assali (ed.), Biology of Gestation, Vol. I. Academic Press, New York, pp. 426-448.

Warkany, J. 1971. Sensitive or critical periods in teratogenesis: Uses and abuses of embryologic timetables. *In* Congenital Malformations: Notes and Comments. Year Book Medical Publishers, Inc., Chicago, pp. 49-52.

Wilson, J. G. 1973. Environment and Birth Defects. Academic Press, New York.

15 The Breast

GROSS ANATOMY

The breast is a secondary sex organ more highly developed in mature women than in men. In both sexes, the pigmented areola surrounds the nipple, but both structures are more highly developed in women. The region immediately below the skin of the nipple and areola is highly organized. It is characterized by a system of mainly circular connective tissue fibers and smooth muscle cells, so arranged that when the latter contract the nipple becomes erect. Light touch evokes this reflex response. The nerves involved in this sensitive response are shown in Figure 15.1. Nipple erection during lactation serves several purposes: 1) The arrangement of the smooth muscle is such that when it contracts, it acts as a sphincter, preventing the escape of milk when suckling is not taking place; 2) The smooth muscle contraction ejects milk by contracting around the milk sinuses (see Figure 15.2); 3) The harder nipple is more efficient for the suckling infant to grasp and retain than would be a softer nipple.

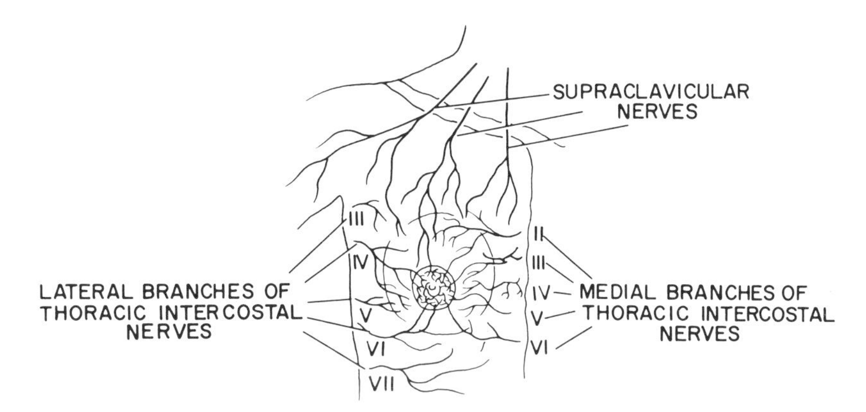

Figure 15.1. The nerve supply of the breast. (After Vorherr, 1974)

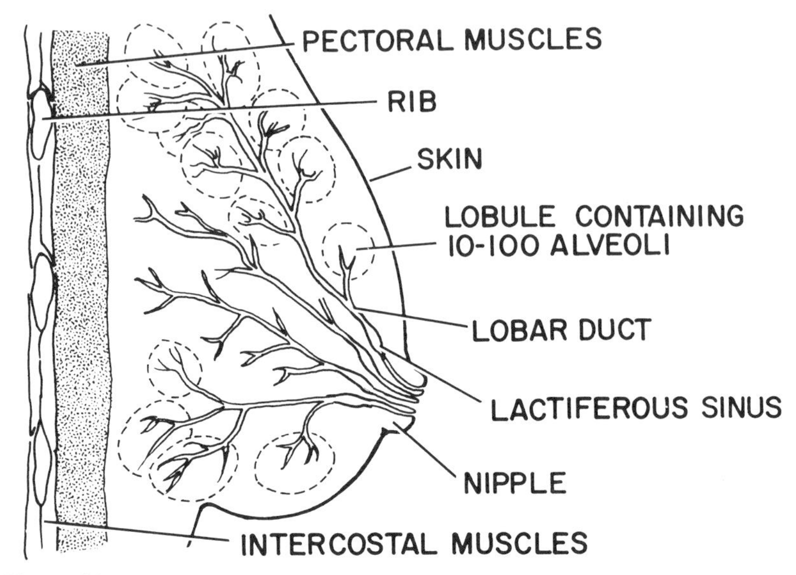

Figure 15.2. The glandular system of the mammary gland. (After Junqueira, Carneiro, and Contopoulos, 1975)

CONSTITUENTS OF THE MAMMARY GLAND

Formation of Milk

At the dome of the nipple are some 15 to 25 openings of the milk ducts. These bear the milk produced by the mammary glands to the surface of the nipple.

Milk is manufactured or synthesized by gland cells which are grouped into grape-like microscopic bodies called *acini* (see later, Figures 15.5A, 15.7, and 15.8). In the lactating state, the cells are arranged in a single layer, and are continuous with the smallest of the conducting ducts that are also of microscopic size. Some 10 to 100 such units or acini are joined by their little ducts to form a structure called a *lobule* which is large enough to be seen with the naked eye or a magnifying glass. The ducts that drain lobules empty into larger ducts that join ducts of other lobules, and the aggregate which these lobules form is known as a *lobe.* (Figure 15.2). There are 15 to 25 lobes in each mammary gland. Each duct dilates just before it enters the base of the nipple. This dilated piece stores milk and is called a *milk sinus.* The sinus narrows again and terminates on the dome of the nipple, where the milk flows on to the surface of the dome.

The acini or secretory portions are surrounded by a fine, microscopic layer of connective tissue, which separates each secretory portion from all others. The connective tissue within each lobule is separated from that of adjacent lobules by a somewhat denser and thicker layer of connective tissue. Also, the aggregate of lobules is surrounded by a denser and thicker layer of connective tissue to form lobes. These lobes are then attached to the surrounding skin and muscles. This is how the mammary gland is anchored in the breast, and how the various ducts and secretory units are kept in suitable relations to each other without tangling, kinking, or tearing. The connective tissue not only supports the mammary gland and all its parts in a functionally useful way, it also acts as a means of supporting the blood vessels (which bring nutrients and oxygen to the gland and remove waste products and carbon dioxide from the gland), lymph vessels, and nerve fibers. In general, regardless of the origin of the arteries and the destination of the veins and lymphatics, they are distributed with the ducts and branch as the ducts branch. Nerve fibers accompany the blood vessels. The most important relations of gland cells of the acini to blood and lymph are in the thin connective tissue layers that enclose each acinus because the finest blood vessels are located here.

The Lymphatic System

The thin layer of connective tissue between capillary and gland cell potentiates two important processes: 1) it allows some components (nutrients, waste products, oxygen, carbon dioxide) to pass to and from the gland cells, while others are restrained, and 2) it serves as the source of the lymph or excess tissue fluid. The lymph enters the lymphatic vessels, which serve as an accessory means, in addition to the veins, for moving fluids from the gland (Figure 15.3).

The smallest lymphatic vessels are capillaries situated between the blood capillaries. Their walls are thinner than those of blood capillaries. They join with others to form larger, interconnected networks that pass along the blood vessels in the larger, denser connective tissue layers of lobules and lobes until they reach the outside of the mammary gland and eventually empty into the large veins of the neck (Figure 15.3).

Three important features about the lymphatic system of the mammary gland are: 1) the walls are very thin and the fluid pressure relations are such that new lymph is being formed continuously in the lactating gland and is being directed always toward the outside of the gland through larger and larger lymphatic vessels; 2) along their course are one-way valves that also direct the lymph toward the outside of the gland; 3) in the course of the lymphatics are interposed fine cellular sieves that are called *lymph nodes* or *glands.* Here particulate material such as broken-down cellular debris is engulfed by cells and thus removed from the lymph stream. Defective cells that enter the lymphatic vessels are also filtered out.

The cells of the lymph node are also a part of the immunity

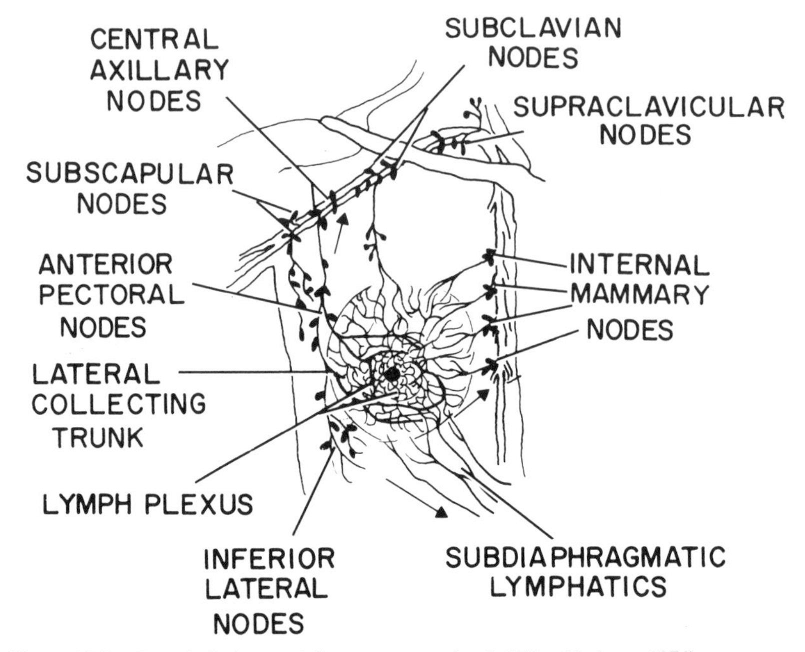

Figure 15.3. Lymph drainage of the mammary gland. (After Vorherr, 1974)

mechanisms of the body. The distribution of cancer cells from a breast carcinoma is a consequence of the three factors listed above, among others. Some cancer cells may separate from the main cancer mass into the thin connective tissue layers or sheets. From here, they move of their own accord or are pushed around until they meet up with the thin-walled lymphatics which they may then penetrate. They then move toward the outside of the gland in the lymph, and may be trapped in the lymph nodes, which are acting as sieves. Some of the cells settle down in the lymph node and multiply; some of these, in turn escape and pass further into other lymph nodes where the same processes take place. Eventually, the cancer cells pass into the large veins and are distributed throughout the body by the circulating blood.

The arterial blood supply is from several sources, indicated in Figure 15.4. By and large, the veins usually accompany the arteries. The lymphatic drainage usually accompanies the blood vessels (Figure 15.3); the positions of the major lymph nodes (deep and superficial) are also given in the same figure.

Myoepithelial Cells

Two additional constituents of the mammary gland are the myoepithelial cells and fat cells. The former are derived developmentally from the duct

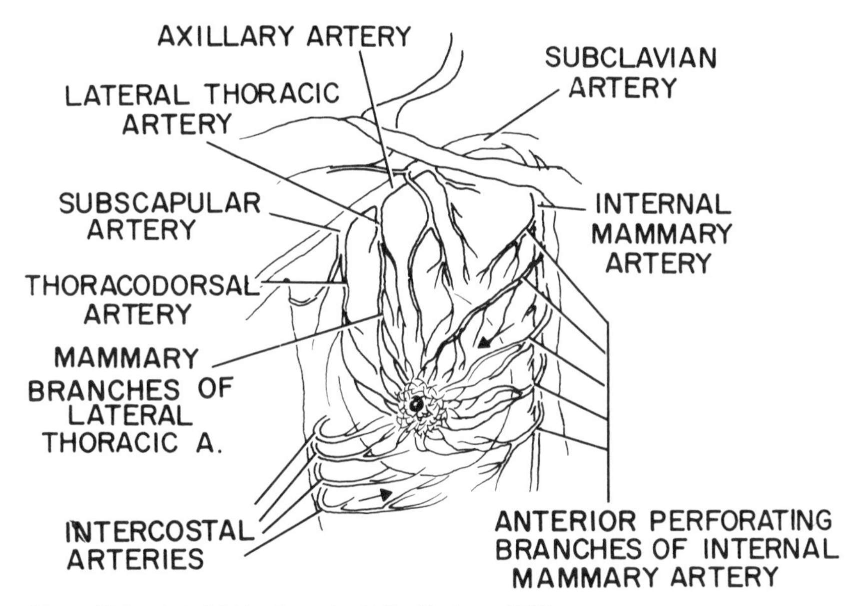

Figure 15.4. Arterial blood supply. (After Vorherr, 1974)

cells and their derivatives, the latter from the connective tissue cells. Myoepithelial cells are very thin cells with numerous cell processes extending around and enclosing gland cells of the secretory acini and duct cells of the smallest dimensions (Figure 15.5). The myoepithelial cells respond to oxytocic hormone by contracting. They aid in expelling milk from the lumen of the secretory acini and finer ducts during lactation. The myoepithelial cells are also present in the wall of the terminal ducts where their simultaneous contraction aids in propelling the milk toward the larger ducts.

Fat Cells

Fat cells of the mammary gland are limited exclusively to the connective tissues, chiefly those just beneath the skin, and between the lobes and lobules of the mammary gland.

DEVELOPMENT AND MATURATION OF THE MAMMARY GLAND

Embryonic Development

The mammary gland is first visible in an embryo about 5 weeks old (8mm), when the mammary streak appears (Figure 15.6A). It consists of a thickening of the outer epithelial cells that stretches from the region of

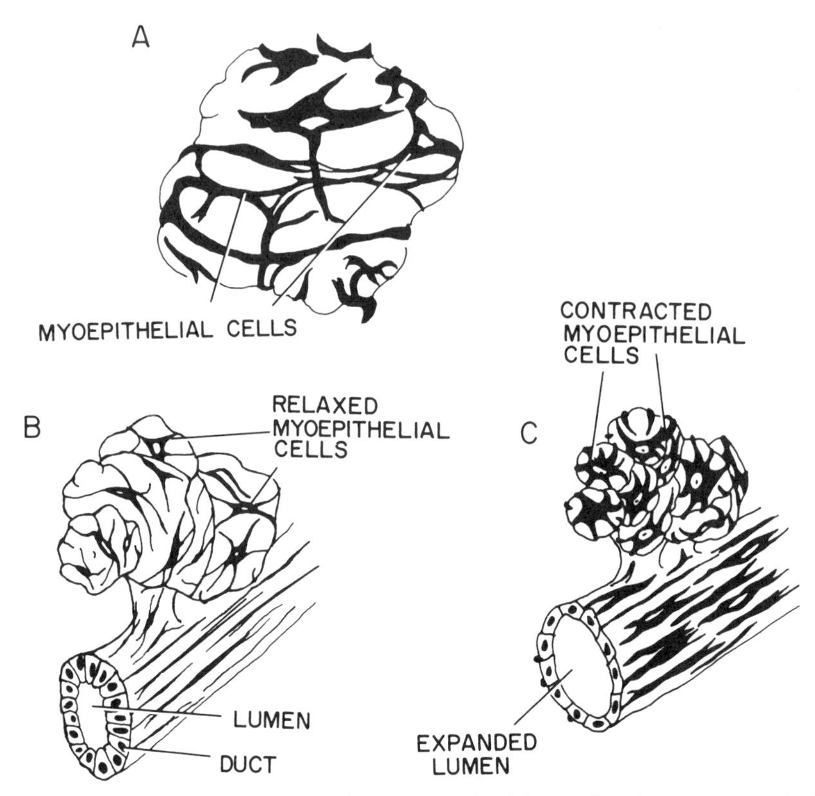

Figure 15.5. A) Outline drawings of a secretory glandular portion show arrangement of myoepithelial cells and their processes. A) Acinus. B) Small duct with relaxed myoepithelial cells. C) Same duct with cells contracted push the milk toward the nipple. The cytoplasm of the myoepithelial cells is rich in myofibrils, as seen in Figure 15.10. (After Vorherr, 1974)

the future axilla to the region of the future groin. This primitive milk streak regresses over most of its length, but in a small proportion of women (up to 6%), some cells are present that may give rise to mammary gland tissue and/or nipples in other parts of the skin, even parts unrelated to the projected course of the mammary streak (Figure 15.6B). At 8 weeks, the persisting part becomes thicker and grows inward, and at 15 weeks sprouts some 15 to 25 solid rootlets, which penetrate the underlying connective tissues. These become canalized during the fifth or sixth month so that they look like ducts. Thereafter growth is accelerated because of the action of sex steroids synthesized in the placenta and in the maternal corpus luteum, chiefly the former. Just before birth, the lobular structure of the mammary gland appears because of the hormonal action on the cells of the terminal ducts and the development of secretory acini. The areola becomes pigmented and a nipple may appear. At birth the se-

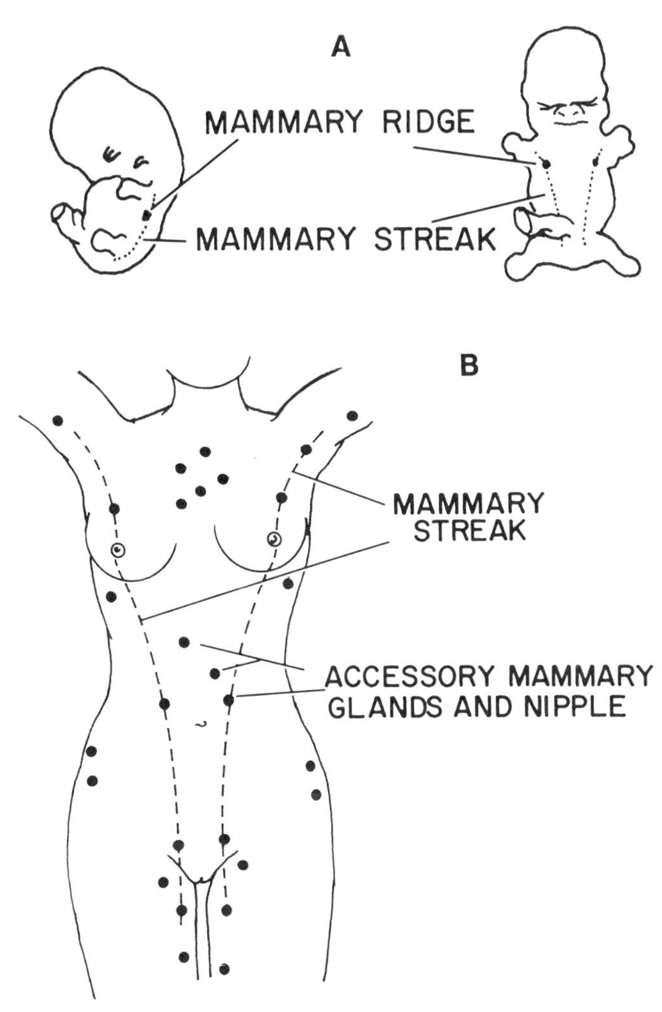

Figure 15.6. Diagram of a very early human embryo shows the location of the mammary streak, which stretches between the developing arm and leg (A). Normally in women the anterior region develops into mammary glands, but in many other species, mammary glands are spaced along the whole streak. In (B), sites of supernumerary nipples are shown along the mammary streak region but also in other regions. (After Vorherr, 1974)

cretory cells of the acini are enlarged because of the prior stimulation by the ovarian and placental steroids, and they secrete a fluid into the lumen of the acini. A yellowish milk-like fluid (colostrum) may be expressed during the next 2 to 3 days and may continue to be formed for a week. Thereafter the secretory activity is reduced and the gland cells regress. From that time on, until puberty, the epithelium resembles that of simple duct cells, without visible traces of glandular or secretory function. Up to this time, the development of the mammary gland, nipple and areola is the same in girls and boys.

Puberty

At puberty, presumably because the releasing hormones of the hypothalamus of girls are being effective, the appropriate basophil cells of the anterior lobe of the hypophysis secrete follicle stimulating and luteinizing hormones. FSH stimulates some primary follicles of the ovary to develop toward Graafian follicles. Even though at first ovulation may not occur, the follicular cells are sufficiently developed to secrete estrogen. This heightened secretion of estrogen causes the terminal ducts of the mammary gland to grow and sprout branches which eventually complete the main lobular divisions. There is an associated increase in the connective tissue including fat and growth of blood vessels. Then when ovulation takes place, and the corpus luteum secretes progesterone, further growth and maturation take place and the breast attains its adolescent size and firmness.

During childhood, the epithelium of the ducts (including the smallest) is 2 cells thick. There are no cytologic signs of secretory activity in these cells. Such signs appear after puberty, when the cells of the terminal portion of the duct develop into superficial dark-staining cells and basal or chief cells. The former slough off into the lumen of the duct and disintegrate. The latter persist and undergo cyclic changes during each menstrual cycle.

During the premenstrual phase, while the corpus luteum is secreting progesterone, the blood flow through the blood vessels of the breast increases, and is associated with swelling of the connective tissues in the mammary gland, which in turn results in heightened turgescence, fullness, and sometimes pain. There is also some sprouting of ducts and probably some appearance of secretory activity in the acinar cells. During the early stages of the menses the swelling and secretory activity persist, but later, the secretory cells become less active or inactive, and the lumen of the acini becomes narrower. During the postmenstrual phase, the swelling of the connective tissue is reduced, and the volume of the breast reaches a minimum at 5 to 7 days of the menstrual cycle. Then as the estrogen level in the blood begins to rise because of the growth and maturation of a succeeding series of follicles, another breast cycle begins. As a result of the succession of growth phases, it is thought that there is a very slight progressive growth of the breast, since there does not seem to be a complete regression of the breast changes during the postmenstrual phase.

Pregnancy

Within the first month of pregnancy, there is increased sprouting of ducts in the mammary gland. During the second month, the breast enlarges noticeably associated with an increased vascularity, and the areola becomes larger and more pigmented. Duct-sprouting continues during the first trimester and is the most prominent feature. During the second trimester, the breast continues to enlarge, probably with further growth of the estrogen-sensitized duct cells. This stage of growth is characterized by increased number, size, and activity of secretory acini (Figures 15.7 and 15.8). Some secretion appears in the lumen of some acini, and both lumen and cell height are enlarged. The prolactin level also rises (Figure 15.9), and probably influences growth further. As the alveoli continue to grow, the connective tissue and fat cells become relatively less prominent. Just before parturition, the blood vessels are dilated, and the blood flow may be double that of the nonpregnant breast. The milk ducts are also dilated with colostrum. The colostrum contains desquamated epithelial cells, phagocytic cells, protein and some fat droplets secreted by the cells of the acini. In addition to these changes, the myoepithelial cells become larger.

Parturition and Lactation

At parturition, great changes in relative hormone ratios take place. Hormones arising from the placenta are lost to the mother; the prolactin level in the blood is at its highest (Figure 15.9). Now free to exert its effect on the gland cells without possibly inhibitory influences arising from the placenta, prolactin induces lactation. Its action is fortified by tactile stimulation of the breast, suckling, and the effects of oxytocin. Ovarian progesterone is not essential for continued lactation (some women who have been ovariectomized nurse babies competently). Synthesis and secretion of milk are fully established at 2 to 5 days postpartum. During nursing, the breasts are enlarged and tender, fully one-third of the breast size being attributed to the storage of milk in the lumen of acini and ducts. The mammary blood vessels remain dilated, and the connective tissue is swollen, partly owing to the venous congestion and lymphatic stasis caused by pressure on the veins and lymphatics by the distended acini and ducts. Prolactin is essential for lactation, and it reaches high levels during suckling, falling again during rest periods (Figure 15.9). Most of the milk is synthesized during nursing and shortly thereafter, when the plasma prolactin levels are higher. In addition to the factors already mentioned (tactile stimuli, suckling, oxytocin) other metabolic hormones are involved in the synthesis of milk, i.e., cortisol, insulin, and thyroid and parathyroid hormone (see Figure 11.14).

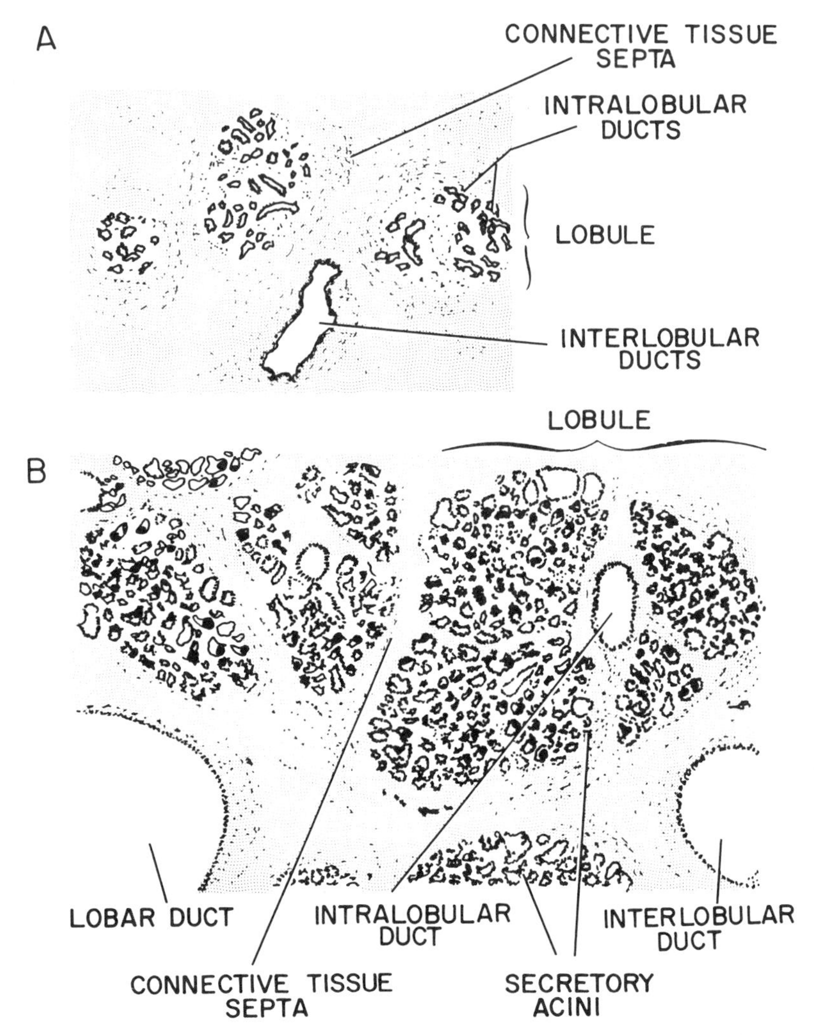

Figure 15.7. Basic microscopic structure of mammary gland tissues. A) Section of a nonlactating gland. Clusters of intralobular ducts are separated by connective tissue into lobules. The interlobular duct is also surrounded by much connective tissue. B) Sections of a lactating mammary gland. The interlobular and intralobular ducts are distended with milk (not shown). Each lobule contains numerous glandular acini whose lumina are also distended with milk. (After Dabelow, 1957)

The acinar cells that synthesize the milk are enlarged during lactation (Figure 15.10). Nuclear and cytoplasmic changes resemble those in other protein-secreting cells (Chapter 8). The secretory granules pass from the cell into the lumen of the acinus and the duct system to be stored until the next nursing period. Fat droplets arise independently of this protein-

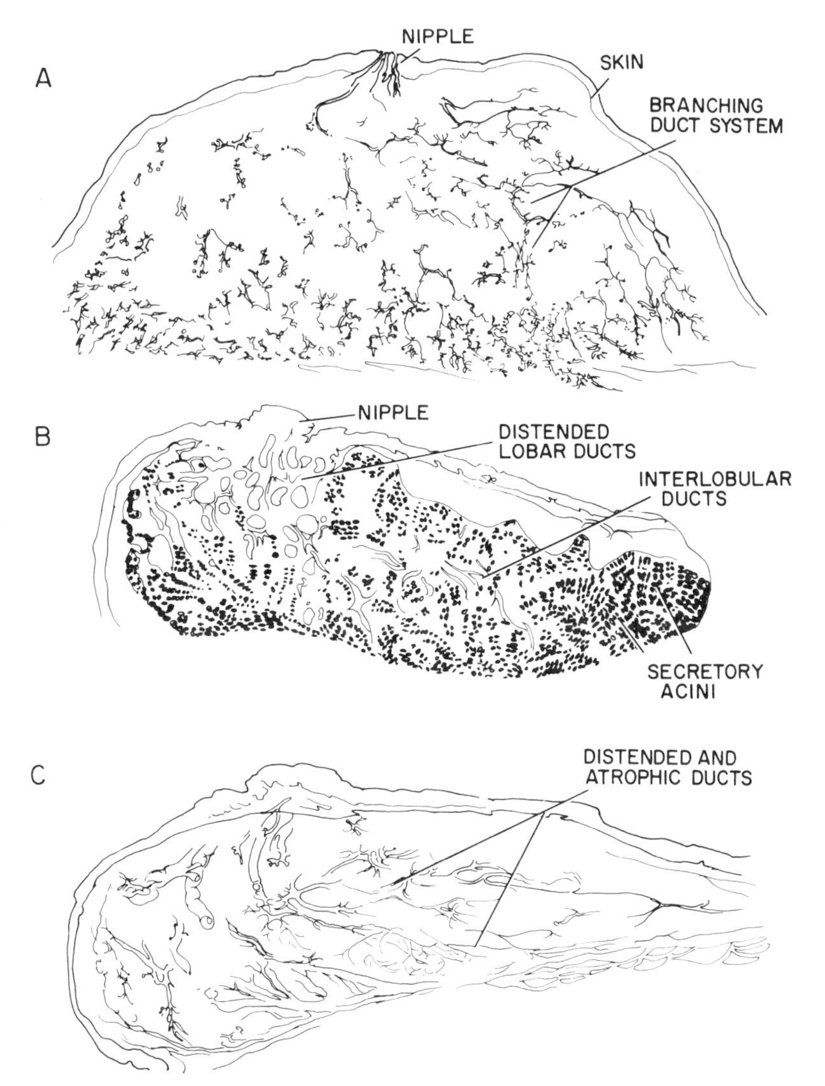

Figure 15.8. Ducts and glandular tissue of thick sections of the breast in different stages of lactogenic activity. A) Nonlactating, from a nonpregnant woman 24 years old. Interlobular and some lobar ducts are stained. The remainder of the breast (unstained and clear in the drawing) is connective tissue, including much fat. B) Lactating, from a 21-year-old woman who died 2 days after childbirth. Interlobular and lobar ducts are distended with milk. C) Atrophic mammary gland from a 76-year-old woman. Ducts are dilated, probably cystic, and atrophic, with loss of apparent lobular structure. (After Dabelow, 1957)

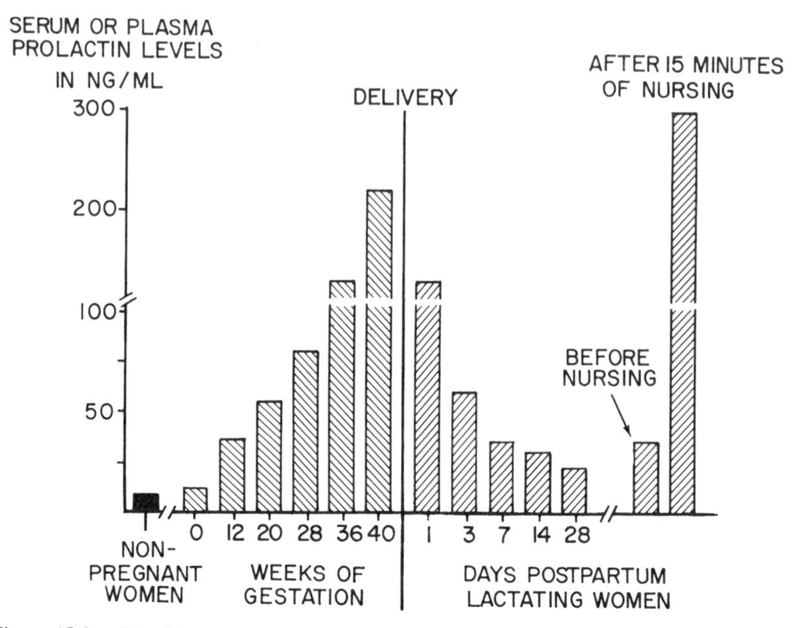

Figure 15.9. Blood levels of prolactin in women before pregnancy, during pregnancy, and after delivery. The level increases during pregnancy and falls to low levels within a month. The rise in the blood level of prolactin after suckling in nursing women is spectacular and rapid. (After Vorherr, 1974)

synthesis system. Fats are synthesized in the luminal (upper) half of the cytoplasm, and accumulate in large droplets that move toward the free surface of the cell. As it leaves the cell, the fat droplet carries with it a thin film of cytoplasm and cell membrane and is stored in this finely dispersed form in the lumen of the acini and the ducts.

This process of formation of milk differs from the origin of colostrum in the newborn, during pregnancy and immediately following parturition. There are chemical differences between colostrum and mature human milk, as shown in Table 15.1. Also included in the table is a comparison of the chemical composition of mature milk of women and cows.

Cessation of Lactation

When nursing is discontinued, the acini are distended by the milk which continues to be produced and stored in the lumen and ducts. This results in pressure on the capillaries around them and other small vessels and thus in their complete or partial closure. Since this means that the cells lack oxygen, nutrients, and water, the gland cells die, disintegrate and lose their organization, separating and floating off into the milk in the lumen. The dead cells are treated like any other necrotic cells anywhere else

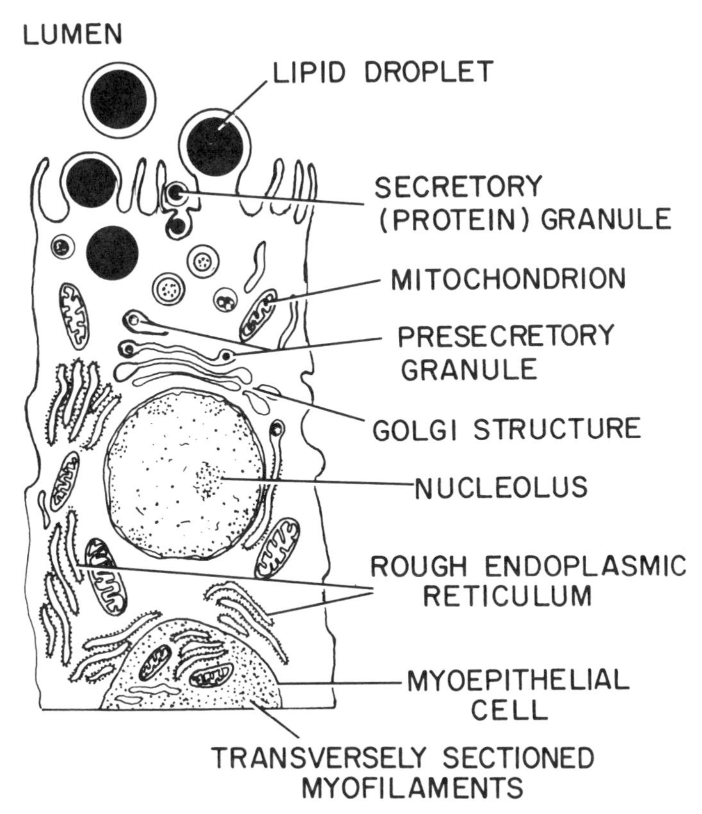

Figure 15.10. A gland cell during lactation. (After Bloom and Fawcett, 1975)

in the body. They are removed by lysosomal enzymes and by phagocytes. The acini gradually regress and disappear, leaving the duct system intact. As a part of the repair process the phagocytes wander off. Some new connective tissue is deposited as the connective tissue framework is reorganized, the capillary network is markedly reduced and the diameter of the larger blood vessels is reduced. Also the water content of the connective tissue returns to the lower prepregnancy levels. Fat cells in the connective tissue increase. This process begins on the outer, peripheral parts of the mammary gland and progresses toward the center. In the end, the breast is larger than before pregnancy, mostly because of a slight increase in the amount of connective tissue and fat cells. The rapidity with which the breast involutes depends on the abruptness of the cessation of lactation, the weaning time, the pressure put on the breast by clothing, the degree of breast manipulation, restriction of fluid intake, and other variables.

Table 15.1. Composition of human and cow's milk

Milk elements	Colostrum	Human milk (mature)	Cow's milk
Water	87.0%	87.5 %	86.0%
Lactose	5.3%	7.0 %	4.8%
Fat	2.9%	3.7 %	4.3%
Total protein	5.8%	1.2 %	3.3%
Casein	1.2%	0.4 %	2.8%
Lactalbumin	1.1%	0.3 %	0.4%
Lactoglobulin	3.5%	0.2 %	0.2%
Ash	0.3%	0.2 %	0.7%
Iron	0.1 mg/100 g	0.15 mg/100 g	0.1 mg/100 g
Sodium	48.0 mg/100 g	15.0 mg/100 g	58.0 mg/100 g
Potassium	74.0 mg/100 g	57.0 mg/100 g	138.0 mg/100 g
Calcium	31.0 mg/100 g	35.0 mg/100 g	125.0 mg/100 g
Magnesium	4.0 mg/100 g	4.0 mg/100 g	12.0 mg/100 g
Chlorine	91.0 mg/100 g	43.0 mg/100 g	103.0 mg/100 g
Phosphorus	14.0 mg/100 g	15.0 mg/100 g	120.0 mg/100 g
Sulfur	22.0 mg/100 g	14.0 mg/100 g	30.0 mg/100 g
Vitamin A	470 IU/100 ml	280 IU/100 ml	180 IU/100 ml
Vitamin D		5 IU/100 ml	2.5 IU/100 ml
Vitamin C	4.5 mg/100 ml	5 mg/100 ml	1.5 mg/100 ml
Specific gravity	1050	1031	1033

From *The Breast*, by H. Vorherr. Courtesy of Academic Press, New York. Copyright © 1974.

Menopause

Before ovulation ceases, during some premenopausal time, the terminal portions of the ducts and the inactive gland cells die. Their remains are removed by phagocytes, as after the cessation of lactation, and the cells are replaced by connective tissue and fat cells. After ovulation ceases, during the menopause, the reduction in gland and duct cells is more marked, the phagocytes disappear, and both are replaced by more connective tissue and fat cells. In the end, some mammary ducts remain intact. The lobular and lobar structure of the mammary gland is lost with the excessive deposition of connective tissue and fat cells (Figure 15.8C). The blood vessels are markedly reduced in diameter and number, and blood flow is correspondingly reduced.

Some topics dealing with the breast are treated in other parts of the book: changes in shape during adolescent maturation (Chapter 17), changes in shape and vascularity during coitus (Chapter 12), and changes in shape during pregnancy and the postmenopause (Figure 13.3).

MILK

Transfer of Toxic Substances

Milk consists of two phases—an aqueous and a lipid phase, each of which has its own solubility characteristics. For example, when the environment contains more than average radioactive iodine or strontium, then this passes into the aqueous phase of the milk, and is transferred to the child. Similarly, when the environment (including food intake) contains fat-soluble toxic substances such as the insecticide DDT or the herbicide Agent Orange (2,4,5-T) contaminated with dioxin, then they pass into the lipid phase of the milk and, again, are transferred to the child. The consequence is that infants, whether breastfed or bottle-fed are exposed to water-soluble and lipid-soluble toxic substances present in the environment.

Detection of Malignancies from Milk Samples

The mammary gland sheds a rich population of cells that appear in breast secretions arising spontaneously or in secretion obtained by means of a breast pump. Following earlier inconclusive studies, there is now increasing interest in the possibility of detecting malignant cells in such secretions at a time earlier than would be possible by other methods.

This possibility is strengthened by the finding of malignant cells in the milk of a cancer-susceptible mouse strain well before the time when detectable tumor nodules normally appear. It has been suggested that milk from all nursing mothers during the child-bearing period could be readily sampled and examined by the PAP technique. The results then would become a part of the mother's permanent health record; any signs of abnormal changes would be a reason for extra-careful watching during subsequent years, since *any* premonitory sign is vital to early detection, and early detection is crucial to treatment and cure of cancers of the breast.

Galactosemia

There are a number of diseases due to mutations that result in an inability to utilize the sugars in milk. Galactosemia, for instance, is due to lack of a specific enzyme necessary for the metabolism of lactose. As a result, the sugar galactose, one of the breakdown products of lactose, is found in high concentrations in the blood. A baby homozygous for the mutant gene will not be able to get enough sugar from the mother's milk, but can thrive on a lactose-free diet that is supplemented with other sugars the infant can metabolize. The presence of this gene in heterozygotes can be detected because they have a measurably reduced enzyme activity. If it is known that both parents carry the mutant gene, a newborn infant can then be tested to see whether it is homozygous, and treated accordingly. For an account of other genetic variants, see Bodmer and Cavalli-Sforza, 1976.

REFERENCES

Bodmer, W. F., and Cavalli-Sforza, L. L. 1976. Genetics, Evolution, and Man. W. H. Freeman, San Francisco.

Bloom, W., and Fawcett, D. W. 1975. A Textbook of Histology. 10th Ed. W. B. Saunders, Company, Philadelphia.

Dabelow, A. 1957. Die Milchdrüse. *In* W. Bargmann (ed.), Handbuch der mikroskopischen Anatomie des Menschen. Vol. III, Part III, Haut and Sinnesorgane. Springer-Verlag, Berlin.

Ensor, D. M. 1978. Comparative Endocrinology of Prolactin. Champan & Hall, London.

Junqueira, L. C., J. Carneiro, and A. Contopoulos, 1975. Basic Histology. Lange Medical Publications, Los Altos, Cal.

Miller-Catchpole, R., McGrew, E. A., and Catchpole, H. R. 1979. The appearance of tumor cells in the milk of C_3H mice in successive pregnancies. Anat. Rec. 193:625.

Tanner, J. M. 1962. Growth at Adolescence. 2nd Ed. Blackwell Scientific Publications, Oxford.

Vorherr, H. 1974. The Breast. Academic Press, New York.

16 Fat Tissue

Because the achievement of slenderness is of great clinical interest and tends in our society to be regarded as an aesthetic goal, it seems desirable to review the biology of fat.

There are two kinds of fat, ordinary white or yellow fat, and brown fat. There is good reason to regard ordinary fat as consisting of many little fat organs, each one centered on blood vessels, or to regard the whole of the ordinary fat in the body as a single organ, dispersed throughout the body as organelles or lobes. The subcutaneous connective tissue, for example, would contain many such unit centers. Other centers would be similarly organized, for example: the omental fat (in the abdominal area), the perirenal fat, the mesenteric fat, the popliteal fat (in the depth behind the knee joint), the inguinal fat (in the groin area), the epididymal fat, and the female genital tract fat.

FAT AND BODY COMPOSITION

The distribution of subcutaneous fat commonly differs in women and men. In women, it is generally prominent in the breasts, the buttocks, and the thighs. In men, it is generally prominent at the nape of the neck, and in the lumbosacral region.

The total amount of fat in people can be very closely estimated, and ranges from 10–20% in men to 15–25% in women. Although women usually have more fat than men, this may not be so in women athletes who are well conditioned; their total body fat approaches that of male athletes trained in the same sport.

Fats, which are chemical substances, should be distinguished from *fat*, which is a tissue. Fats are esters of fatty acids and glycerol, chiefly triglycerides, and can be hydrolyzed into their components:

$$
\begin{array}{lll}
R-COO-CH_2 & & R-COOH \quad HOCH_2 \\
\quad\quad\quad\quad | & & \quad\quad + \quad\quad\quad\quad | \\
R^1-COO-CH + 3H_2O \leftrightarrows & & R^1-COOH + HOCH \\
\quad\quad\quad\quad | & & \quad\quad + \quad\quad\quad\quad | \\
R^2-COO-CH_2 & & R^2-COOH \quad HOCH_2
\end{array}
$$

Neutral fat — Free fatty acids — Glycerol

Fat tissue may differ chemically from one part of the body to another. For example the iodine number (which is a measure of unsaturated bonds of fats) is nearly twice as great in liver fat as in subcutaneous fat. But the composition is nearly the same in all fat depots in the sense that more than 99% of lipids are neutral fat, with only traces of fatty acids, mono- and diglycerides, cholesterol, cholesterol esters, and phospholipids. Neutral fat consists primarily of triglyceride esters of fatty acids, which are chiefly long-chain and saturated. Rarely do fatty acids of fat depots have fewer than 14 carbons or more than 22; most have 16 (palmitic acid) or 18 (stearic acid). The structure is represented as follows:

$$CH_3—(CH_2)_{16}—COOH$$

Stearic acid

Some fatty acids have one or more unsaturated bonds, as:

$$CH_3—(CH_2)_7—CH = CH—(CH_2)_7—COOH$$

Oleic acid

$$CH_3(CH_2)_4—CH = CH—CH_2—CH = CH—(CH_2)_7—COOH$$

Linoleic acid

$$CH_3—(CH_2)_7—CH=CH—CH_2—CH=CH—CH_2—CH=CH—(CH_2)_7—COOH$$

Linolenic acid

Functions

Fat serves essentially as a storage depot for readily available energy. Subcutaneous fat insulates against heat loss from the surface of the body. Fat also protects against mechanical trauma and shock, in such areas as the orbit of the eye, major joints, and the palm of the hand and the sole of the foot. Fat also is an integral part of a mass response of the sympathetic nervous system, such as in "panic" flight or rage, since fat can be readily mobilized in the form of fatty acids to produce energy. It is possible that the extensive sympathetic innervation of fat cells and the blood vessels may underly at least in part this "total" reaction.

ORDINARY WHITE OR YELLOW FAT

Nearly all the fat in the body is inside cells, and of all the cells of the body, fat cells are the richest in fat. Fat cells are among the largest cells in the body, reaching to more than 0.1 mm in diameter. The cells are nearly spherical.

Metabolism

Depot fat may seem to be relatively inert most of the time because the amount does not vary greatly over long periods. Yet it is known that it has a very high metabolic turnover rate. In the rat, it is estimated that nearly 10% of the fatty acids of fat depots are replaced per day by others coming from other sites. In mice the percentage of turnover is greater. The rate of turnover in people is unknown, but is assumed to be appreciable. It is no surprise, then, that this high protoplasmic activity of assimilation and mobilization of fat is matched by a very rich capillary vascularity in relation to the volume of protoplasm of fat cells.

Each fat cell is enclosed in a fine network of microscopic connective tissue fibers that may reach a degree of fineness appreciable only with the electron microscope. Also lying in this fibrillar network is a rich network of blood capillaries. Between the fat cells, bathing the whole space where the fibrils are, is the tissue fluid. All exchanges between the blood plasma of the capillaries and the outside surface of the fat cells must take place through this tissue fluid space. The tissue fluid space here contains collagen, chondroitin sulfate-protein complexes and other similar substances that constitute the ground substance.

Only a small portion (about 5% of the total volume) of the fat cell consists of protoplasm—a thin shell that may be less than $1/2$ μm thick, and which contains the nucleus, the cell membrane, mitochondria, enzymes and other essential cell components present in all other cells of the body. These are controlled by genes in the chromatin that are like those in all other cells. In addition, genes are active in the nucleus that are inactive in those of other cells, for their functions are in the control of a very marked fat and carbohydrate metabolism. This is not inert, but is in a constant state of flux; some fat molecules are removed and transferred to cells elsewhere while others enter from other cells or are synthesized by the protoplasm from sugars or fatty acids reaching it from the blood. Ignoring for the moment the volume occupied by the fat droplet, the capillary surface per unit of protoplasm of fat cells is as great as in the most active muscle in the body, the heart. All of these factors are in line with the idea that fat cells are extremely active, and respond readily to environmental factors, including food intake, temperature, work performed, and response to hormones. Excess body fat is a detriment in that it puts an extra load on the heart, which must pump blood through the blood vessels of the fat tissue. At the same time, it constitutes essentially an extra burden that each overweight person must support and carry around.

The constancy of fat depots over long times is attributed to two factors: 1) the relative constancy of the rate of assimilation of fat from carbohydrate intermediates as well as from lipids from other sites, and 2) the relative constancy over time of the rate of mobilization of lipids as fatty acids. In turn, these factors are influenced by dietary intake and metabolic oxidation for the supply of energy. When there is an excess of dietary intake, the lipid depots become larger; when there is an excess of metabolic oxidation from excessive exercise, the fat depots become smaller. There is, in most people, a balance between the feelings of hunger and satiation, thought to be to a degree controlled by centers in the hypothalamus, that determines dietary intake. There are of course other factors in addition, such as olfaction, taste, and psychologic motivations. While this formulation accounts for most changes in fat tissue, there are enough clinical cases of overweight that are not accommodated to justify the suggestion that "... in many patients with obesity there is some metabolic defect and that often obesity in man has a metabolic component" and that "... the balance of evidence is in favor of the existence of metabolic differences between the lean and the obese...." (Kekwick, 1965).

Anorexia Nervosa

Some young people, chiefly female, ranging from 9 to 25 years of age, reduce their food intake to such an extent that it may lead to starvation and death. The wish to become slim may be one of several precipitating factors of the condition known as *anorexia nervosa*. This calls for psychiatric intervention and reeducation. In some clinics, not only the patient's psychologic orientation, but also the family's behavior pattern are treated as if the clinical condition were the outcome of family maladjustments.

FAT CELLS AND BLOOD VESSELS

The close relations of fat cells and blood vessels are prominent already in the embryonic origin of fat. The first fat cells appear in relation to small blood vessels, and are strictly limited to the site where numerous branching vessels form a close network. The earliest cells look like mesenchymal cells or fibroblasts whose cytoplasm contains a few small fat droplets. In later stages, these increase in number as the cell increases in size. Later they fuse into a large central fat drop surrounded by cytoplasm. The fused drop increases in size and attains its final size at about 0.1 mm. When fat is lost, as in starvation, the fat cells become smaller in a reverse manner, and eventually the cells come to resemble mesenchyme or fibroblast-like cells. In such reduced circumstances, the close relation between an extremely close network of blood vessels and the "fat" cells become prominent. At the same time that the cells lose their fat droplets, the amount of connective tissue is markedly reduced. When the starved animal is re-fed, fat re-accumulates in the same parts of the body and in the same relative amounts as in the original cycle. This phenomenon probably applies to people also. This indicates that sites where fat accumulates preferentially are determined, as if they were multiple parts of a single, dispersed organ. There is also some evidence that although the cells look like mesenchymal cells or fibroblasts, they may actually be irreversibly differentiated as fat cells. The number of such cells may be fixed early in life. The evidence there is shows that fat babies tend to become fat adults, and slender babies tend to become slender adults, and that when an adult person accumulates moderate amounts of fat, the number of fat cells remains the same, but the fat cells (and their enclosed fat drop) become larger. The documentation for these general statements is not convincing and an actual demonstration of early irreversible differentiation of fat cells from mesenchymal cells or fibroblasts is lacking.

FAT METABOLISM

Fat metabolism is relatively simple: the fatty acids are built up two carbons at a time from CO_2 and from acetate, with esterification by glycerol when the chain length is sufficiently long. Fat is oxidized by a similar mechanism: it is first hydrolyzed, then the free fatty acids are subject to oxidation, two carbons at each step. The process is tied in with glucose metabolism through the citric acid cycle. This is the basis for the intimate relations of dietary sugar and the deposition or depletion of fat. All the metabolic processes are performed in the cytoplasm of fat cells, as well as of other cells, and account for the transfer of fat, its absorption, mobilization, and storage. The following steps are involved in fat transport: glucose from the blood enters the cell and may be stored as glycogen (for later conversion to glucose) in the cytoplasm or metabolized to CO_2. The CO_2 or some intermediates are converted to fatty acids and then to fat, which then coalesces with the central fat drop where it is stored; or ingested fat, which has been absorbed from the intestinal tract as minute droplets 1 μm or less in diameter, passes through the endothelial cell wall of blood capillaries, where it is broken down to fatty acids and glycerol. The fatty acids pass into the intercellular space and into the cytoplasm of the fat cell where they are reconverted to fat, which is added to the central storage fat drop. When fat is removed from the cell, fat goes through the reverse stages: neutral fat leaves the fat droplet, and enters the cytoplasmic rim. There it is digested to form fatty acids and glycerol. These pass through to the tissue fluid space, and enter the cytoplasm of the endothelial cell of the blood capillary, supposedly through pinocytosis. When ejected into the lumen of the capillary the fatty acids combine with a plasma protein. The resulting lipoprotein is carried in the blood plasma to other parts of the body, where the lipid component may be deposited.

The conversions of glucose to fatty acids and fat, and the interconversion of fat and fatty acids require the action of several cytoplasmic enzymes. These are under the control of several hormones, chiefly insulin, epinephrine, and norepinephrine. But numerous other hormones may be involved: adrenal steroids; pancreatic glucagon; and the following pituitary hormones—vasopressin, thyrotropin, adrenocorticotropin, luteotropin, and growth hormone.

FAT SOLUBILITY

Atmospheric Nitrogen

Because of its solubility properties, fat is responsible for two biologic maladjustments to technology. It dissolves certain substances selectively, and accumulates them selectively on exposure. Two different kinds of examples are cited. The first example is atmospheric nitrogen (which is in the air we breathe). It is more than five times more soluble in fat than is oxygen in the tissue fluid and blood plasma. For this reason, when divers are decompressed too rapidly on returning to the surface from a dive, the excess nitrogen dissolved in fat cannot be carried away sufficiently rapidly in the dissolved form in the blood plasma. When nitrogen breaks out of solution as gas bubbles in fat tissues and veins leading from the fat depot, the expanding gas bubbles impinge on adjacent nerve fibers and give rise to pain by pulling, tearing, stretching, and compressing them. Also, the gas bubbles in the veins may enter the general blood circulation and create damage through the gas emboli that block the circulation of blood in the region.

Toxic Substances

The second example is the accumulation in fat of certain toxic substances such as DDT and 2,4,5-T (Agent Orange). The latter is used as a herbicide, and contains an extremely toxic impurity called *dioxin*. Stored in fat because of their differential solubility in fat, they are only slowly eliminated from the body. Thus the exposure of the body to possibly toxic levels is prolonged. In addition, when such pollutants are ingested by animals in the food chain, they are of course concentrated in their fat. Hence, the toxic substances are at a higher concentration than their original environmental concentration. When ingested, dissolved in the fat of meat, the pollutant is stored in the person's fat. Then it is slowly released from the fat into the circulating blood of fat tissue, whence it is distributed generally throughout the body.

BROWN FAT

Brown fat is common in the near term human fetus and in the newborn. It appears most commonly along the vertebrae and certain other sites. In rodents and hibernating mammals it is more conspicuous. Brown fat is involved in the control of body temperature because it is readily oxidized and generates heat. In people, brown fat sometimes persists into adulthood, but is difficult to recognize and in any case is in such small quantity as to be negligible as compared with the amount of white or yellow fat. The cells of brown fat are smaller than those of yellow fat, and contain a brown pigment. The fat occurs as many fat droplets which are crowded in the cytoplasm. Brown fat is very highly vascularized, and some of its functional value in infants derives from the fact that it is readily available to generate heat metabolically and to protect the infant from changes in body temperature, at a time when the temperature control system is not yet fully developed.

REFERENCES

Bray, G. A. 1973. Obesity in Perspective. DHEW Publication No. (NIH) 75-708. Superintendent of Documents, U.S. Government Printing Office, Washington, D.C.

Brodie, B. B., R. P. Maickel, and D. N. Stern. 1965. Autonomic nervous system and adipose tissue. *In* A. E. Renold and G. F. Cahill, Jr. (ed.), Handbook of Physiology. Section V, Adipose Tissue. American Physiological Society, Washington, D.C., pp. 583-600.

Brooks, C. McC., and K. Koizumi. 1974. The hypothalamus and control of integrative process. *In* V. B. Mountcastle (ed.), Medical Physiology, Vol. I, 13th Ed. C. V. Mosby Co., St. Louis, pp. 813-836.

Bulfer, J. M., and C. E. Allen. 1979. Fat cells and obesity. BioScience 29:736-741.

Buskirk, E. R. 1974. Obesity: A brief overview with emphasis on exercise. Fed. Proc. 33:1948-1951.

Catchpole, H. R., and I. Gersh. 1947. Pathogenetic factors and pathological consequences of decompression sickness. Physiol. Rev. 27:360-397.

Galton, D. J. 1971. The Human Adipose Cell. Butterworths, London.

Gersh, I., G. E. Hawkinson, and E. N. Rathbun. 1944. Tissue and vascular

bubbles after decompression from high pressure atmospheres—correlation of specific gravity with morphological changes. J. Cell. Comp. Physiol. 24:35-70.
Gersh, I., and M. A. Still. 1945. Blood vessels in fat tissue. Relation to problems of gas exchange. J. Exp. Med. 81:219-232.
Greenwood, M. R. C., and P. R. Johnson. 1977. Adipose tissue cellularity and its relationship to the development of obesity in females. *In* M. Winick (ed.), Current Concepts in Nutrition, Vol. V, pp. 119-135.
Hirsch, J. 1965. Fatty acid patterns in human adipose tissue. *In* A. E. Renold and G. F. Cahill, Jr. (eds.), Handbook of Physiology. Section V, Adipose Tissue. American Physiological Society, Washington, D.C., pp. 181-189.
Keesey, R. E., and T. L. Powley. 1975. Hypothalamic regulation of body weight. Am. Sci. 63:558-565.
Kekwick, A. 1965. Adiposity. *In* A. E. Renold and G. F. Cahill, Jr. (eds.), Handbook of Physiology. Section V, Adipose Tissue. American Physiological Society, Washington, D.C., pp. 617-624.
Koizumi, K., and C. McC. Brooks. 1974. The autonomic nervous system and its role in controlling visceral activities. *In* V. B. Mountcastle (ed.), Medical Physiology, Vol. I. 13th Ed., C. V. Mosby Co., St. Louis, pp. 783-812.
Lindberg, O. (ed.). 1970. Brown Adipose Tissue. American Elsevier Publication Co., Inc. New York.
Minuchin, S., B. L. Rosman and L. Baker. 1978. Psychosomatic Families. Anorexia Nervosa in Context. Harvard University Press, Cambridge.
Schoenheimer, R. 1942. The Dynamic State of Body Constituents. Harvard University Press, Cambridge.
Shock, N. W. 1966. Physiological growth. *In* F. Falkner (ed.), Human Development. W. B. Saunders Co., Philadelphia.
Sprynarová, S., and J. Parízková. 1969. Comparison of the functional, circulatory and respiratory capacity in girl gymnasts and swimmers. J. Sports Med. Phys. Fitness 9:165-172.
Stern, J. S., and M. R. C. Greenwood. 1974. A review of adipose cellularity in man and animals. Fed. Proc. 33:1952-1955.
Wassermann, F. 1965. The development of adipose tissue. *In* A. E. Renold and G. F. Cahill, Jr. (eds.), Handbook of Physiology, Section V, Adipose Tissue. American Physiological Society, Washington, D.C., pp. 87-107.

17 Sex Differences of Women and Men

Many biologic differences exist between women and men, apart from the obvious primary and secondary sex differences. The former are statistical and do not necessarily apply to any specific person. Moreover, while some characteristics may apply to a specific person, others may not. That is, there is a range of variation in the expression of characters in women and men and the distribution may differ for women and men, with some overlap.

Such differences between the sexes may be physical, physiologic, biochemical, metabolic, or behavioral, i.e., psychologic or social. These differences may arise or become manifest during embryonic or fetal life, during childhood, adolescence, and adulthood, or during middle or old age. It is not clear to what extent sex differences are genetically determined or influenced by environmental (including cultural) factors. Nor is it clear how these two factors are expressed in the observed characteristics.

CHILDHOOD

Typically, at birth, boy babies are slightly larger than girl babies. Girls and boys grow at approximately the same rate throughout childhood, but there are a few exceptions, as for example the attainment of wider external dimensions of the pelvis in boys and of the larger inner region of the pelvis in girls, and the earlier skeletal and dental development in girls.

The time of eruption of milk teeth is the same in girl and boy babies. Data from identical twins suggest that the sequence and eruption of these teeth is markedly influenced genetically. By contrast, there is a notable sex difference in the time of shedding of milk teeth and the time of eruption of permanent teeth; girls are more advanced than boys by two to eleven months, depending on the tooth. Studies of identical twins suggest that in these stages also there is a marked genetic control of the processes involved.

ADOLESCENCE

Most changes take place during adolescence, and it is then that the patterns of change differ in the two sexes. About the time of puberty, girls enter on a growth spurt, and for some years they are taller and weigh more than boys of the same age. Two or more years after the girls begin their growth spurt, the boys enter on theirs, so that, while the growth of girls slows down and levels off, the boys continue to grow. The adolescent spurt with respect to weight begins for girls when they are 10–12 years old, while that for boys begins at 12–14 years of age; the corresponding times for height are 10–14 years and 11–16 years. In general, the girls with an early menarche have an early growth spurt, and their growth period is shorter than that of girls with a late menarche. The sexual dimorphisms in height and weight are illustrated graphically in Figures 17.1–17.3. These particular growth curves apply to middle-class populations of English and American whites. The curves are similar but not identical for American and African blacks.

DEFINITION OF TERMS

The following general terms are used in the rest of the chapter. *Puberty* is the beginning of sexual maturity. In girls it is marked by the *menarche,*

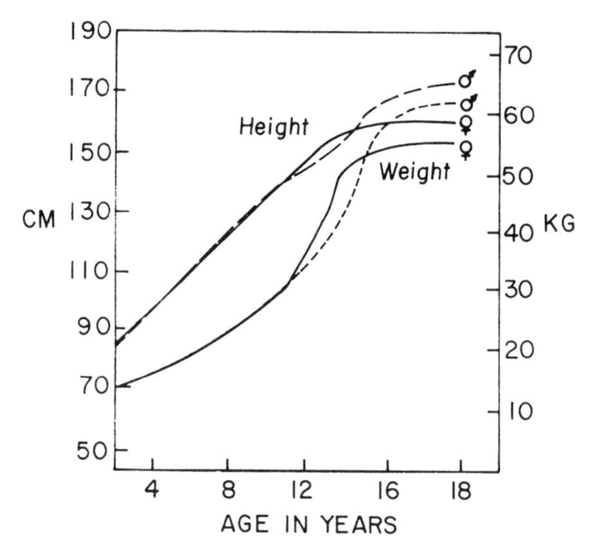

Figure 17.1. Typical individual height and weight curves for a girl and a boy. This kind of longitudinal curve is obtained by making measurements on the same persons at regular intervals. When combined with similar curves from many subjects, they serve as a basis for percentile curves. (After Tanner, Whitehouse, and Takaishi, 1966)

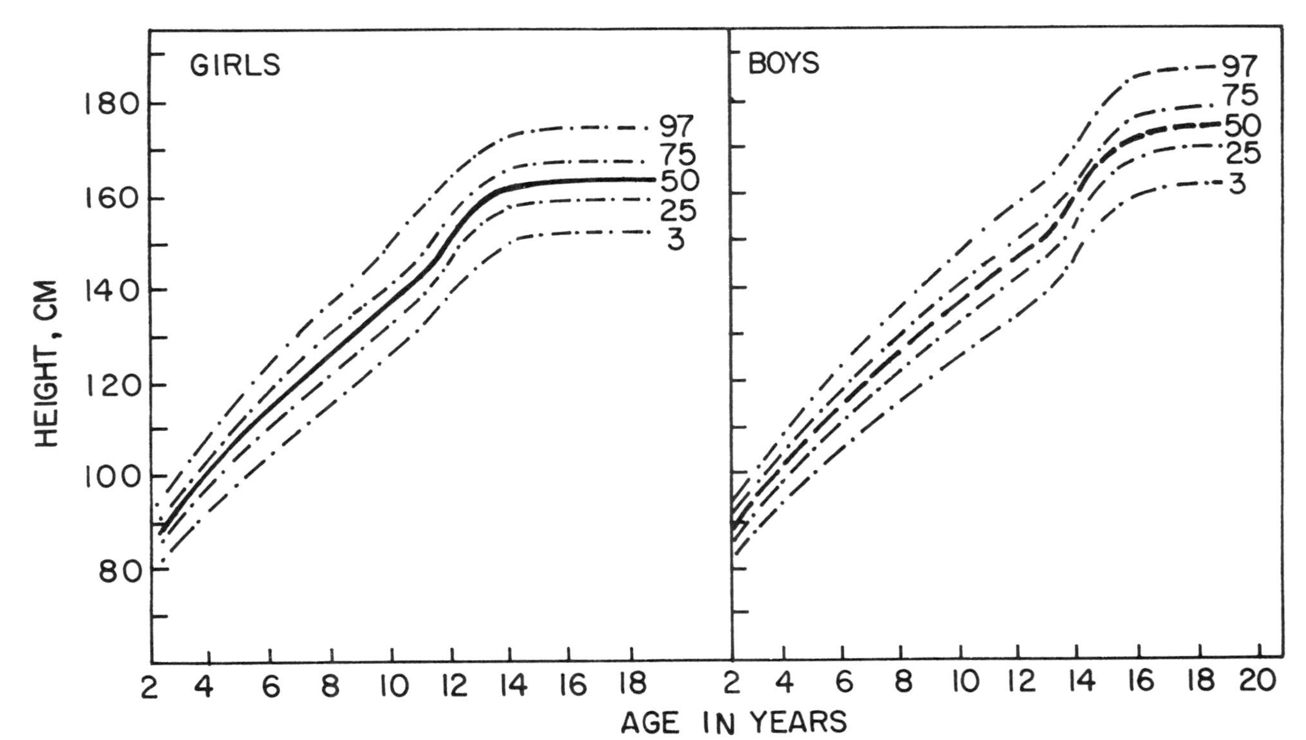

Figure 17.2. Longitudinal height curves for English girls and boys show the variation in height at yearly intervals in a relatively homogeneous cultural class. Repeated measurements were made on the same individuals, and the percentile curves are based on the proportion of subjects who fell below the median percentile group and those above the median percentile group. (After Marshall, 1977)

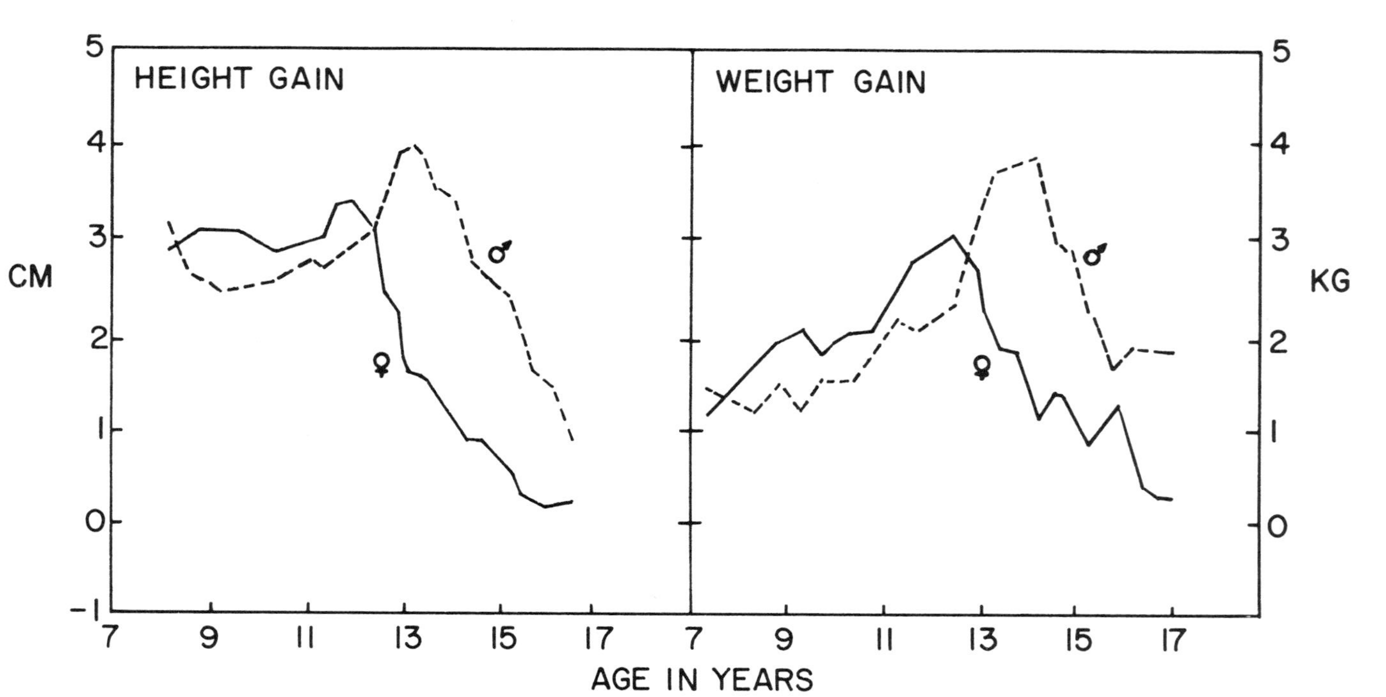

Figure 17.3. Curves of differences in mean heights and weights between successive age groups of American white youths. These curves are based on cross-sectional measurements, when successive age groups of youths were subjects. The subjects were measured once, in contrast with the subjects of longitudinal studies in which measurements of each subject were made at repeated intervals over years. Gains in height and weight reflect differences in rate of gain or loss at different ages. (After Hamill, Johnston, and Lemeshow, 1973)

the time the first menstruation occurs (Chapter 11). In boys the onset of puberty is not so clear; but it has certainly occurred by the time of the first ejaculation. *Adolescence* is the period of transition from childhood to adulthood, and begins before puberty. The *adolescent spurt* is the increase in rate of growth or activity of various parts of the body that takes place during adolescence. The maximum rate of change occurs at the spurt peak. *Sexual dimorphisms* (differences in form and physiology characteristic of females and males) are the net result of these changes in growth and growth rate or activity. This term includes the totality of the coincidental changes that take place in girls and boys during adolescence, superimposed on those developed in embryonic and fetal life and in childhood.

SKELETAL GROWTH

Adolescent skeletal growth is characterized by marked sexual dimorphism in all bones with respect to the time of the first appearance of bone in the embryo, the order of appearance of ossification centers, maturation of bones including the cessation of bone growth, and the robustness of bones, including the density of bone, the calcium content of bone, and the rate of remodeling of bone. Girls are at least one year more advanced than boys, not only in regard to individual bones, but also in regard to skeletal regions:

Growth in height ends earlier in girls (Figure 17.2).
Bones are thinner and less dense in women (Figure 17.4).
Skull is lighter and more rounded in women (Figure 17.5, Table 17.1).
Pelvis is smaller and rounder in women (Figure 17.6, Table 17.2).

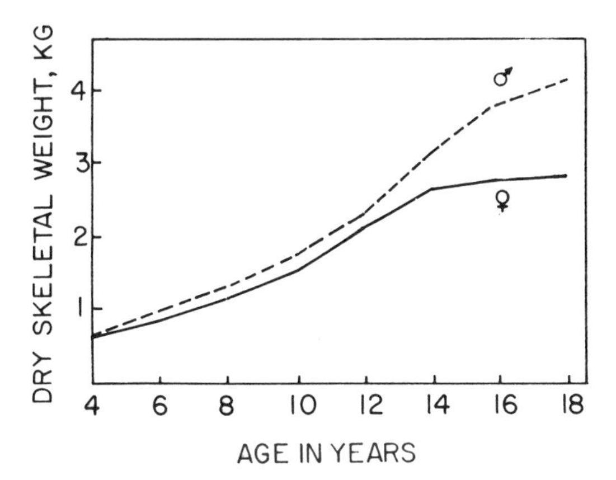

Figure 17.4. Changes in dry skeletal weight of youths. (After Cheek, 1974)

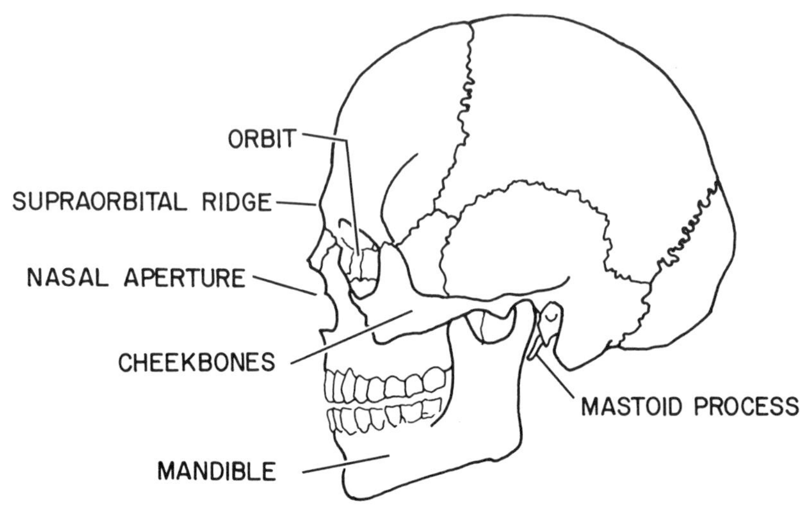

Figure 17.5. Lateral view of skull for identification of parts of the skull compared in Table 17.1. (After Krogman, 1962)

Table 17.1. Comparison of bony structures of the skull in women and men (see Figure 17.5)

Skull size	Generally smaller, rounded in women
Volume of cranium	Smaller than in men, but not in proportion to body size
Mandible	Smaller, thinner, in women
Nasal aperture	Lower, wider in women
Orbit	Higher in face, sharper margins, larger and more rounded in women
Supraorbital ridges	Less prominent in women
Forehead contour	Higher, smoother, more vertical in women
Mastoid process	Less bulky in women
Cheek bones	Higher and less prominent in women

From *The Human Skeleton in Forensic Medicine*, by W. M. Krogman. Courtesy of Charles C Thomas, Publisher, Springfield, Illinois. Copyright © 1962.

Trunk (sitting height)—adolescent spurt earlier, briefer, and smaller in girls (Figure 17.7).

Growth of long bones—adolescent spurt earlier, shorter and less marked in girls (Figure 17.8).

Growth of hands and wrist—adolescent spurt begins and ends earlier in girls (Table 17.3).

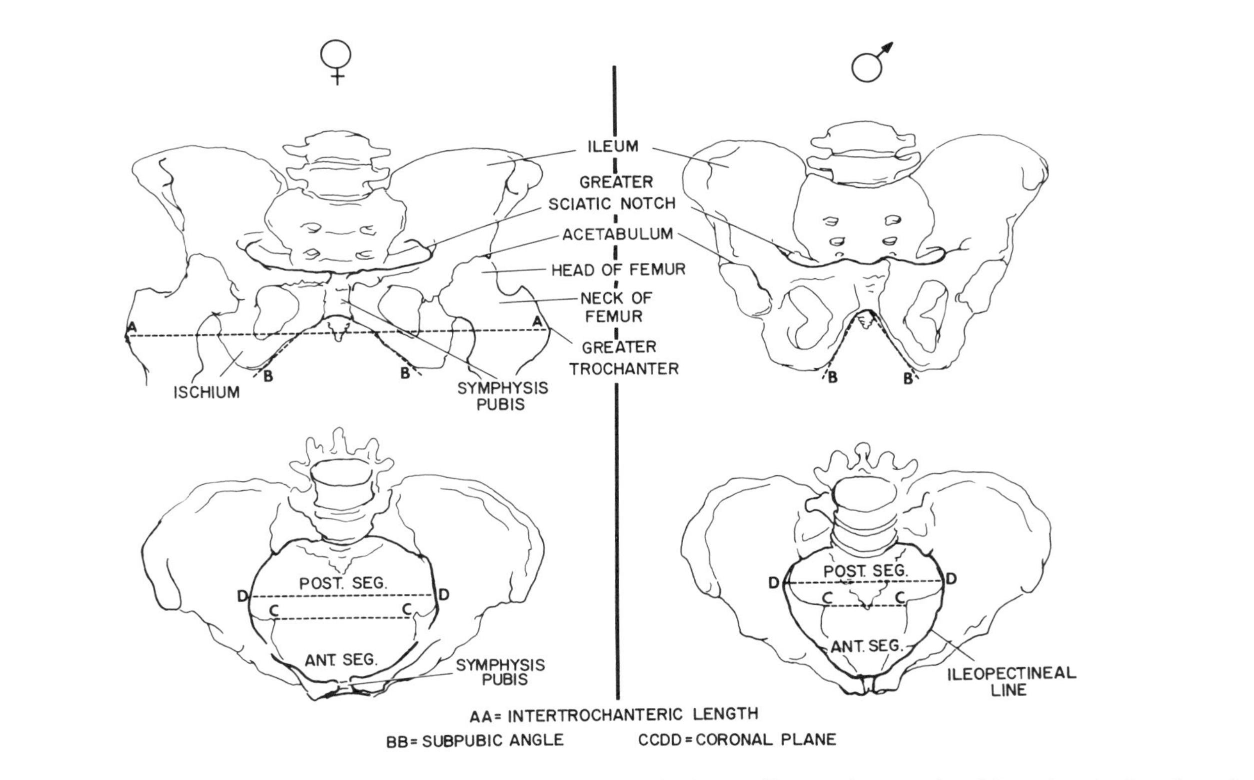

Figure 17.6. Top, frontal view of pelvis for identification of the parts mentioned in the text; Bottom, the same viewed from above to show the major differences in the pelvic rim or iliopectineal line in the two sexes. Ant. Seg., the portion anterior to the coronal plane (CCDD). Post. Seg., the portion posterior to the coronal plane. See Table 17.3. (After Krogman, 1962; Steer, 1959)

Table 17.2. Comparison of bony structures of the pelvis and trunk in women and men (see Figure 17.6).

Pelvis	Generally smaller, rounder in women
Symphysis	Lower in women
Subpubic angle	U-shaped, rounded, broader in women
Acetabulum	Small, directed anterolaterally in women; laterally in men
Greater sciatic notch	Larger, wider, shallower in women
Ischiopubic rami	More strongly everted in women
Ilium	Lower, laterally divergent in women; vertically in men
Sacrum	Shorter, broader in women
Pelvic brim	Circular or elliptical in women; heart-shaped in men
True pelvis	Shallower and more spacious in women

From *The Human Skeleton in Forensic Medicine*, by W. M. Krogman. Courtesy of Charles C Thomas, Publisher, Springfield, Illinois. Copyright © 1962.

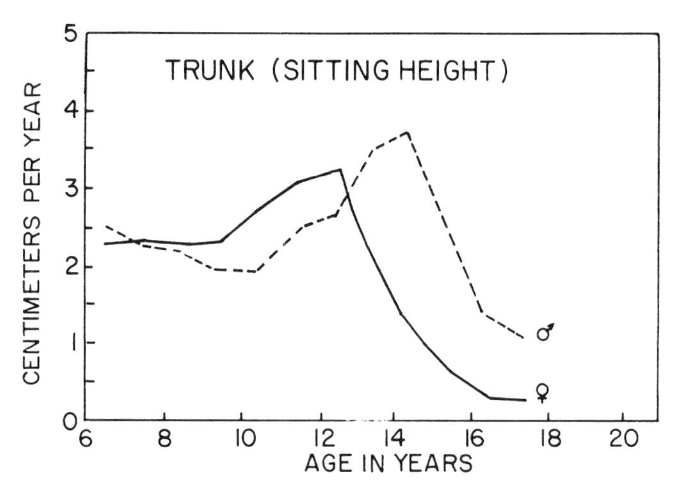

Figure 17.7. Rate of trunk growth (sitting height) of girls and boys. (After Anderson, Hwang, and Green, 1965)

Ossification of Bone

Bone is one of the connective tissues of the body. It is differentiated from other connective tissues by the characteristic arrangement of bundles of collagen fibers and by the high concentration of calcium in the form of apatite crystals. The cells that lay down these intercellular materials are called *osteoblasts.* Other cells, which are responsible for removing bone, are called *osteoclasts.*

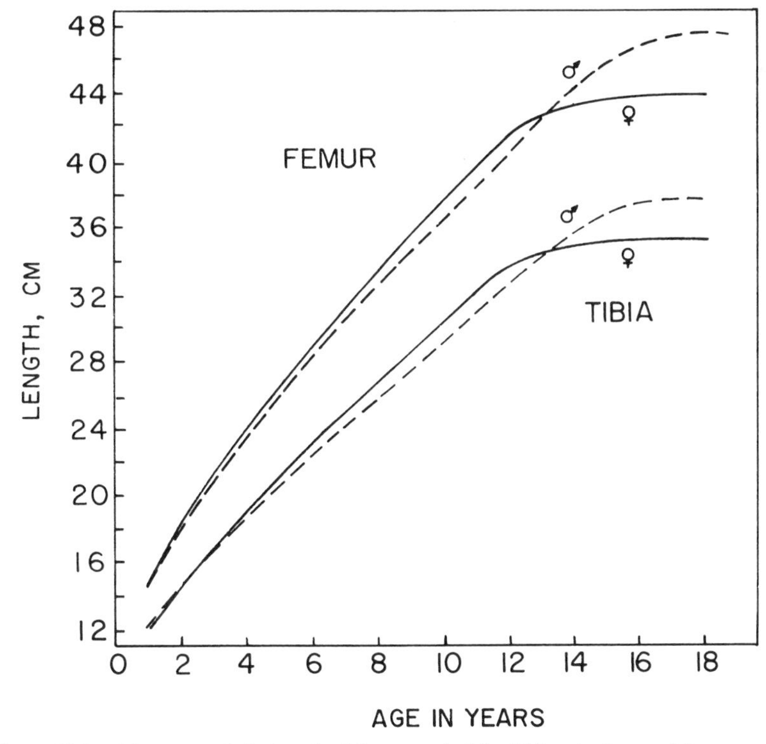

Figure 17.8. Mean growth in length of femur and tibia. (After Anderson, Messner, and Green, 1964)

Table 17.3. Chronological age in months for the two sexes, of equivalent states of skeletal development of the hands and wrist

Girls	Boys
10	12
14	19
21	28
27	36
37	48
47	60
58	72
113	132
154	186
175	204

Bone may be formed in relation to a cartilage model that is laid down first and eventually replaced by the new bone; or bone may be formed from membranes or sheets of connective tissue whose cells develop into bone-forming and bone-destroying cells. Bones formed from cartilage include all the long bones of the body (e.g., the humerus, femur, radius, ulna, tibia, and fibula). Bones formed directly from membranes include the flat bones of the head (e.g., frontal, parietal, and occipital). The long bones grow in length by going through the cartilage stage, and in diameter by going through the membrane stage. The sites where bone formation begins in each bone whether in cartilage models or in membranes, are called *centers of ossification*. The size and general shape of individual bones as well as of skeletal regions are probably largely genetically determined.

Centers of ossification based on cartilage models appear in the long bones (of the arm and leg) and in the bones of the vertebral column, the thorax, shoulder, hip and elsewhere. Long bones have at least three centers of ossification separated by cartilage—one for the shaft, and one for each end. The cartilage continues to grow until the bone reaches the mature state, and is continuously replaced by the expanding bony region. In this way, bones grow in length until intervening cartilage disappears, and the adjacent ossification centers fuse. From this time on, the *length* of a bone is fixed. While this process is taking place, the *diameter* of the bone increases by the continual deposition of membrane bone on the outside of the growing cartilage and bone model. But such a bone would be "solid," structurally weak, and much too heavy; and besides, there would not be space enough for bone marrow, which is essential for the formation of red blood cells and other lines of cells involved in immunologic processes. This is prevented by a resorption of bone from the inside, resorption lagging somewhat behind the deposition of new bone on the outside. The end process of bone deposition and resorption is a hollow shaft with various internal strengthening buttresses and external processes. The shaft is better able to withstand weights and strains by bending, and the hollow central portion can accommodate the bone marrow. The two processes of bone deposition and bone resorption take place also in membrane bone, bone being added to the outer surfaces and being removed from the inside or central cavity. These processes together account for the process of bone remodeling that takes place not only during development but also throughout life in response to physical strains and stresses by muscles and tendons, and also for bone repair following bruises, fractures, changes in work habits, or disuse. The mechanisms by which these intrinsic (genic) and extrinsic (environmental) factors are expressed by the bone cells and their extracellular products as they elegantly control the growth and remodeling processes of bone remain unknown.

In general, the sequence of appearance of centers of ossification during early life is the same in girls and boys, but the time of appearance differs in the two sexes. The centers of ossification appear earlier in girls, sometimes by as much as two years. In general, the order of fusion of the centers of ossification (when bone growth in length ceases) is the same in both sexes, but fusion takes place earlier in girls. The adolescent spurt also takes place earlier in girls than in boys, and terminates earlier also. Thus skeletal maturation is more precocious in girls, with the consequence that, statistically speaking, most women are shorter than most men. This is attributed to the effect of estrogen secreted at a higher rate after menarche. It is illustrated by the ratio of various indices in girls versus boys: height, 0.94; sitting height, 0.95; arm length, 0.8; leg length, 0.93; circumference of head, 0.96; face height, 0.90; face breadth, 0.90.

BODY SHAPE

The shape of the body is determined chiefly by the skeleton and the muscle masses attached to it, as well as by the fat tissue deposited in and around the muscles, in the subcutaneous connective tissue and in the abdominal cavity. The following terms are used in the discussion of these components of the body:

Total Fat	= Subcutaneous fat + inter- and intramuscular fat + visceral fat
Lean body mass	= Total body weight − total fat
Muscle mass	= Total body weight − (fat + bones + viscera)
Total body water	= Extracellular water + intracellular water + plasma water
Intracellular water	= Total body water − (extracellular + plasma water)
Extracellular water	= Total body water − (intracellular + plasma water)

All estimates of all body components listed are based on certain unproved assumptions, and are to this extent empirical or arbitrary.

The following five general observations are based on the use of these indirect methods:

The fat content of preadolescent girls and boys is nearly the same (Figure 17.9).

Early in adolescence, the fat content of girls increases, and thereafter remains higher than that of boys. By contrast, the fat content of boys falls during adolescence (Figures 17.9 and 17.10).

The fat content increases slowly during aging, and the sex difference persists (Figure 17.11).

Lean body mass increases sharply in boys during adolescence, and remains higher than in adolescent girls. The lean body mass in adolescent girls shows a lesser spurt but at an earlier time (Figures 17.12 and 17.13).

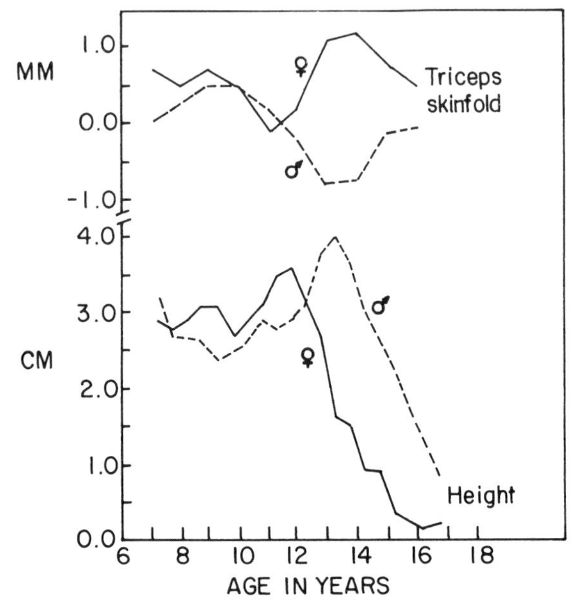

Figure 17.9. Smoothed curves represent rate of changes of triceps skinfold thickness in girls and boys, in relation to their rate of change in height. This depicts the relative growth of fat in adolescent girls and boys. (After Johnston, Hamill, and Lemeshow, 1974)

Muscle strength increases in girls and boys during adolescence, but the increase is greater (and begins later) in boys (Figure 17.14).

Some pertinent examples of how such estimates clarify some problems will be given. For example, some physically fit athletes, females or males, may weigh as much as or more than less active young persons. But analysis shows that the lean body mass was larger in the athletes, than in the non-athletes, and the opposite was true of estimates of fat. After the athletes finished training and were less active, their lean body mass fell, though not to the same low level of the pretraining period. The increase in lean body mass of trained athletes is greater in some sports than in others. For example, it is higher in gymnasts and runners than in swimmers or weight lifters. Lean body mass is related to muscle mass and to a number of other criteria; for example, heart volume, maximum oxygen consumption, and oxygen consumption at rest.

Connective Tissue

It is impossible to estimate the amount of muscle mass of the body without also measuring the amount of connective tissue in muscle. The chief components of a voluntary muscle are the muscle cells, which constitute about 80%, and the extracellular components of the connective tissue. This

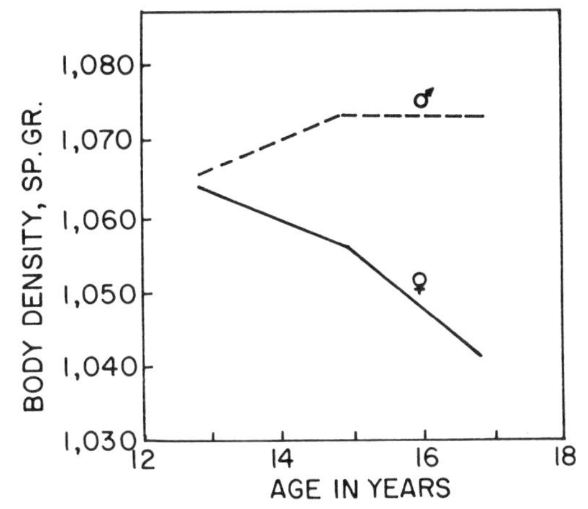

Figure 17.10. Changes in specific gravity of girls and boys during their adolescence. (After Haak, 1966)

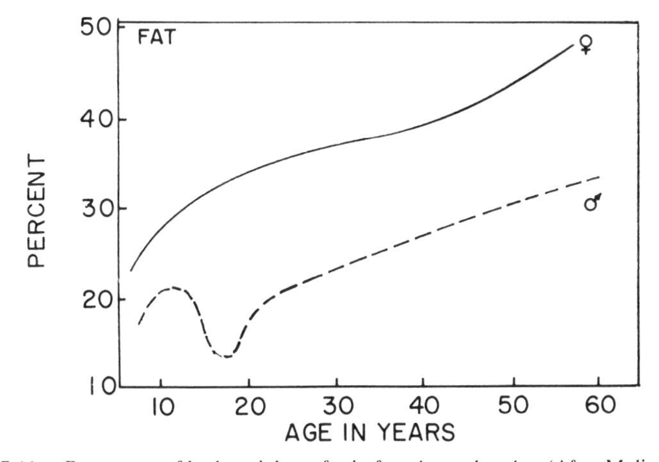

Figure 17.11. Percentage of body weight as fat in females and males. (After Malina, 1969)

binds the muscle cells to each other and to their origins and insertions, and binds the blood vessels, lymphatics, and nerves of muscle. The connective tissue in turn consists of cells (which for our present purpose can be ignored because they constitute a small percentage of the connective tissue) and extracellular components. The latter comprise collagen or collagen-like fibrils visible with the light and/or electron microscope, and the nonfibrillar components that constitute the ground substance. Only one report has attempted to characterize changes in the collagen content of the body during adolescence. According to this report, there may be a

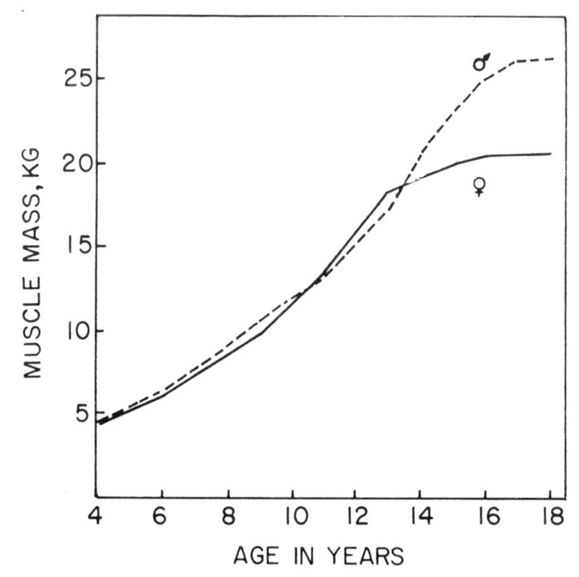

Figure 17.12. Changes in adolescent girls and boys in muscle mass as calculated from daily creatinine excretion. (After Cheek, 1974)

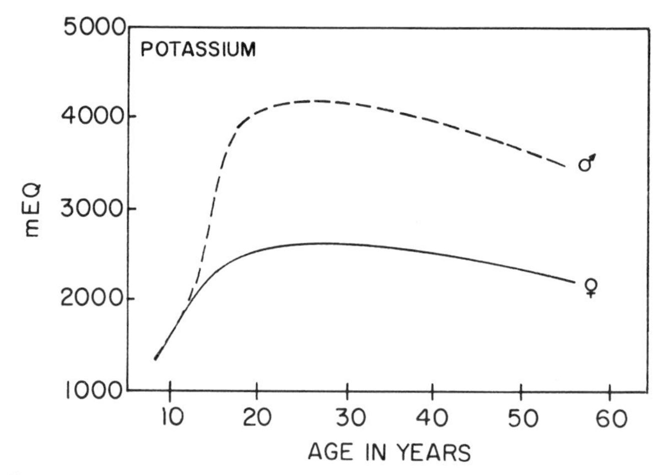

Figure 17.13. Smoothed curves of average potassium concentration of males and females, which can be roughly equated with muscle mass. (After Malina, 1969)

possible correlation of a urinary metabolic product and collagen with increasing height, weight, and age of children until puberty. At this time, a sex difference appears and boys have markedly elevated levels as compared with girls. This may be due in large part to the adolescent muscle spurt (Cheek, 1974).

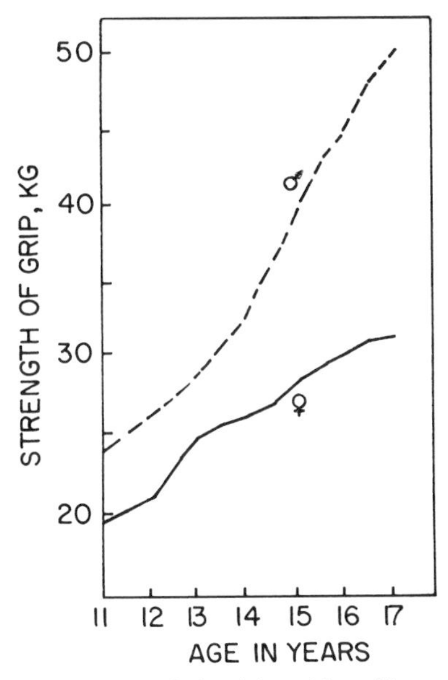

Figure 17.14. Strength of left hand grip in girls and boys illustrates the small adolescent spurt in the former and the more marked and longer-lasting spurt in the latter. (After Tanner, 1962)

As yet no estimates have been made of the solid components of the nonfibrillar ground substance, or its constituents (such as chondroitin sulfate, etc.) during growth and development of children. Nor have attempts been reported to ascertain the composition of collagen of sexually maturing young people undergoing the process of the growth spurt.

BODY COMPOSITION

Water

The simplest summary of estimates of water in various compartments of the body is given in Figure 17.15. This graph shows that total body water, intracellular water, extracellular water, and plasma volume are lower in women than in men.

Average Body Composition and Metabolism

Some of the data on differences in average body composition in women and men are summarized in Table 17.4. These values are estimated, and of course represent the particular group and class sampled. The data also show changes that take place (on the average) within each sex with age. In

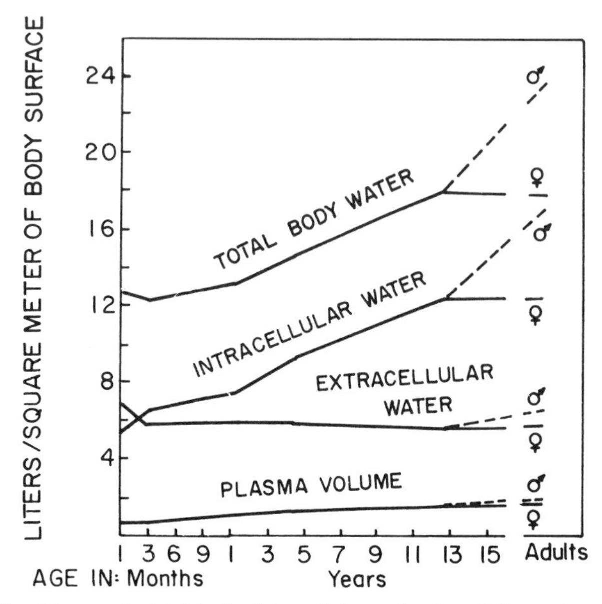

Figure 17.15. Changes in total body, intracellular and extracellular water, and plasma volume expressed in relation to square meters of body surface, which is used here as a measure of size. (After Friis-Hansen, 1965)

addition to sex differences in form, size, relations, and chemical composition, there are also physiologic and biochemical differences:

Systolic blood pressure is lower in women beginning with adolescence (Figure 17.16).

Lower hemoglobin content and red blood cell count occurs in women, beginning with adolescence (Figure 17.17).

Heart rate and mouth temperature is higher in women beginning with adolescence (Figure 17.18).

Basal metabolic rate is lower throughout life in women (Figure 17.19).

Numerous respiratory measurements are smaller in women beginning with adolescence.

Alkaline phosphatase in blood of postmenopausal women exceeds that of men (Figure 17.20).

Enzyme Levels

Sex differences in the level of certain enzymes in the blood plasma or serum have been found. The alkaline phosphatase level in men is reported to be higher than in women up to about 55 years of age, after which the level in women is higher than in men (Figure 17.20). This may account for the heightened resorption of calcium from bones in postmenopausal

Table 17.4. Body composition in average normal adults at age 25 and age 65 to show numerical values of various body components as estimated by a single method[a]

	Women				Men			
	Absolute		Relative (% Body Weight)		Absolute		Relative (% Body Weight)	
	Age 25	Age 65	Age 25	Age 65	Age 25	Age 65	Age 25	Age 65
Body weight	69.0 kg	60.0 kg			70.0 kg	70.0 kg		
Body water	30.8 liters	28.0 liters	51.3%	46.7%	41.2 liters	37.0 liters	58.9%	52.9%
Intracellular	16.6 liters	14.3 liters	27.7%	23.8%	24.0 liters	19.2 liters	34.3%	27.4%
Extracellular	14.2 liters	13.7 liters	23.6%	22.9%	17.2 liters	17.8 liters	24.6%	25.5%
Plasma volume	2760 ml	2462 ml	4.6%	4.1%	3302 ml	2940 ml	4.7%	4.2%
Lean body weight	42 kg	38.2 kg	70.2%	63.7%	56.3 kg	50.5 kg	80.4%	72.1%
Body fat	17.9 kg	21.8 kg	29.8%	36.3%	13.7 kg	19.5 kg	19.6%	27.9%
Skeletal weight	4.4 kg	4.2 kg	7.3%	7.0%	5.8 kg	5.7 kg	8.3%	8.1%

[a] Adapted from Oleson (1965).

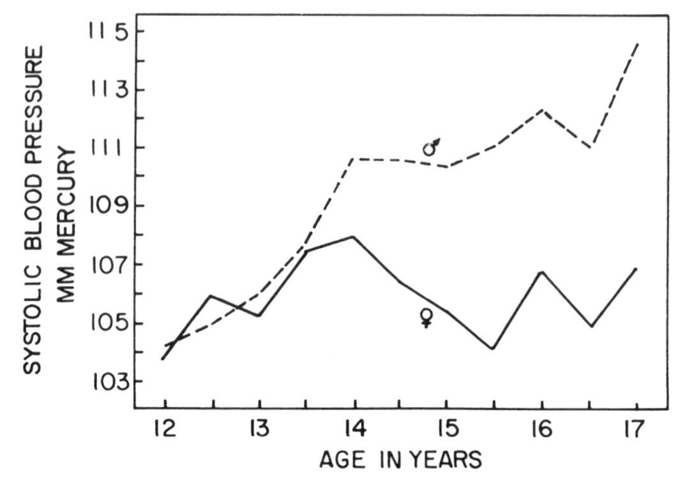

Figure 17.16. Comparison of systolic blood pressure during the adolescent spurt of girls and boys. The blood pressure rise is modest and of short duration compared to that of the boys. (After Tanner, 1962)

women, and perhaps also for the reduction in the thickness of cortical bone that takes place in these women. Curves for serum cholesterol show the same sex differences in adults (Figure. 17.20).

LARYNX

Changes in the larynx occurring late in the adolescence of boys result in lowering of the voice. No corresponding change takes place in girls.

HAIR

Hair is more important cosmetically and psychosocially than biologically. Hair is one example of many sex differences whose biologic origin and function are not known. Fetal (or lanugo) hair is shed during the last quarter of gestation and during the first two years of infancy. It is replaced postnatally by fine hair called *vellus,* or coarse hair called *terminal.* When the latter become separated from their roots and cease to grow they are called *transitional,* and may be pulled, brushed or washed out without pain. The transitional hair is replaced in time by new terminal hair. In the scalp, more than four-fifths of the hair is terminal, and the remainder is transitional. There are no marked sexual differences in hair density, i. e., the number of hairs per unit area. The sexual differences are in hair pattern and in the relative proportion of vellus and terminal hair. While hair on some parts of the body shows no significant sexual differences, other

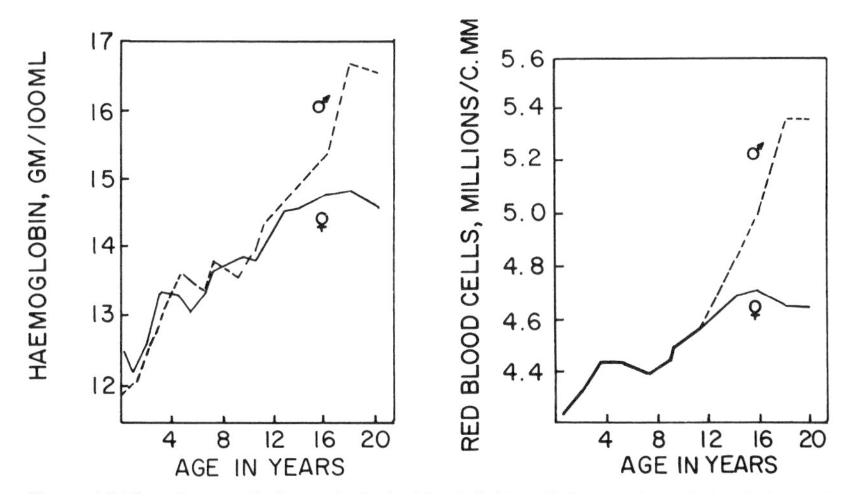

Figure 17.17. Changes in hemoglobin in blood (left) and the number of circulating red blood corpuscles (right) of girls and boys. (After Tanner, 1962)

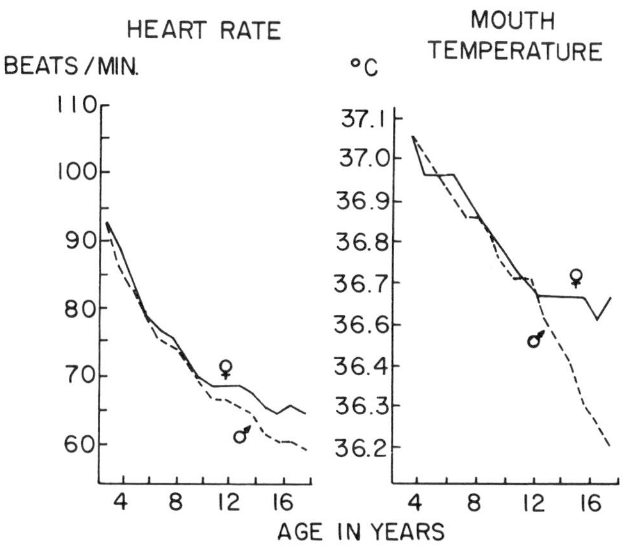

Figure 17.18. Changes in basal heart rate (left) and mouth temperature (right) of girls and boys. (After Tanner, 1962)

parts do show sexual differences. Hair patterns are genetically determined, but those of some parts of the body (such as hair of the pubis, chest, armpit, face, neck and scalp) may be markedly affected by sex hormones, especially androgens. There is little doubt that environmental factors such as infections, cleanliness, and manual manipulation also in-

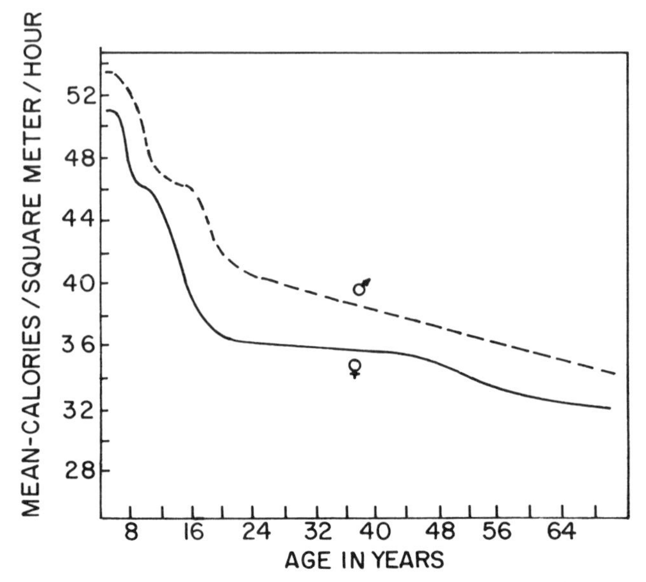

Figure 17.19. Differences in basal metabolism of females and males throughout most of life. (After Shock, 1966)

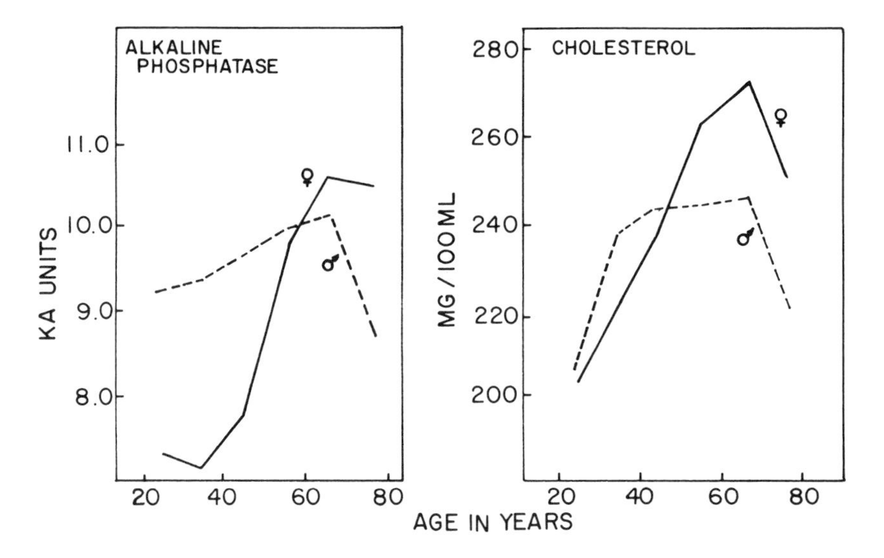

Figure 17.20. Serum alkaline phosphatase activity and cholesterol concentration in women and men. (After deWijn, 1966)

fluence the hair pattern. Comparisons of hair patterns in identical twins and of siblings emphasize the importance of genetic factors. Targets of androgenic hormones (hair of the beard, pubis, armpit, chest, and scalp) respond more readily to hormones. It is thought that androgens of the circulating blood combine with receptors on the surface of the sensitive hair bulb, and directly and indirectly activate protein synthesis (which goes primarily toward the synthesis of keratin). The source of androgen is thought to be chiefly from the adrenal cortex and ovary (chiefly the former) in girls and women, and from the testis in boys and men.

The main sex differences in hair are that:

1. Hair patterns differ in certain sites such as the face and chest.
2. The proportion of terminal (coarse) hair is higher in men than in women. In adult women, about 35% of hair on the chest, trunk, legs, and arms are terminal, as against about 90% in adult men.
3. The terminal hair is coarser in men than the terminal hair of women.
4. The pubic hair pattern in women is that of an inverted triangle, while that of men is that of a rhomboid, with the anterior apex reaching to the umbilicus. The appearance of pubic hair is one of the earliest signs of puberty (Figure 17.21).

Changes During Menopause and Pregnancy

The hair pattern of adult women is rather constant with two exceptions: 1) During the menopause the proportion of terminal hair in the scalp decreases (giving the appearance of a thinning of the hair), and some coarse hair may appear on the face, probably stimulated by androgens from the adrenal cortex. 2) In the last two-thirds of pregnancy, the percentage of coarse hair increases while the growth rate decreases; after parturition, there may be an excessive general loss of hair all over the body, but the hair is replaced in the course of time. Axillary hair goes through the same changes as scalp hair during and after pregnancy.

Hirsutism

Hirsutism (the excessive growth of hair in the male pattern) occurs in some women. It may be caused by excessive secretion of androgens by the adrenal cortex and the ovaries. The target cells in the hair follicle may be unduly sensitive to circulating androgen. If this appearance is psychologically troublesome, it may be alleviated by suitable medical care.

Baldness

Baldness (alopecia) is almost always limited to men. It begins with the receding forehead line, where there is a progressive reduction in the diameter, length, and pigmentation of the hair. In other words, the coarse hair of the scalp is replaced by finer and finer hair resembling vellus hair

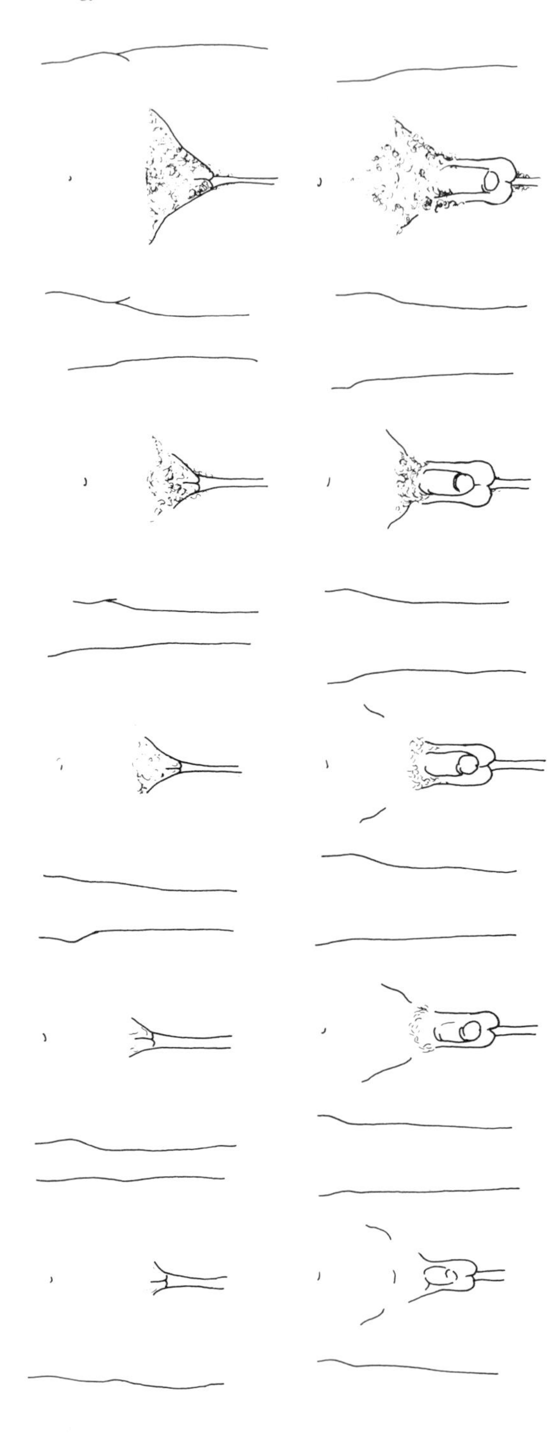

Figure 17.21. Stages in the growth of pubic hair in girls and boys. The stages are described on p. 279. (After Tanner, 1962)

of the infant. In old people, this process is supplemented by loss of hair through atrophy of hair follicles. In women, there may be a slight recession of the hair line, and a slight general reduction in the percentage of coarse hair, with some atrophy of hair follicles supervening in older women.

Pubic Hair

The development of pubic hair during adolescence has been empirically separated into five stages for purposes of comparison in different individuals (Figure 17.21). The accompanying descriptions for the two sexes are:

Stage 1: Infantile or young child stage, when there is no true pubic hair (not shown).
Stage 2: Sparse growth of slightly pigmented hair.
Stage 3: Hair darker and coarser, spreads.
Stage 4: Hair adult in character, confined to a smaller area than in adult.
Stage 5: In girls, hair distributed as inverse triangle, and spreads to medial side of thighs. In boys, hair distributed as inverse triangle, which serves as the base of another triangle whose apex reaches up to the umbilicus.
Stage 6: Adult state.

SKIN GLANDS

Sebaceous Glands

Sebaceous glands secrete a fatty fluid (sebum) on the surface of the skin. Sebum is secreted over most of the body through small ducts that empty into hair follicles and pass between the follicles and the hairs to reach the surface. In a few small parts of the body the ducts empty directly on the surface of the skin—the nipples and areolae, the labia minora and the prepuce. The glands seem to be smaller in women than in men except for those of the nipple and areola. In postmenopausal women, the amount secreted is reduced. The glands are highly sensitive to androgens, and, in boys, secrete more sebum when they reach puberty.

Sweat Glands

Sweat glands are present all over the body, with some exceptions—the glans clitoridis and the labia minora in women, and the glans penis and inner surface of the prepuce in men. The coiled tubular glands empty directly on the surface of the skin through very fine ducts between hair follicles. The ducts may be plugged and cause acne, or they may become leaky and cause rashes. When these conditions are bothersome, a physician should be consulted for treatment. The secretion of sweat is part of the mechanism of lowering body temperature through evaporation of the water. It is claimed that the sweating mechanism of women athletes is less efficient than that of men, with the result that female athletes tend to overheat.

Apocrine Glands

Apocrine glands are modified sweat glands that are especially common in the armpit (axilla). Like sebaceous glands, nearly all empty their secretion on the surface of the skin through the hair follicles. They are said to go through cyclic periods, enlarging during the premenstrual period, and regressing during menstruation.

BREASTS

The breast was discussed in detail in Chapter 15. Concerning the maturation of the breast during adolescence, the five developmental stages (illustrated in Figure 17.22) have been described as follows:

Stage 1: (Pre-adolescent): Elevation of papilla.
Stage 2: (Bud stage): Elevation of breast and papilla as small mound. Enlargement of areola diameter.
Stage 3: Continued enlargement of breast and areola.
Stage 4: Projection of the areola and papilla beyond the primary breast mound.
Stage 5: Projection of papilla only, with recession of the areola to the plane of the primary mound.

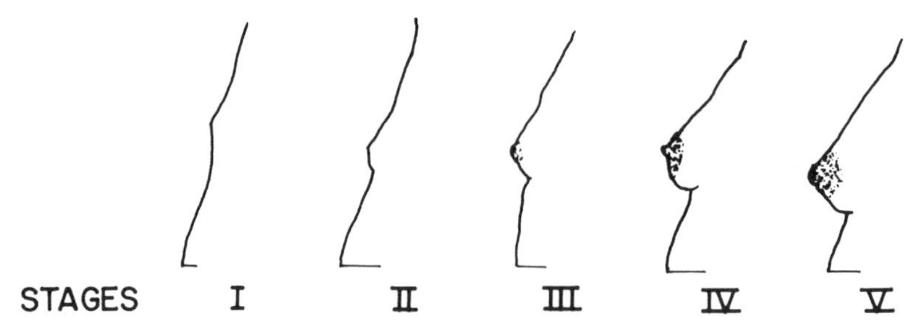

Figure 17.22. Stages in the maturation of the breast during adolescence. For description, see above. (After Tanner, 1962)

REFERENCES

Anderson, M., S.-C. Hwang, and W. T. Green, 1965. Growth of the normal trunk in boys and girls during the second decade of life. J. Bone Joint Surg. 47A:1554-1564.

Anderson, M., M. B. Messner, and W. T. Green. 1964. Distribution of lengths of the nomal femur and tibia in children from one to eighteen years of age. J. Bone Joint Surg. 46A:1197-1202.

Avers, C. J. 1974. Biology of Sex. John Wiley & Sons, Inc., New York.

Cheek, D. B. 1974. Body composition, hormones, nutrition, and adolescent growth. *In* M. M. Grumbach, G. D. Grave, and E. E. Mayer (eds.), Control of the Onset of Puberty. John Wiley & Sons, New York, pp. 424-440.

Coleman, W. H. 1969. Sex differences in the growth of the human bony pelvis. Am. J. Phys. Anthropology 31:125-152.

Conger, P. R., and R. B. J. Macnab. 1967. Strength, body composition, and work capacity of participants and nonparticipants in women's intercollegiate sports. Research Quart. 38:184-192.

Corbitt, R. W., D. J. Erickson, and M. L. Thornton. 1974. Female athletics. JAMA 228:1266-1267.

deGaray, A. L., L. Levine. J. E. L. Carter (eds.). 1974. Genetic and Anthropological Studies of Olympic Athletes. Academic Press, New York.

deWijn, J. F. 1966. Changing levels of blood constituents during growth. *In* J. J. van der Werff ten Bosch and A. Haak (eds.), Somatic Growth of the Child. H. E. Stenfert Kroese, Leiden.

Drinkwater, B. L. 1973. Physiological responses of women to exercise. Exerc. Sport Sci. Rev. 1:125-153.

Enlow, D. H. 1968. The Human Face. Harper & Row, Publishers, New York.

Eveleth, P. B., and J. M. Tanner, 1976. Worldwide Variation in Human Growth. Cambridge University Press, Cambridge.

Ferriman, D. 1971. Hair Growth in Health and Disease. Charles C Thomas, Springfield.

Filer, L. J., Jr. 1966. Developmental aspects of biochemistry. *In* F. Falkner (ed.), Human Development, W. B. Saunders, Co., Philadelphia.

Friis-Hansen, B. 1965. Hydrometry of growth and aging. *In* J. Brožek (ed.), Human Body Composition. Symposia of the Society for the Study of Human Body Composition, VII, Pergamon Press, New York, pp. 191-209.

Frisch, R. E. 1974. Critical weight at menarche, initiation of the adolescent growth spurt, and control of puberty. *In* M. M. Grumbach, G. D. Grave and E. E. Mayer (eds.), Control of the Onset of Puberty. John Wiley & Sons, New York, pp. 403-422.

Frisch, R. E. 1975. Critical weights, a critical body composition, menarche, and the maintenance of menstrual cycles. *In* Symposium on Biosocial Interrelations in Population Adaptation. Mouton Publications, The Hague, pp. 319-352.

Garn, S. M. 1963. Human biology and research in body composition. *In* J. Brozek (ed.), Body Composition. Monograph of a 6-day conference on this subject held by N.Y. Acad. Sci., Ann. N.Y. Acad. Sci., Vol. 110, pp. 429-446.

Haak, A. 1966. Body composition. *In* J. J. van der Werff ten Bosch and A. Haak (eds.), Somatic Growth of the Child. H. E. Stenfert Kroese, Leiden, pp. 89-98.

Hamil, P. V. V., F. E. Johnston, and S. Lemeshow. 1973. Height and Weight of Youths 12-17 Years. DHEW Publication No. (HSM) 73-1606. National Center

for Health Statistics, Rockville, Md. Vital and Health Statistics, Series 11, No. 124. U.S. Government Printing Office, Washington, D.C.
Hamilton, M. E. 1975. Variation among five groups of amerindians in the magnitude of sexual dimorphism of skeletal size. Unpublished doctoral thesis, University of Michigan.
Harris, D. V. 1973. Conditioning for stress in sports. American Association of Health, Physical Education and Recreation. Division for Girls and Women's Sports. DGWS Research Reports: Women in Sports, pp. 73-83.
Johnston, F. E., P. V. V. Hamill, and S. Lemeshow. 1974. Skinfold Thickness of Youths 12-17 Years. DHEW Publication No. (HRA) 74-1614, Vital and Health Statistics Series 11, No. 132. U.S. Government Printing Office, Washington, D.C.
Jokl, E. 1964. Physiology of Exercise. Charles C Thomas, Springfield.
Jokl, E., and P. Jokl. 1968. The Physiological Basis of Athletic Records. Charles C Thomas, Springfield.
Jokl, E., and E. Simon. 1964. International Research in Sport and Physical Education. Charles C Thomas, Springfield (see especially: pp. 238-253, 262-285).
Katchadourian, H. 1977. The Biology of Adolescence. W. H. Freeman & Co., San Francisco.
Krogman, W. M. 1962. The Human Skeleton in Forensic Medicine. Charles C Thomas, Springfield.
Malina, R. M. 1969. Quantification of fat, muscle, and bone in man. Clin. Orthop. 65:9-38.
Marshall, W. A. 1977. Human Growth and Its Disorders. Academic Press, New York.
Milvy, P. (ed.). 1977. The Marathon: Physiological, Medical, Epidemiological and Psychological Studies. Ann. N.Y. Acad. Sci. 301:726-733, 764-776, 793-807, 808-815.
Montagna, W., and R. L. Dobson (eds.). 1969. Hair Growth. Pergamon Press, New York.
Montagna, W., and P. F. Parakkal. 1974. The Structure and Function of Skin. Academic Press, New York.
Oleson, K. H. 1965. Body composition in normal adults. *In* J. Brožek (ed.), Human Body Composition. Symposia of the Society for the Study of Human Body Composition, Vol. 7, Pergamon Press, New York, pp. 177-190.
Parizkova, J. 1968. Body composition and physical fitness. Curr. Anthropol. 9:273-287.
Pyle, S. I., A. M. Waterhouse, and W. W. Greulich. 1971. A Radiographic Standard of Reference for the Growing Hand and Wrist. The Press of Case Western Reserve University; Year Book Medical Publishers, Inc., Chicago.
Scammon, R. E. 1930. The measurement of the body in childhood. *In* J. A. Harris, C. M. Jackson, D. G. Paterson, and R. E. Scammon (eds.), The Measurement of Man. University of Minnesota Press, Minneapolis, pp. 171-215.
Shock, N. W. 1966. Physiological growth. *In* F. Falkner (ed.), Human Development. W. B. Saunders Co., Philadelphia.
Sinclair, D. 1978. Human Growth after Birth. 3rd Ed. Oxford University Press, New York.
Steer, C. M. 1959. Evaluation of Pelvis in Obstetrics. W. B. Saunders Co., Philadelphia.
Tanner, J. M. 1962. Growth at Adolescence. 2nd Ed. Blackwell Scientific Press, Oxford.
Tanner, J. M. 1974. Sequence and tempo in the somatic changes in puberty. *In*

M. M. Grumbach, G. D. Grave, and E. E. Mayer (eds.), Control of the Onset of Puberty. John Wiley & Sons, New York, pp. 448-467.

Tanner, J. M. 1978. Fetus in Man. Harvard University Press, Cambridge.

Tanner, J. M., R. H. Whitehouse, and M. Takaishi 1966. Standards from birth to maturity for height, weight, height velocity and weight velocity: British children, 1965. Arch. Dis. Child. 41:454-471, 613-635.

Van der Linden, F. P. G. M., and H. S. Duterloo 1976. Development of the Human Dentition. Harper & Row Publishers, New York.

Van Gerven, D. P. 1972. The contribution of size and shape variation to patterns of sexual dimorphism of the human femur. Am. J. Phys. Anthrop. 37:49-60.

Young, J. Z. 1971. An Introduction to the Study of Man. Oxford University Press, Oxford.

18 Comments on Sex Differences

The sequence of biologic events during adolescence is shown for some traits in Figure 18.1. These figures remind us of the tremendous variability in the time of onset and rate of maturation that characterize the development of various functions. The times given apply to certain specific groups of English girls and boys measured in a specific decade. The times could be different for another group of young people in England, Wales, Europe, or the United States, even during the same decade. In addition, the degree of development of some traits at maturity would vary among the individuals in this specific group, as well as in all other groups that might be measured at that time or any other time. The heavy bars represent the average period of maturation of the respective trait, which takes place in several stages not shown in the figure. The range of variation is indicated by the associated thin line.

Changes in other organs also take place. Shortly before menarche, the stratified squamous epithelium of the vagina becomes thicker and begins to accumulate cytoplasmic glycogen. When these cells are shed into the lumen of the vagina and die, the glycogen is digested by the bacilli normally resident in the vagina, and lactic acid is generated. This begins a change in the acidity of the vagina that normally persists to menopause. A second feature occurs during the growth spurt in adolescent girls: more fat is accumulated in certain regions of the body, especially the thighs, buttocks, chest, lower abdomen, and the mons veneris. A third feature is that after the menarche the modified sweat glands of the axilla (armpit) and labia majora may undergo cyclical changes in activity in phase with the menstrual cycle.

CAUSE OF EARLY ADOLESCENT SPURT IN GIRLS

The adolescent spurt must be considered a unity, despite the variation in the degree of development of one or more factors. That is, the accelerated

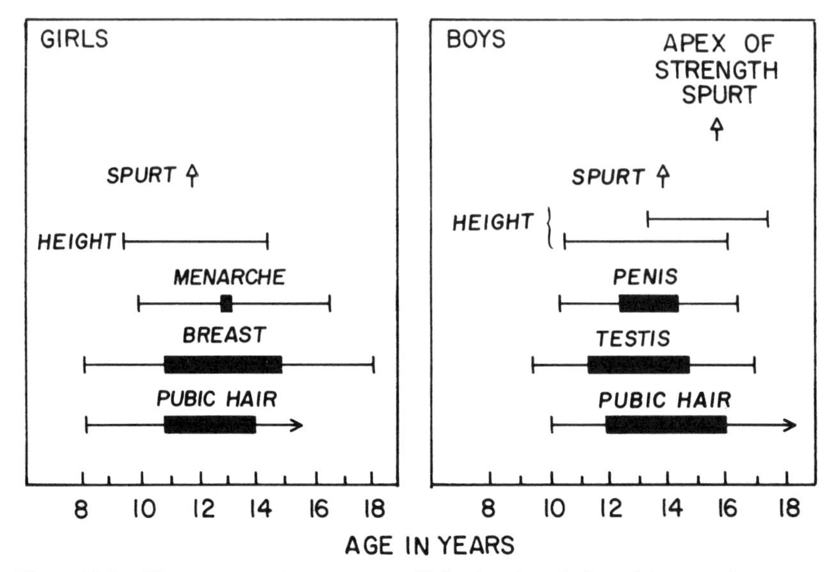

Figure 18.1. The sequence of some events which take place during adolescence in average English girls and boys. Age is shown in the horizontal axis. The thick bars show the peak period when the empirically defined maturational stages take place. The thin bars show the range of variation of the traits, which is considerable. The mean of the growth spurt in height is shown by the arrows labeled "spurt." The apex of the strength spurt of boys is similarly marked. The time of onset and end of the height spurt for girls is indicated by the thin bar. The range of time of onset of the height spurt for boys is indicated by the lower height bar, and the end of the spurt by the upper height bar. The arrow for pubic hair indicates that the final stages may be reached at some indefinite time in adulthood. The range of variation as to time of onset, degree of development, and time of termination is very great for all traits, and all values are "normal." (After Tanner, 1962; Young, 1971)

rate of growth in height and weight, the skeleton, the muscle mass and fat occur together whether the menarche is early or late. Permeating all are the genetic effects, as manifested by comparative studies of age and rates of development of various structures or functions of identical and non-identical twins, mother-daughter data, and data from familially unrelated girls. Such studies were the basis for earlier discussions of the genetic role in menarche, skeletal growth spurts and attainment of maturity, and loss of milk teeth and eruption of permanent teeth. In these respects, girls are more advanced than boys. It has been suggested that this relative retardation of boys may be caused by genes in the Y chromosome, or alternatively that the double dose of X chromosomes in girls inhibits muscle development and growth. But the evidence is not at all clear. Whatever the genetic basis might be, it seems that, in nearly all traits considered, the genetic effects probably depend on the combined action of genes located at several different loci and probably also on different chromosomes, interacting with the environment.

HORMONAL EFFECTS

It is also clear that hormones are part of the mechanism of genic control, in the sense that in the postadolescent or adult (at least before menopause) the synthesis and secretion of the hormones of all the endocrine gland cells are under the control of the genes of the endocrine cells. Probably the receptors of hormones in the target organs or cells are also genically controlled. These statements are especially pertinent to the action of estrogens, adrenal corticosteroids, testosterone and other androgens, and to the sequence of hormones from the hypophysis and hypothalamus that antecede these. But all hormones concerned with known cellular or organismic activities are automatically involved in the processes of growth and maturation. This includes the adrenal hormones, the pituitary hormones, the ovarian and testicular hormones, the hormones of the islets of Langerhans, and the parathyroid hormones. All of these are necessary for the continued health of the individual, but replacement therapy or excessive doses have not usually affected the growth rate of the child in the past. This is in marked contrast with growth hormone (and growth hormone-releasing hormone), iodinated thyroglobulin (or thyroxine or triiodotyrosine), and thyroid stimulating hormone as well as thyroid stimulating-releasing hormone. Absence of growth hormone results in dwarfed stature, which is corrected by the administration of growth hormone. Similarly, absence of iodinated thyroid hormone results in cretinism, with limited growth and retarded IQ, symptoms that are corrected by early administration of the iodinated compounds.

DARWINIAN SELECTION

It has been proposed that the particular pattern of the exaggeratedly delayed turning on of cellular (and hormonal) mechanisms associated with adolescence has been selected for, during the evolution of the human species. The exaggeratedly prolonged infancy, childhood, and adolescence of the young would tend to favor their learning in a protected environment under parental guidance and their subsequent survival in conditions where learned skills were at a premium. This hypothesis, however, can not be proved.

Statistically, women are shorter, their pelvic brim more circular, their true (lower) pelvis roomier, and their bodies fatter and less muscular than men. These differences are associated with corresponding differences in the circulatory system, the respiratory system, and the basal metabolic rate, as well as some sex differences of lesser importance such as those involving the texture of the skin, hair, and glands of the skin.

The explanation now favored for at least some of the sex differences enumerated above is that they arose early in the history of the ancestors of *Homo sapiens* and evolved through natural selection as adaptations to survival and successful procreation. According to this hypothesis, the smaller height, weight and muscle mass, and the larger fat depots of women are related to the storage of energy in the form of fat in amounts sufficient for the support of pregnancy and lactation in a hostile, sometimes unfavorable environment. In times of nutritional shortage, less food would be required to maintain the smaller muscle mass and more could be stored as fat. This interpretation accommodates also the observations that women tend to survive nutritional stress and the stress of illness more readily than men. Furthermore, it is consistent with the findings that the fat content must exceed a certain level for the onset of menarche and to overcome amenorrhea. (The association of percentage body fat with menarche and menstruation may, however, be fortuitous in that it reflects the close association in the hypothalamus of the centers for accumulation of fat and the neuroendocrine nuclei of the hypothalamus; this association is either by close physical relationship or promoted by way of the afferent tracts to these nuclei.) But now this adaptation is superfluous in advanced and industrialized societies, at least for a certain portion of the population for whom prenatal and pediatric care are more readily available than in earlier times.

The desire for slimness (especially in women of childbearing age) and the emphasis on physical fitness, both culturally influenced goals, deviate from a pattern that might have developed eons past because fat provided a degree of protection from famine that favored survival of the species under the harsh and uncertain conditions of the time. This hypothesis, too, can not be proved.

GROWTH RECORDS AND PUBLIC HEALTH

It would seem that the anthropologic studies on which most of these reports are based depend on certain standardized methods, and one would suppose that some agreement can be reached on standardized, routine, reliable, and rapid tests at least for lean body mass, muscle mass, fat, and various cardiac and respiratory traits. The importance of such determinations for evaluating the health of growing children cannot be overestimated, for these growth curves collectively can be used as an index of normal versus abnormal development. This becomes feasible now on a national scale with the present state of the art of computer usage. Such a program would make it possible to identify abnormalities of development at an early age in individual children, at least in certain growth disorders. As Eveleth and Tanner (1976) put it in a most remarkable contribution: "A careful assessment of the growth of children remains one of the best guides to the healthy development of present populations." Their stupendous book, a collection of data on child growth, was compiled as a part of the contributions of the Human Adaptability Section of the International Biology Program (1964–1974). Data show that there are differences in growth of ethnic groups in this country that may be largely cultural in nature. Some racial differences seem to have a genetic basis, at least in part. Whether we are dealing with racial, sexual or individual differences, even when we know there is a genetic component, the prudent course is to improve the environmental factors as much as possible for all children, and to enlarge our understanding of the factors responsible for the onset of puberty, various growth spurts, sexual dimorphisms, and other sex differences.

SPORTS AND EXERCISE

The current upsurge of physical activity of women must be attributed largely to the present cultural views of women—the drive toward the full expression of the total range of women's potential and ability in all fields, including in law, politics, industry, and education. This movement is still in the process of developing, however haltingly. The great explosion of physical energy of women spills over into the expansion of sports for both the professional and the amateur. This is a far cry from two generations ago, when sports for women were commonly frowned on as tending to masculinize "young ladies." Then, girls were supposed to reach their peak of activity in sports at 15 or 16 years of age, at which time they were prepared to settle down physically to become patient, passive, nurturant (though emotionally sensitive and unstable) ladies who left the vigorous activities to the more aggressive men. Exercise and sports were supposed to make women musclebound, and to interfere with menstruation and procreation. The rate of improvement in women's records in sports has accelerated without realization of any of these dire predictions.

As Jokl (1964) wrote: ". . . advancement of athletic records for women reflects the same social, economic, and technologic changes that underlie the evolution of athletic records for men . . ." and it may be true that "Current rates of improvement of women's athletic records are greater than those of men. This differential tendency will continue for some time. . . ." Social approval of the current expansion of sports in schools and colleges will lead to more challenges to male supremacy in more sports in the future. In the end, when the curves of rate of improvement of women's records flatten out, the prediction is that in many sports "performance equality will never be reached." This is especially so in the "heavy" sports, e.g., wrestling, weight lifting, football, and ice hockey, where we can expect few women to be competitive with the exceptional men. These predictions on the limits of women in sports are based on differences between trained men and women, which seem to be biologically determined: trained women have a smaller lean body mass and muscle mass, and a higher proportion of fat than the trained man. Also, women are shorter, have smaller legs and strides, have smaller hearts and lungs, and less circulating hemoglobin. For example, among female and male distance runners, the woman is about 10 cm shorter, 13 kg lighter in total body weight, 10 kg lighter in lean body weight, and has 2 kg more fat.

These physical differences are not necessarily important in industry, where women are now employed as stevedores, truck drivers, and miners, for example, and in the various trades as painters, carpenters, and electricians, where they can earn good salaries, especially when unionized.

Women have been improving their sports records at a great rate in recent years. This must be attributed to improved training, better environmental conditions in the affluent countries (at least for the sports participants), higher rewards (in finances and status), stronger motivation, and the exuberance of an exploding social movement based on a burgeoning sense of feminism.

Trained women are more physically fit than untrained people because usually their lean body mass and muscle mass are greater and the relative proportion of fat is reduced. The physical improvements during training are achieved gradually during serious training; the increased muscle mass is caused primarily by enlargement of the size of the muscle fibers. The decrease in fat content and in the proportion of fat is probably primarily owing to the decrease in the size of individual fat cells. The mechanism by which fat is differentially used up is unknown. It may be related to the preferential shunting of blood into fatty regions, or the opening up of the rich capillary bed of fat. The physiologic mechanisms concerned in the adaptation to strenuous training are largely unknown, and need further elucidation, especially in women.

For most athletes in most sports, fat is a burden or weight handicap to be borne by the competitor. It is said to be a good insulator against cold in such sports as swimming. But it has yet to be proved that such insulation does in fact protect significantly against the cold. Women are believed to be less efficient in dissipating heat through sweating, although the evidence is not convincing. In earlier times, menstruation and procreation were claimed to be adversely affected in athletes. There is no good evidence to support this claim, and plenty of evidence to the contrary. Sportswomen tolerate pregnancy better than untrained women: they have more complication-free pregnancies and greater ease of delivery than normal but less active women. After a prolonged period of training, women tend to become amenorrheic. But after the period of training and competition is completed, menstrual cycles resume. Athletic women tend to develop (or accentuate preexisting) conditions of anemia or iron deficiency; both are readily correctable by slowing the pace of training and proper nutritional supplements.

The beneficial effects of exercise and sports have been recognized by the Committee on the Medical Aspects of Sports of the American Medical Association, which gave an almost unqualified approval to sports when engaged in by girls' or women's teams. The Committee also endorsed mixed-sex sports where body contact is minimal, providing there is an annual medical examination, and proper training, coaching, officiating, equipment, and physical facilities. They opposed the participation of women in mixed-sex teams of sports played with great vigor or violence, such as football and ice hockey. The reasons given were not prudish, but were based on the sex differential in height, weight, and strength, that could endanger the health and safety of women.

REFERENCES

Eveleth, P. V., and J. M. Tanner. 1976. Worldwide Variation in Human Growth. Cambridge University Press, Cambridge.

Hamburg, B. A. 1974. Psychobiology of sex differences: an evolutionary perspective. *In* R. C. Friedman, R. M. Richart and R. L. Vande Wiele (eds.), Sex Differences in Behavior. John Wiley & Sons, New York, pp. 373-392.

Jokl, E. 1964. Physiology of Exercise. Charles C Thomas, Springfield.

Lennane, K. J., and J. R. Lennane. 1973. Alleged psychogenic disorders in women—a possible manifestation of sexual prejudice. N. Engl. J. Med. 288:288-292.

Lowe, M. 1978. Sociobiology and sex differences. Signs 4:118-125.

Marx, J. 1979. Dysmenorrhea: basic research leads to a rational therapy. Science 205:175-176.

Tanner, J. M. 1962. Growth at Adolescence. 2nd Ed. Blackwell Scientific Press, Oxford.

Young, J. Z. 1971. An Introduction to the Study of Man. Oxford University Press, Oxford.

Epilogue

This book shows that there are manifold ways in which male and female, in our society, are biologically different. The differences range from those affecting the process of development, to those involving the skeletal structure and body tissues of adults; from the anatomy and function of the reproductive system and levels of hormone production, to differences in behavior and cognitive skills, possibly influenced by hormonal effects on the brain.

To say this is not to make value judgments, to prefer one sex over the other, or to say that biology predisposes women and men for different occupations. This becomes especially clear when it is observed that, apart from the reproductive systems, the differences between human males and females are differences in quantitative, measurable characteristics, and in most cases the range of measurements for males and females overlaps. This means that one cannot say flatly that women's biology predisposes them for different occupations (outside of childbearing) than men, because, regarding any given characteristic, within the overlapping part of the range, there are men and women who are similarly and equally predisposed. The existence of nonoverlapping parts of the range indicates that some men may excel in one particular respect and some women in another; but there is room for the good, as well as the excellent, in any kind of occupation.

Lack of basic information should also prevent the making of categorical statements. Clinicians have discussed whether dysmenorrhea is psychologically or somatically determined (Lennane and Lennane, 1973) for decades until at last a probable physiologic basis has been identified and the possibility of an effective treatment is opened up, at least for some cases (Marx, 1979). There are still differences of opinion about the causes of premenstrual tension: research is needed in this area too. Some of the symptoms associated with the climacteric (such as osteoporosis) are not obviously due to estrogen deficiency: it is important to know what their causes are and find ways to treat them. As to the effects of hormones on the brain, we are only at the beginning of an understanding, and the results of different studies frequently show contradictions (Bleier, 1979). Far more studies and measurements must be made, including identification of regions of the brain where hormone receptors are present and the times during which these regions are receptive to hormones. This is a tremendously complicated field in which many new findings are being made on the catecholamines and other synaptic transmitters. The functional relations between these substances and the steroid hormones must be established. When this has been done in a variety of experimental animals, the question of which, if any, of the findings are valid for people will remain. Whether measurable sex differences in different parts of the brain are present prenatally in humans, and the extent to which different cultural conditions contribute to their development postnatally are other questions that must be answered.

Another major question is how, in human cases, can the contribution of genes and environment, or nature and nurture, to the development of individual traits be evaluated? It is fairly easy to set up controlled experiments with plants. With animals it is more difficult and requires such refinements as double-blind experiments to eliminate biased treatments, because animals respond to the people who handle them. When it comes to human studies, adequately controlled experiments are virtually impossible (Lowe, 1978). Reports of experiments claiming to show genetically based or hormonally induced differences must be received sceptically. Some differences between the sexes (skeletal measurements, for instance, body composition, and times and rates of developmental processes) seem more or less universal, fixed, and unchangeable. But many other gender-related differences are not common to all cultures and their genetic basis is therefore much more dubious. Even where some degree of genetic determination seems probable (e.g., in spatial skills), the genes can only manifest themselves through interaction with environmental factors, many of which can be controlled. One of the most striking examples of gender differences induced culturally is provided by the case studied by Money and Ehrhardt (1972) involving twin male infants, one of whom was brought up as a boy, while the other was raised as a girl and acquired a culturally feminine gender identity. Biology may seem like destiny, but only when one fails to consider the interaction between genes and environment.

Hamburg (1974) puts forward an interesting paradigm based on language, which may be relevant to the development of gender differences. The basic equipment necessary for speech comprises the human larynx, tongue and lips, and appropriate structures in the central and peripheral nervous systems. These are gene-determined and any of the components may be affected by a mutation (such as, for instance, hare lip), which affects the subject's speech and response to speech. On the other hand, the character of the language that each human infant develops (whether it be English, French, Swahili, or Chinese) depends entirely on the culture to which the infant is exposed. It is also possible for an infant possessing the necessary physical equipment to grow up without any language skills, if the child is not exposed to the use of language at an early age. It is reasonable to expect that the range of variation in many human skills and aptitudes is as much culturally determined as the range of language and linguistic aptitude. By analogy, this may also apply to the gender differences of different cultures.

As behavior can be modified and skills learned, the most important question is not whether anatomy is or is not destiny. There are more useful questions that are amenable to research and may have practical implications. How do different cultural conditions contribute to the development or suppression of a given type of behavior, or a specific skill? To what extent should we try to equalize the development of male and female, on the one hand, or to permit or even encourage the development of sex-related differences, on the other? In our discussion of human variation (Chapter 6) we take the view that maximizing the permissible variation for both women and men permits a greater choice for both, and is more desirable than seeking conformity to prescribed gender roles.

We hope that more women, as well as more men, will be motivated to do research on the biology of sex differences, focusing on areas (some of which we have touched on) where our knowledge is clearly inadequate or where current interpretations are biased. In this century, physicists have recognized that the observer is a factor in a scientific experiment. A recent article (Bleier, 1979) applies this insight to a discussion of bias in biological research on sex differences. The input into research on sex and gender should include different kinds of bias, including that of the feminist. In the end, a bias that is not consonant with reality will have to deal with contradictory data. From that point there are two alternatives: to ignore the data and cling to the myth, or to respect the data and change our bias. Even more important is the fact that the empathy of women (which might be expected to be less inhibited than that of men) provides a better climate for research in this area.

While we await answers to some of the questions posed above, women and men alike are in unexplored territory. If one sex can be good, but not excellent, in a particular area, is it worth trying to compete? Can training develop excellence, or is self-confidence all that is necessary? But if there are many uncertainties, there are also tremendous opportunities for pioneering, and there are some things that we can look forward to. One is that, as scarce as role models are now, there will be a greater variety and a greater number for the next generation. Also, as more women rebel against sexism and discrimination, and as more men become dissatisfied with circumscribed sex roles, a new milieu for childrearing will develop. Then we shall be able to compare the new generation of children of increasingly nonsexist parents with those of previous generations. Perhaps the curves will shift and the degree of overlap will change.

We hope that those who have read this book have acquired a sense of the wonderful intricacy of the systems of genes, of hormones, of nerve cells, and of tissues and organs, which are integrated in our bodies. Both women and men should take pride in their bodies as well as in their minds.

REFERENCES

Bleier, R. 1979. Social and political bias in science: an examination of animal studies and their generalizations to human behavior and evolution. In R. Hubbard and M. Lowe (eds.) Genes and Gender II. Gordian Press, Staten Island, New York, pp. 49-69.

Freud, S. 1925. Some psychological consequences of the anatomical distinction between the sexes. *In* J. Strachey (ed.), The Standard Edition of the Complete Psychological Works of Sigmund Freud. pp. 248-258, Vol. 19 Hogarth Press, London.

Hamburg, B. A. 1974. Psychobiology of sex differences: an evolutionary perspective. In R. C. Friedman, R. M. Richart and R. L. Vande Wiele (eds.) Sex Differences in Behavior. John Wiley & Sons, New York, pp. 373-392.

Lennane, K. J. and Lennane, J. R. 1973. Alleged psychogenic disorders in women—a possible manifestation of sexual prejudice. N. Eng. J. Med. 288:288-292.

Lowe, M. 1978. Sociobiology and sex differences. Signs 4:118-125.

Marx, J. 1979. Dysmenorrhea: basic research leads to a rational therapy. Science 206:175-176.

Money, J. and Ehrhardt, A. A. 1972. Man and Woman, Boy and Girl. Johns Hopkins University Press, Baltimore, pp. 118-123.

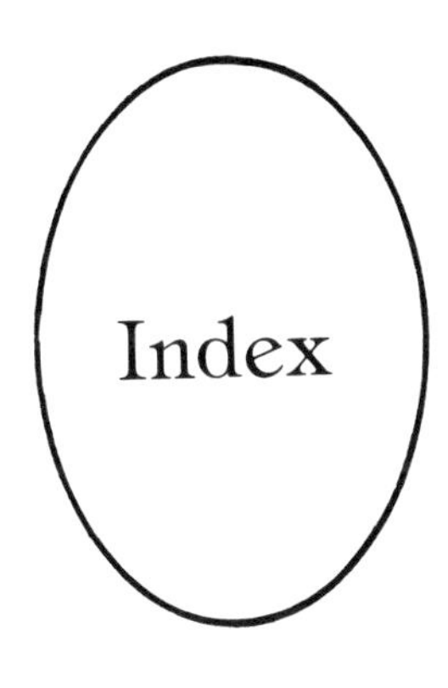
Index